STOCHASTIC VERSUS FUZZY APPROACHES TO MULTIOBJECTIVE MATHEMATICAL PROGRAMMING UNDER UNDERTAINTY

THEORY AND DECISION LIBRARY

General Editors: W. Leinfellner and G. Eberlein

Series A: Philosophy and Methodology of the Social Sciences
Editors: W. Leinfellner (Technical University of Vienna)
G. Eberlein (Technical University of Munich)

Series B: Mathematical and Statistical Methods
Editor: H. Skala (University of Paderborn)

Series C: Game Theory, Mathematical Programming and Operations Research
Editor: S. H. Tijs (University of Nijmegen)

Series D: System Theory, Knowledge Engineering and Problem Solving
Editor: W. Janko (University of Economics, Vienna)

SERIES D: SYSTEM THEORY, KNOWLEDGE ENGINEERING AND PROBLEM SOLVING

Editor: W. Janko (Vienna)

Volume 6

Scope

This series focuses on the design and description of organisations and systems with application to the social sciences. Formal treatment of the subjects is encouraged. Systems theory, information systems, system analysis, interrelated structures, program systems and expert systems are considered to be a theme within the series. The fundamental basics of such concepts including computational and algorithmic aspects and the investigation of the empirical behaviour of systems and organisations will be an essential part of this library. The study of problems related to the interface of systems and organisations to their environment is supported. Interdisciplinary considerations are welcome. The publication of recent and original results will be favoured.

For a list of titles published in this series, see final page.

STOCHASTIC VERSUS FUZZY APPROACHES TO MULTIOBJECTIVE MATHEMATICAL PROGRAMMING UNDER UNCERTAINTY

edited by

ROMAN SLOWINSKI
Technical University of Poznan, Poland

and

JACQUES TEGHEM
Faculté Polytechnique de Mons, Belgium

KLUWER ACADEMIC PUBLISHERS
DORDRECHT / BOSTON / LONDON

Library of Congress Cataloging-in-Publication Data

Stochastic versus fuzzy approaches to multiobjective mathematical programming under uncertainty / edited by Roman Slowinski, Jacques Teghem.
p. cm. -- (Theory and decision library. Series D, System theory, knowledge engineering, and problem solving ; v. 6)
ISBN-13: 978-94-010-7449-0 e-ISBN-13: 978-94-009-2111-5
DOI: 10.1007/978-94-009-2111-5

1. Programming (Mathematics) 2. Fuzzy systems. 3. Stochastic processes. I. Słowiński, Roman. II. Teghem, Jacques, 1948- . III. Series.
QA402.5.S74 1991
519.7--dc20 90-42735

ISBN-13: 978-94-010-7449-0

Published by Kluwer Academic Publishers,
P.O. Box 17, 3300 AA Dordrecht, The Netherlands.

Kluwer Academic Publishers incorporates
the publishing programmes of
D. Reidel, Martinus Nijhoff, Dr W. Junk and MTP Press.

Sold and distributed in the U.S.A. and Canada
by Kluwer Academic Publishers,
101 Philip Drive, Norwell, MA 02061, U.S.A.

In all other countries, sold and distributed
by Kluwer Academic Publishers Group,
P.O. Box 322, 3300 AH Dordrecht, The Netherlands.

Printed on acid-free paper

CONTENTS

FOREWORD vii

D. DUBOIS (F), H. PRADE (F)

Part I. THE GENERAL FRAMEWORK.

Chapter 1. Multiobjective programming under uncertainty : scope and goals of the book. 3

R. SLOWINSKI (PL), J. TEGHEM (B)

Chapter 2. Multiobjective programming : basic concepts and approaches. 7

D. VANDERPOOTEN (F)

Chapter 3. Stochastic programming : numerical solution techniques by semi-stochastic approximation methods. 23

K. MARTI (D)

Chapter 4. Fuzzy programming : a survey of recent developments. 45

M. INUIGUCHI (J), H. ICHIHASHI (J), H. TANAKA (J)

Part II. THE STOCHASTIC APPROACH.

Chapter 1. Overview of different approaches for solving stochastic programming problems with multiple objective functions. 71

I.M. STANCU-MINASIAN (R)

Chapter 2. "STRANGE" : an interactive method for multiobjective stochastic linear programming, and "STRANGE-MOMIX" : its extension to integer variables. 103

J. TEGHEM (B)

Chapter 3. Application of STRANGE to energy studies. 117

P. L. KUNSCH (B)

Chapter 4. Multiobjective stochastic linear programming with incomplete information : a general methodology. 131

B. URLI (CDN), R. NADEAU (CDN)

Chapter 5. Computation of efficient solutions of stochastic optimization problems with applications to regression and scenario analysis. 163

K. MARTI (D)

Part III. THE FUZZY APPROACH.

Chapter 1. Interactive decision-making for multiobjective programming problems with fuzzy parameters. 191
M. SAKAWA (J), H. YANO (J)

Chapter 2. A possibilistic approach for multiobjective programming problems. Efficiency of solutions. 229
M. DELGADO (E), J.L. VERDEGAY (E), M.A. VILA (E)

Chapter 3. "FLIP" : an interactive method for multiobjective linear programming with fuzzy coefficients. 249
R. SŁOWIŃSKI (PL)

Chapter 4. Application of "FLIP" method to farm structure optimization under uncertainty. 263
P. CZYŻAK (PL)

Chapter 5. "FULPAL" : an interactive method for solving (multiobjective) fuzzy linear programming problems. 279
H. ROMMELFANGER (D)

Chapter 6. Multiple objective linear programming problems in the presence of fuzzy coefficients. 301
M.K. LUHANDJULA (DZ), M. SAKAWA (J)

Chapter 7. Inequality constraints between fuzzy numbers and their use in mathematical programming. 321
M. ROUBENS (B)

Chapter 8. Using fuzzy logic with linguistic quantifiers in multiobjective decision making and optimization: A step towards more human-consistent models. 331
J. KACPRZYK (PL), R. R. YAGER (USA)

Part IV. STOCHASTIC VERSUS FUZZY APPROACHES AND RELATED ISSUES.

Chapter 1. Stochastic versus possibilistic multiobjective programming. 353
J.J. BUCKLEY,(USA)

Chapter 2. A comparison study of "STRANGE" and "FLIP". 365
R. SŁOWIŃSKI (PL), J. TEGHEM (B)

Chapter 3. Multiobjective mathematical programming with inexact data. 395
I.M. STANCU-MINASIAN (R), ST. TIGAN (R)

INDEX 419

FOREWORD

Operations Research is a field whose major contribution has been to propose a rigorous formulation of often ill-defined problems pertaining to the organization or the design of large scale systems, such as resource allocation problems, scheduling and the like. While this effort did help a lot in understanding the nature of these problems, the mathematical models have proved only partially satisfactory due to the difficulty in gathering precise data, and in formulating objective functions that reflect the multi-faceted notion of optimal solution according to human experts. In this respect linear programming is a typical example of impressive achievement of Operations Research, that in its deterministic form is not always adapted to real world decision-making : everything must be expressed in terms of linear constraints ; yet the coefficients that appear in these constraints may not be so well-defined, either because their value depends upon other parameters (not accounted for in the model) or because they cannot be precisely assessed, and only qualitative estimates of these coefficients are available. Similarly the best solution to a linear programming problem may be more a matter of compromise between various criteria rather than just minimizing or maximizing a linear objective function. Lastly the constraints, expressed by equalities or inequalities between linear expressions, are often softer in reality that what their mathematical expression might let us believe, and infeasibility as detected by the linear programming techniques can often been coped with by making trade-offs with the real world.

This situation has motivated a search for more flexible formulations of optimisation problems, that, although remaining rigorous, may help bridging the gap between the mathematical models and the real decision-making situation, through the introduction of uncertainty and fuzziness. This book has the great merit to bring together two distinct lines of research that try to address some of these issues : stochastic linear programming and fuzzy linear programming that have developed independently. It may be interesting to understand why this is so and what are their respective areas of applications. This is what the Editors of this volume have tried to do, not only by inviting leading contributors of each approach to present their methodologies, as well as examples of applications, but also by carrying out a comparative analysis of stochastic and fuzzy linear programming on the same problems.

From a more basic point of view, what underlies the debate that is the topic of the book, is the difference, in meaning and aims, between probability, understood as the models of uncertainty relying on Kolmogorov axioms, and fuzzy sets, including the approach to uncertainty analysis called possibility theory. Historically the additive model of probability has received full development with the emergence of statistics, and it is interesting to note that several contributors to stochastic programming justify the presence of ill-know coefficients in linear expressions by statistical variability. The use of subjective probabilities is often resorted to, so as to cope with the lack of statistical data, and subjective probabilities often turn out to be subjectively assessed frequencies in practice, despite the existence of theoretical works justifying the probabilistic model of uncertainty outside any frequentist interpretation.

By contrast, the —now 25-years old— fuzzy tradition first intended to soften the notion of constraint, thus acknowledging that objective functions and constraints were of the same nature, and that the distinction between them is gradual rather than abrupt. In the first stand, fuzzy linear programming was a methodology for multiple objective optimisation under soft constraints, the equality and inequality between left and right-hand side of linear constraints becoming graded. At this point the problems addressed by stochastic programming (coefficient variability) and fuzzy programming (soft (in)equalities) are quite distinct. However, the full-fledged fuzzy linear programming approach also involves ill-definition of coefficients modelled by fuzzy numbers. A fuzzy number models an ill-bounded set of possible values, quite in the spirit of linear programming with interval-valued coefficients. In other words fuzzy linear programming combines soft constraints and graded sensitivity analysis with incompletely known coefficients.

While the above analysis is about what can be said "in principle" regarding the difference between stochastic and fuzzy linear programming, this difference must be understood at the level of the operational researcher, and of the user of optimisation packages. Here, there is an open field of investigation, and this book is a major step in this direction. Not only will it be useful to specialists of optimisation for straightening up their ideas about fuzzy versus stochastic approaches ; but this book must be pointed out as one of the very first examples of interdisciplinary cooperation between proponents of fuzzy set theory and advocates of probability theory, and as such, deserves a much wider audience. Let us hope that similar projects will be carried out in other fields where the antagonism between fuzzy set theory and probability theory is still raging. No doubt that this type of books, and particularly this one, will help triggering a high level scientific dialogue between both schools of uncertainty, a crucial condition for the proper recognition and perenniality of fuzzy set and possibility theory.

Toulouse, April 6, 1990

Didier DUBOIS   Henri PRADE

Part I.
THE GENERAL FRAMEWORK

Part I
Chapter 1

MULTIOBJECTIVE PROGRAMMING UNDER UNCERTAINTY : SCOPE AND GOALS OF THE BOOK

ROMAN SLOWINSKI
Technical University of Poznan
60-965 Poznan, Poland

JACQUES TEGHEM
Faculté Polytechnique de Mons
7000 Mons, Belgium

In recent years we have been able to observe that a classical mathematical programming model is insufficient in many real-world situations, particularly in long-term planning problems and programming of development strategies. The nature of these problems requires taking into account multiple objectives on the one hand, and various kinds of uncertainty, on the other hand. During the past decade, the development of multiobjective mathematical programming has been particularly fruitful. At the same time many authors dealt with modelling of various kinds of uncertainty in decision problems. Two different ways of handling uncertainty were at the origin of stochastic mathematical programming and fuzzy mathematical programming. Recently, these two kinds of generalization have been combined in the framework of multiobjective stochastic mathematical programming and multiobjective fuzzy mathematical programming.

The purpose of this monograph is to present a state-of-the-art in this field, advantages and disadvantages of both approaches as well as various relationships existing among them.

The book is composed of four parts. The first part contains three survey chapters related to three key components of multiobjective mathematical programming, separately.

VANDERPOOTEN introduces the basic concepts of multiobjective optimization and describes the different existing approaches to solving multiobjective mathematical programming problems. He gives a special attention to the interactive methods and points out their practical importance.

MARTI, after a recall of the two basic approaches (the penalty function approach and the chance-constrained approach) to stochastic programming, devotes its contribution to the methods related to the first one. He describes the different numerical techniques using stochastic approximation procedures and shows that their convergence can be accelerated by the use of descent directions. He discusses in detail the construction of improved step directions.

R. Slowinski and J. Teghem (eds.),
Stochastic versus Fuzzy Approaches to Multiobjective Mathematical Programming under Uncertainty, 3–6.

NUIGUCHI, ICHIHASHI and TANAKA classify the major types of fuzzy programming into three categories : mathematical programming with vagueness, mathematical programming with ambiguity, mathematical programming with vagueness and ambiguity. For each of them, they describe the different formulations which can be derived depending on the interpretation of the uncertainty and review the main methods proposed in the literature.

Three remaining parts of the book are devoted to combination of multiobjective programming and uncertainty.

The second part groups chapters representing stochastic approach to modelling of uncertainty.

First, STANCU-MINASIAN analyses the different multiobjective stochastic programming problems : the Chebyshev problem; the fractional programming problem; the stochastic goal programming; the group decision making in stochastic programming; the multiple minimum-risk problem. He gives an overview of the existing approaches to solving them. He shows moreover that these methods are extensions of methods used before for solving multiobjective linear programming (MOLP) and stochastic programming problems, and points out the difficulties of these generalizations.

TEGHEM presents an interactive method, called STRANGE, for a particular case of stochastic MOLP in which the uncertainty is modelled with scenarios. To take into account the possible violation of the uncertain constraints, an additional risk criterion is introduced. An extension to the case of integer variables is proposed with the STRANGE-MOMIX method, using an interactive branch and bound.

KUNSCH describes two real-world applications, in the field of energy planning, which have been successfully treated by STRANGE and its extension. These case studies concern the optimal use of nuclear fuel and a power-mix problem in which contenders are classical Diesel engins and solar power systems.

URLI and NADEAU analyse a stochastic MOLP problem for which only an incomplete information is available for the uncertain parameters. For such situations, they propose a general methodology for transforming the problem into an equivalent deterministic MOLP problem. Their methodology is illustrated through a didactical example.

MARTI considers an LP problem with a general stochastic objective function defined by a loss function. This loss function is often not exactly known and there is some uncertainty in defining it. To treat this situation, several numerical procedures are proposed and discussed.

The third part collects chapters representing fuzzy approach to modelling of uncertainty.

SAKAWA and YANO present interactive methods for multiobjective linear, linear fractional and nonlinear programming problems with fuzzy parameters. The key concept of these methods is α-multiobjective programming and M-α-Pareto optimality, based on the α-level sets of the fuzzy numbers.

DELGADO, VERDEGAY and VILA give a unified framework for using fuzzy sets or possibility theory in multiobjective programming. They explore its general properties focusing on efficiency of the solutions.

SLOWINSKI presents an interactive method, called FLIP, for multiobjective linear programming with fuzzy coefficients. A proposed principle for comparison of fuzzy numbers enables transformation of the fuzzy MOLP problem into a multiobjective linear fractional program.

CZYŻAK shows one of agricultural applications of FLIP. He solves a real-world example of a farm structure optimization problem using the FLIP software on a micro.

ROMMELFANGER presents another interactive method for solving MOLP problems with fuzzy coefficients. Its name is FULPAL and it is based on a new interpretation of the inequality relation in fuzzy constraints.

LUHANDJULA and SAKAWA continue considering MOLP problems with fuzzy coefficients. Basing on the principle of the bounded rationality, they derive a concept of satisfying solutions and establish necessary and sufficient conditions for a potential action to be a satisfying one.

ROUBENS considers a fundamental problem of fuzzy programming which is comparison of fuzzy numbers. In the case of L-R fuzzy numbers, the comparison based on compensation of areas determined by the membership functions is reduced to the comparison of upper and lower bounds of α-cuts.

KACPRZYK and YAGER discuss some nonconventional models of multiobjective optimization in which an "optimal" alternative is sought that best satisfies, say, most (or any other linguistic quantifier as almost all, much more than a half, ...) of the important objectives. Fuzzy-logic-based calculi of linguistically quantified propositions are employed.

The last part is composed of chapters in which stochastic and fuzzy approaches are considered together.

BUCKLEY deals with MOLP problems where uncertain parameters are modelled using random variables (stochastic programming) or fuzzy variables (possibilistic programming). He explores both the similarities, and differences between stochastic and possibilistic programming with respect to his solution procedure.

SLOWINSKI and TEGHEM compare two methods for solving MOLP problems under uncertainty on a didactic example. The first method is STRANGE representing the stochastic approach, and the second is FLIP representing the fuzzy approach. For each stage of problem solving, possible limitations, advantages and inconveniencies of both methods are discussed.

STANCU-MINASIAN and TIGAN complete the consideration of uncertainty in multiobjective programs with two approaches handling inexact coefficients in the objective functions. They correspond to the conservative and nonconservative approaches to mathematical programming with set coefficients.

Throughout the book, many possible sources of uncertainty in problems formulated as multiobjective mathematical programs are handled using stochastic or/and fuzzy (possibilistic) approach. Many solution methods propose an interactive way of modelling user's preferences and uncertainty of parameters. The best compromise solution is conditioned by these two models, usually strongly correlated. Presenting different aspects of these complex problems was possible due to a large participation of outstanding specialists in this field. We wish therefore to thank very much all the Contributors for their valuable contributions and active cooperation in this collective work. We wish also to extend our thanks to D. DUBOIS and H. PRADE (I.R.I.T., Université Paul Sabatier, Toulouse, France) who kindly accepted to write the Foreword to the present book.

Part I
Chapter 2

MULTIOBJECTIVE PROGRAMMING: BASIC CONCEPTS AND APPROACHES

D. VANDERPOOTEN
LAMSADE, Université Paris Dauphine
Place du Maréchal De Lattre de Tassigny
75775 Paris Cedex 16
France

ABSTRACT. This paper aims at introducing, in a tutorial style, the basic concepts used in Multiple Criteria Decision Aid (MCDA), with an emphasis on multiobjective programming (MOP). Rather than precisely discussing some of the techniques and methods used in MOP (many references are provided), we present and comment the basic approaches underlying these methods.

1. Introduction

When considering a methodological approach to deal with a decision problem, analysts and/or decision makers are faced with two basic options:

- using a single criterion which is supposed to capture the essence of the decision,
- resorting to multiple criteria, each of which reflects a specific viewpoint relevant to the problem.

The adoption of a single criterion approach amounts to considering that the decision maker's preferences are conveniently rendered through a single measure. Such an approach is certainly prevailing when these preferences are based on a unique or predominant viewpoint. The construction of a prescription (recommendation, suggestion) is then straightforward. If a best solution is looked for, we are faced with a, sometimes technically difficult but clearly stated, problem of detecting the solution(s) optimizing the criterion.

In many decision contexts however, several, usually conflicting, viewpoints are to be considered. A single criterion approach is still possible by synthesizing these viewpoints into a unique measure. The resulting criterion, which is expressed in a more or less fictitious reference unit (monetary unit, utility,...) is often difficult to interpret. Moreover, its elaboration requires the aggregation of heterogeneous aspects using conversion rates which are often very questionable (the evaluation of the classical "shortage penalty cost" in inventory models or the estimation of the cost of a human life in health studies are crucial examples). Finally qualitative

R. Slowinski and J. Teghem (eds.),
Stochastic versus Fuzzy Approaches to Multiobjective Mathematical Programming under Uncertainty, 7–22.

aspects, which are difficult to evaluate are often, more or less consciously, discarded.

Considering explicitly the relevant viewpoints by using several criteria allows to circumvent most of these difficulties. However, the construction of a prescription is no longer straightforward, since each criterion will usually favour different groups of solutions. What could be perceived as a severe drawback actually stresses the fact that a decision involving conflicting viewpoints is intrinsically difficult and should be made in close collaboration with the decision maker (DM) and his preference structure (in the following, the DM is considered as a more or less identified person whose preferences must be taken into account in the decision process).

The expansion of MCDA and MOP is indicated by the publication of many papers (see *e.g.* the bibliographies by Achilles *et al.* (1979) and Stadler (1984)) and books which emphasize either a methodological (see *e.g.* Zeleny (1982), Roy (1985), Schärlig (1985), Vincke (1989)) or a technical (see *e.g.* Cohon (1978), Hwang and Masud (1979), Goicoechea *et al.* (1982), Chankong and Haimes (1983), Sawaragi *et al.* (1985), Yu (1985), Steuer (1986)) perspective.

Some preliminary definitions and concepts are introduced in Section 2. Section 3 discusses the role of preference parameters and scalarizing functions in the construction of prescriptions; some classical results are also presented. We introduce and comment the basic approaches used in MOP in Section 4. Section 5 contains some final remarks.

2. Some preliminary definitions and concepts

2.1 GENERAL STATEMENTS

When a decision problem, involving multiple criteria, consists in determining a best compromise solution, it may be stated as follows:

$$\begin{array}{ll} \text{Max} & [g_1(a),\dots,g_p(a)] \\ \text{s.t.} & a \in A \end{array} \qquad (2.1)$$

where:

- A denotes the set of potential *alternatives* (or feasible solutions),
- $\{g_1,\dots,g_p\}$ represents a set of p ($p \geq 2$) real-valued functions called *criteria*.

Mathematically, statement (2.1) is not well-defined since, usually, there is no alternative which optimizes each criterion simultaneously. Thus, the notation Max indicates here that we are looking for the *best compromise solution* according to the DM's preference structure, taking into account each of the p criteria (assuming, without loss of generality, that each criterion is to be maximized).

Multiobjective programming (MOP) deals with multiple criteria decision problems such that:

- each alternative is characterized by a vector $(x_1,\dots,x_n) = \mathbf{x}$ ($\mathbf{x} \in \mathbb{R}^n$) of *decision variables,*
- the set A is defined in terms of the decision variables :

$$A = \{\mathbf{x} \in \mathbb{R}^n \ / \ h_i(\mathbf{x}) \leq 0 \ \ (i = 1,...,m) \ \text{ and } \ \mathbf{x} \in S\}$$

where $h_i(\mathbf{x})$ $(i = 1,...,m)$ are constraint functions and $S \subset \mathbb{R}^n$ is used to represent additional constraints which cannot be expressed through functions (for instance, $S = Z^n$ or $S = \{0, 1\}^n$),

- each criterion g_j may be expressed as a function of the decision variables and is called *objective function.*

The general statement of a MOP problem is then:

$$\begin{aligned} \text{Max} \quad & [g_1(\mathbf{x}),...,g_p(\mathbf{x})] \\ \text{s.t.} \quad & h_i(\mathbf{x}) \leq 0 \qquad (i = 1,...,m) \\ & \mathbf{x} \in S \end{aligned} \tag{2.2}$$

This field has received considerable attention as a natural extension of classical (single objective) mathematical programming. General books about MOP are Chankong and Haimes (1983), Jahn and Krabs (1987), Sawaragi *et al.* (1985), Serafini (1985)... Previous general surveys are Buchanan (1986), Chankong *et al.* (1985), Evans (1984), MacCrimmon (1973)...

A well-studied special case of (2.2) is the multiobjective linear programming (MOLP) problem:

$$\begin{aligned} \text{Max} \quad & [\mathbf{c}^1\mathbf{x},...,\mathbf{c}^p\mathbf{x}] \\ \text{s.t.} \quad & D\,\mathbf{x} \leq \mathbf{b} \\ & \mathbf{x} \geq 0 \end{aligned}$$

where $\mathbf{c}^j$ are n-dimensional coefficient vectors, D is a m×n constraint coefficient matrix and $\mathbf{b}$ is a m-dimensional vector of right hand side coefficients. General overviews emphasizing the MOLP case are Cohon (1978), Hwang and Masud (1979), Steuer (1986), Zionts (1988).

Surveys dedicated to special cases of multiobjective programming are also available: multiobjective nonlinear programming (Narula and Weistroffer (1989)), multiobjective integer programming (Teghem and Kunsch (1986)), multiobjective zero-one programming (Rasmussen (1986)).

2.2 DECISION SPACE AND CRITERION SPACE

The above statements are represented in the *decision space*, which is the classical representation in mathematical programming. Although some specific concepts (*e.g.* related to duality theory) and the practical techniques used to "solve" multiobjective programming problems are based on classical mathematical programming, most of the basic concepts and definitions may be introduced independently using the *criterion space*. In this space each potential alternative a (or $\mathbf{x}$) is represented by its vector of criterion values $(g_1(a),...,g_p(a))$ (or $(g_1(\mathbf{x}),...,g_p(\mathbf{x}))$ - see Fig. 2.1). We should notice that the criterion space is more significant than the decision space, in a multiobjective context, since the interest of each alternative should be appreciated only regarding its values on the criteria. Let Z ($\subset \mathbb{R}^p$) be the criterion space and Z_A the image of A, *i.e.* the set of feasible outcomes. We have then:

$$Z_A = \{\ \mathbf{z} \in Z \ / \ z_j = g_j(a) \ (j = 1,...,p) \ \text{ where } a \in A\ \}$$

Definition 2.1: $\mathbf{z}' \in Z_A$ is *nondominated* iff there is no $\mathbf{z} \in Z_A$ such that $\mathbf{z} > \mathbf{z}'$.

In other words, a nondominated point is such that any other point in Z_A which increases the value of one criterion also decreases the value of at least one other criterion.

A restricted definition of nondominated points was introduced so as to eliminate undesirable nondominated points with unbounded trade-offs between the various criteria. The corresponding concept of properly nondominated point may be defined as follows (Geoffrion (1968)):

Definition 2.2: $\mathbf{z}' \in Z_A$ is *properly nondominated* iff it is nondominated and there exists a scalar $M > 0$ such that for each $i \in \{1,...,p\}$ and each $\mathbf{z} \in Z_A$ with $z_i > z'_i$, there exists at least one $j \in \{1,...,p\}$ with $z'_j > z_j$ and $(z_i - z'_i)/(z'_j - z_j) \leq M$.

In the MOLP case, the set of properly nondominated points corresponds to the set of nondominated points.

A relaxed definition of nondominated points is as follows:

Definition 2.3: $\mathbf{z}' \in Z_A$ is *weakly nondominated* iff there is no $\mathbf{z} \in Z_A$ such that $\mathbf{z} \gg \mathbf{z}'$.

Weakly nondominated points are irrelevant from a practical viewpoint. However, from a technical viewpoint, it is sometimes only possible to detect such points.

These three definitions are illustrated below (see Fig. 2.2).

Fig. 2.2 : Nondominated points (in the criterion space)

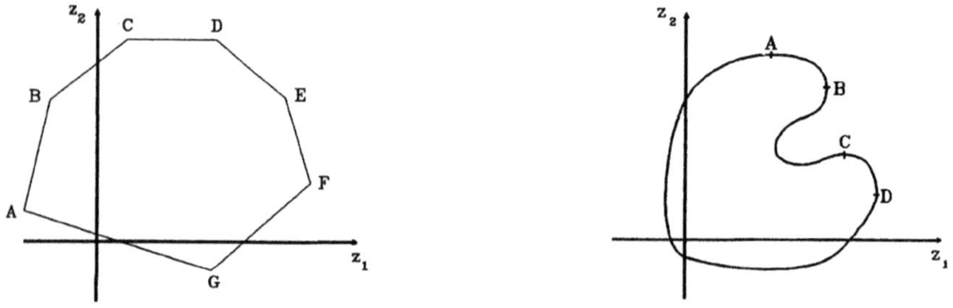

The set of nondominated points corresponds to line segments DE and EF. The set of weakly nondominated points corresponds to line segments CD, DE and EF (MOLP case).

The set of nondominated points corresponds to boundary segments AB and CD. Points A, B, C and D are improperly nondominated.

Fig. 2.1 : Decision space and criterion space

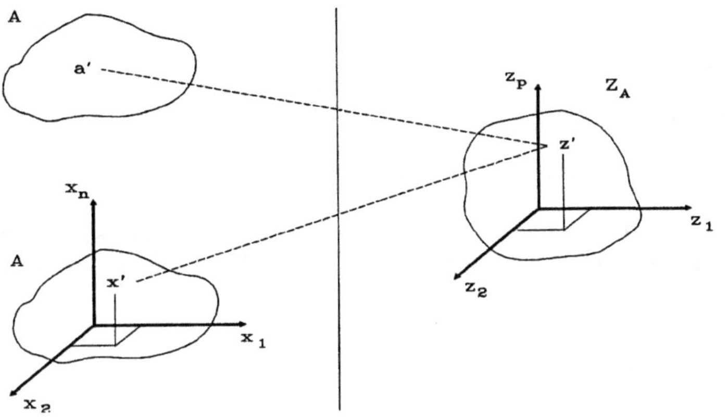

A MOP problem, and more generally a multiple criteria decision problem, may then be considered as the selection of a best compromise *criterion vector* $\mathbf{z}$ from Z_A, which may be stated as:

$$\begin{array}{ll} \text{Max} & [z_1,...,z_p] = \mathbf{z} \\ \text{s.t.} & \mathbf{z} \in Z_A \end{array}$$

2.3 NONDOMINATED POINTS AND EFFICIENT SOLUTIONS

Without any information regarding the DM's preference structure, the comparison between criterion vectors can be based on the natural partial orders defined on Z:

$\forall\ \mathbf{z} \in Z,\ \forall\ \mathbf{z}' \in Z$ we have :

- $\mathbf{z} \geq \mathbf{z}'$ iff $z_j \geq z'_j \quad \forall\ j \in \{1,...,p\}$,
- $\mathbf{z} > \mathbf{z}'$ iff $\mathbf{z} \geq \mathbf{z}'$ and $\mathbf{z} \neq \mathbf{z}'$,
- $\mathbf{z} \gg \mathbf{z}'$ iff $z_j > z'_j \quad \forall\ j \in \{1,...,p\}$.

Ideally, we should look for a point (criterion vector) $\mathbf{z}' \in Z_A$ such that $\mathbf{z}' \geq \mathbf{z}$ for all $\mathbf{z} \in Z_A$. Such a point usually does not exist. However, we may restrict our attention to nondominated points which constitute potential candidates for the best compromise criterion vector.

Obviously, these concepts may also be defined in the decision space. The solutions corresponding to (properly, weakly) nondominated points are called (properly, weakly) *efficient* solutions. Efficient solutions are also known as Pareto optimal or noninferior solutions.

As clearly indicated by the above examples, the set of nondominated points (and the corresponding efficient set) is usually infinite, which does not allow to characterize a prescription. However, such concepts are useful to guide the construction of this prescription.

2.4 IDEAL POINT, NADIR POINT, PAYOFF MATRIX

A natural way of obtaining initial information about the decision problem consists in optimizing each criterion separately over A. Let $\tilde{a}^j$ $(j=1,...,p)$ be a solution to the following problem:

$$\begin{aligned} &\max \; g_j(a) \\ &\text{s.t. } a \in A \end{aligned} \tag{2.3}$$

We shall denote $\tilde{z}_j^k = g_j(\tilde{a}^k)$ and $z_j^* = \tilde{z}_j^j = g_j(\tilde{a}^j)$.

Definition 2.4: The vector $\mathbf{z^*} = (z_1^*,...,z_p^*)$ is called *ideal point.*

$\tilde{z}_j^k$ are classically gathered into a table called the *payoff matrix*:

	g_1		g_j		g_p
$\tilde{a}^1$	z_1^*	...	$\tilde{z}_j^1$	...	$\tilde{z}_p^1$
$\tilde{a}^j$	$\tilde{z}_1^j$	...	z_j^*	...	$\tilde{z}_p^j$
$\tilde{a}^p$	$\tilde{z}_1^p$	...	$\tilde{z}_j^p$	...	z_p^*

The ideal point is on the diagonal of the payoff matrix.

Let n_j be the minimum value in the jth column of the pay-off matrix:

$$n_j = \min_{k=1,...,p} \{ \tilde{z}_j^k \} \qquad (j = 1,...,p).$$

Definition 2.5 : The vector $\mathbf{n} = (n_1,...,n_p)$ such that is called *nadir point.*

In case of alternative optima for (2.3), the payoff matrix and consequently the nadir point are not uniquely defined.

The above concepts are interesting to obtain an insight into the decision problem. The ideal point is usually not attainable because of the conflicts between criteria.

However, it may be regarded and used as a realistic approximation of (the image of) the best conceivable alternative (since it is defined from the current set of alternatives and thus corresponds to optimistic, but realistic criterion values - see *e.g.* Zeleny (1976)).

The payoff matrix which provides rather representative points, also gives an estimate of the ranges of the potential outcomes. However, it should be noticed that, in general, the components of the nadir point do not correspond to the minimum values of the criteria over the efficient set (see *e.g.* Isermann and Steuer (1987)).

3. Preference parameters and scalarizing functions

The concepts introduced in the previous section are only derived from the decision problem statement. They provide an insight in the problem and in the type of solution required, but do not lead to the construction of a prescription. In order to progress towards the definition of a best compromise solution, it is necessary to use additional information so as to aggregate the various criteria.

3.1 PREFERENCE PARAMETERS

Any nondominated point is a candidate for representing the best compromise criterion vector. The selection of a specific point requires information about the DM's preference structure. Such a preference information is usually obtained from the DM in a more or less direct way so as to reduce the cognitive strain. Prefernce information is then translated into *preference parameters.*

Let Ω be the set of preference parameters. Classical basic preference parameters are:

- *intercriteria parameters* which reflect the relative importance of each criterion (trade-offs, weighting vectors),
- *reference points* which represent desirable (or undesirable) levels on each criterion,
- *reservation levels* which represent minimal requirements on each criterion.

A preference parameter ω may consist of one or several basic parameters of the above type, according to the information required by the method.
Preference parameters, when integrated into a solution scheme, are used to:

- aggregate the various criteria (intercriteria parameters, reference points),
- delimit a subregion of interest $Z'_A \subset Z_A$ (reservation levels).

3.2 SCALARIZING FUNCTIONS

A *scalarizing function* is a function $s : Z_A \times \Omega \rightarrow \mathbb{R}$ which aggregates the criterion values of any alternative according to a given preference parameter.

It should be emphasized that such a function must be considered as a mere technical tool used for generating solutions. This generation process is simply achieved by selecting a specific parameter $\omega' \in \Omega$ and optimizing the resulting function over Z_A (or $Z'_A \subset Z_A$ if delimiting information is contained in ω'). MOP methods based on a scalarizing function will then solve one or several (depending on their underlying approach - see § 4) problems of the following type:

$$\underset{\mathbf{z} \in Z'_A}{\text{opt}} \quad s(\mathbf{z}, \omega') \tag{3.1}$$

Examples of classical scalarizing functions are:

- *weighted sum*:
$$s(\mathbf{z}, \boldsymbol{\lambda}) = \sum_{j=1}^{p} \lambda_j z_j$$
- *weighted Tchebychev norm*:
$$s(\mathbf{z}, (\boldsymbol{\lambda}, \tilde{\mathbf{z}})) = \max_{j=1,\ldots,p} \{|\lambda_j(\tilde{z}_j - z_j)|\}$$
- *augmented weighted Tchebychev norm*:
$$s(\mathbf{z}, (\boldsymbol{\lambda}, \tilde{\mathbf{z}})) = \max_{j=1,\ldots,p} \{|\lambda_j(\tilde{z}_j - z_j)|\} + \rho \sum_{j=1}^{p} |\lambda_j(\tilde{z}_j - z_j)| \quad \text{with } \rho > 0.$$

The choice of a specific scalarizing function depends on the type of preference parameter required by the method (weighting vectors, reference points,...). This function should also satisfy some desirable properties. Two basic requirements are:

- s only generates nondominated points (R.1)
- any nondominated point may be generated by s (R.2)

As a practical requirement, s should not involve a large computational burden (considering the original problem: if a MOLP is to be solved, s should preferably involves the resolution of linear programming problem). Other requirements are suggested by Henig and Ritz (1986) and Wierzbicki (1986).

3.3 SOME RESULTS

Scalarizing functions have been extensively studied in the literature regarding their ability to characterize the set of nondominated points (*i.e.* with respect to requirements (R.1) and (R.2)). The interested reader is referred to *e.g.* Chankong and Haimes (1983, § 4.3), Gal (1986), Jahn (1984), Sawaragi *et al.* (1985, § 3.4), Wierzbicki (1986).

We just indicate a general and powerful result with respect to requirement (R.1), whose proof is given here because of its simplicity. Let us first give the following definitions.

Considering any $\mathbf{z} \in Z$ and any $\mathbf{z}' \in Z$, a function $f : Z \to \mathbb{R}$ is:

- *monotonically increasing* iff $\mathbf{z} > \mathbf{z}' \Rightarrow f(\mathbf{z}) \geq f(\mathbf{z}')$,
- *strongly monotonically increasing* iff $\mathbf{z} > \mathbf{z}' \Rightarrow f(\mathbf{z}) > f(\mathbf{z}')$,
- *strictly monotonically increasing* iff $\mathbf{z} \gg \mathbf{z}' \Rightarrow f(\mathbf{z}) > f(\mathbf{z}')$.

We have then:

Theorem 3.1: Let a function $f : Z_A \to \mathbb{R}$ and $\mathbf{z}'$ an optimal solution to

$$\max_{z \in Z_A} \quad f(z) \tag{3.2}$$

(a) if f is *strongly monotonicallly* increasing on Z_A, then $\mathbf{z}'$ is nondominated,

(b) if f is *monotonically* increasing on Z_A and if $\mathbf{z}'$ is the unique optimal solution to (3.2), then $\mathbf{z}'$ is nondominated,

(c) if f is *strictly monotonically* increasing on Z_A, then $\mathbf{z}'$ is weakly nondominated.

Proof :
(a) and (b): Suppose $\mathbf{z}'$ is dominated. Then, there exists $\mathbf{z}'' \in Z_A$ such that $\mathbf{z}'' > \mathbf{z}'$.
- (a): we have $f(\mathbf{z}'') > f(\mathbf{z}')$, which contradicts the optimality of $\mathbf{z}'$.
- (b): we have $f(\mathbf{z}'') \geq f(\mathbf{z}')$, which contradicts the fact that $\mathbf{z}'$ is the unique optimal solution to (3.2).

(c): Suppose $\mathbf{z}'$ is not weakly nondominated. Then, there exists $\mathbf{z}'' \in Z_A$ such that $\mathbf{z}'' \gg \mathbf{z}'$. Thus, we have $f(\mathbf{z}'') > f(\mathbf{z}')$, which contradicts the optimality of $\mathbf{z}'$. □

Considering the scalarizing functions introduced in the previous section, we have the following theorems (whose parts (a) may be directly derived from theorem 3.1).

Theorem 3.2: Considering the following problem:

$$\max_{z \in Z_A} \quad \sum_{j=1}^{p} \lambda_j z_j \tag{3.3}$$

(a.1) if $\mathbf{z}'$ is optimal solution to (3.3) with $\boldsymbol{\lambda} \gg \mathbf{0}_p$, then $\mathbf{z}'$ is nondominated.

(a.2) if $\mathbf{z}'$ is the unique optimal solution to (3.3) with $\boldsymbol{\lambda} > \mathbf{0}_p$, then $\mathbf{z}'$ is nondominated.

(a.3) if $\mathbf{z}'$ is optimal solution to (3.3) with $\boldsymbol{\lambda} > \mathbf{0}_p$, then $\mathbf{z}'$ is weakly nondominated.

(b) if $\mathbf{z}'$ is nondominated and Z_A is *convex*, then there exists $\boldsymbol{\lambda} > \mathbf{0}_p$ such that $\mathbf{z}'$ is optimal solution to (3.3).

Theorem 3.3: Considering the following problem with $\bar{\mathbf{z}} \gg \mathbf{z}^*$:

$$\min_{z \in Z_A} \quad \max_{j = 1,\dots,p} \{|\lambda_j(\bar{z}_j - z_j)|\} \tag{3.4}$$

(a.1) if $\mathbf{z}'$ is the unique optimal solution to (3.4) with $\boldsymbol{\lambda} > \mathbf{0}_p$, then $\mathbf{z}'$ is nondominated.

(a.2) if $\mathbf{z}'$ is optimal solution to (3.4) with $\boldsymbol{\lambda} \gg \mathbf{0}_p$, then $\mathbf{z}'$ is weakly nondominated.

(b) if $\mathbf{z}'$ is nondominated then there exists $\boldsymbol{\lambda} \gg \mathbf{0}_p$ such that $\mathbf{z}'$ is the unique optimal solution to (3.4).

Theorem 3.4: Considering the following problem with $\bar{\mathbf{z}} \gg \mathbf{z}^*$:

$$\min_{\mathbf{z} \in Z_A} \left(\max_{j=1,\ldots,p} \{|\lambda_j(\bar{z}_j - z_j)|\} + \rho \sum_{j=1}^{p} |\lambda_j(\bar{z}_j - z_j)| \right) \tag{3.5}$$

(a.1) if $\mathbf{z}'$ is optimal solution to (3.5) with $\boldsymbol{\lambda} \gg \mathbf{0}_p$, then $\mathbf{z}'$ is nondominated.

(a.2) if $\mathbf{z}'$ is the unique optimal solution to (3.5) with $\boldsymbol{\lambda} > \mathbf{0}_p$, then $\mathbf{z}'$ is nondominated.

(a.3) if $\mathbf{z}'$ is optimal solution to (3.5) with $\boldsymbol{\lambda} > \mathbf{0}_p$, then $\mathbf{z}'$ is weakly nondominated.

(b) if $\mathbf{z}'$ is nondominated then there exists $\boldsymbol{\lambda} \gg \mathbf{0}_p$ and a sufficiently small positive value ρ such that $\mathbf{z}'$ is the unique optimal solution to (3.5).

As to the computational burden involved by the above functions, we can notice that, since we impose $\bar{\mathbf{z}} \gg \mathbf{z}^*$ in theorem 3.4, problem (3.5) is equivalent to:

$$\min \left(\mu - \rho \sum_{j=1}^{p} \lambda_j z_j\right)$$

$$\text{s.t.} \quad \mu \geq \lambda_j (\bar{z}_j - z_j) \qquad (j = 1,\ldots,p)$$

$$\mathbf{z} \in Z_A$$

A very similar transformation is clearly possible for problem (3.4). This shows that, if the original problem is a MOLP, problems (3.4) and (3.5), as well as problem (3.3), are linear programming problems.

4. Basic approaches used in MOP

When dealing with a MCDA problem, two fundamental approaches are available so as to model the DM's preferences and to derive a prescription:

(1) aggregate the various criteria into a scalarizing function and exploit the resulting function,

(2) define one or several global preference relations on A and exploit the resulting relation(s).

The second approach aims at modelling the DM's preferences using binary preference relations (see Roubens and Vincke (1985), Roy (1985)). It is widely used in discrete problems involving a limited number of alternatives (see *e.g.* the outranking methods (Roy (1989), Vanderpooten (1989a)). This approach is not discussed here since it is not used in MOP.

MOP methods are optimization-based procedures resorting to scalarizing functions. A possible classification results from the optimization scheme used by the methods. We can distinguish methods based on:

- a unique optimization,
- a parametric optimization,
- iterative optimizations.

The previous distinction emphasizes a technical perspective. In order to get a real understanding of each approach, it is also necessary to distinguish the role devoted to the DM. Hwang and Masud (1979) propose a classification according to the stage at which information is required from the DM. The different classes respectively corresponding to the above optimization schemes are:

- a priori articulation of preference information,
- a posteriori articulation of preference information,
- progressive articulation of preference information.

We briefly introduce and comment the basic approaches used in MOP. Techniques and methods are not discussed here.

4.1 UNIQUE OPTIMIZATION WITH PRIOR ARTICULATION OF PREFERENCES

This is the most classical approach which consists in directly replacing the original MOP problem with a single objective problem. This is achieved by constructing or selecting a specific scalarizing function and determining relevant preference parameters. A unique problem (3.1) is then solved to get the prescription.

One obvious practical advantage of this approach is that it requires a rather limited computational effort. However, a large cognitive strain may be imposed on the DM.

Different classes of methods may be distinguished according to the meaning devoted to the scalarizing function.

In a first class of methods, the scalarizing function is not considered as a mere technical tool for generating solutions, but as an analytical formulation of the DM's preference structure. This kind of approach actually derives from classical utility theory where DMs are assumed to behave according to a *value* (or *utility*) function. The value function v of a specific DM is such that, for all $\mathbf{z} \in Z_A$, $\mathbf{z}' \in Z_A$:

$$v(\mathbf{z}') > v(\mathbf{z}) \Leftrightarrow \text{the DM prefers } \mathbf{z}' \text{ to } \mathbf{z} \qquad (4.1)$$
$$v(\mathbf{z}') = v(\mathbf{z}) \Leftrightarrow \text{the DM is indifferent between } \mathbf{z}' \text{ and } \mathbf{z}.$$

Expliciting this value function requires rather complex elicitation techniques which both aims at determining the form of the function and the required parameters (see Keeney and Raiffa (1976)). The validity of the resulting function, and even the fact the existence of such value functions, may be questioned. However, once this is accepted, any optimal solution of v gains an additional legitimacy.

In a second class of methods, the scalarizing function is imposed beforehand. In this case, the scalarizing function is not assumed to represent the DM's preference structure (and to verify relations (4.1)). It is just conceived such that its optimal solutions have a desirable and often easily interpretable property (like being feasible solutions which achieve desirable pre-specified aspiration levels or goals as "closely" as possible). A very well-known group of such methods are different forms of goal programming (see Charnes and Cooper (1961), Ignizio (1976), Lee (1972)). Compromise programming (where solutions minimizing a distance to the ideal point are looked for - see Zeleny (1973)) is a similar approach. Lexicographic methods (whose resulting solution optimizes the most important criterion and possibly, if there is any tie, the other criteria according to an order of importance pre-specified by the DM) are conceived in the same spirit.

It should be noticed that this classical approach has given rise to many practical application (in particular, methods based on goal programming and value functions).

4.2 PARAMETRIC OPTIMIZATION WITH A POSTERIORI ARTICULATION OF PREFERENCES

In this approach, the scalarizing function is conceived as a tool for generating the set or a subset of efficient solutions. In general MOP problems, this is achieved by parametrically solving problem (3.1). In the MOLP case, this parametric optimization is performed implicitly using Simplex-based procedures. A very large number of procedures were proposed in this case (*e.g.* Ecker and Kouada (1978), Evans and Steuer (1973), Isermann (1977), Yu and Zeleny (1975)).

The basic purpose of this approach is as follows. It aims at exploiting the MOP formulation of the decision problem, without any additional assumption about the DM's preference structure, so as to provide him with as exhaustive as possible information about the potential candidate solutions. Once this information is supplied, the final choice is to be made by the DM.

A first difficulty with this approach is the extensive computational burden which is required (even in the MOLP case).

Another severe limitation results from the very large number of generated solutions. The DM is then unable to analyze this information. Some techniques for presenting representative samples of solutions have been devised (*e.g.* the filtering technique of Steuer and Harris (1980)).

Apart from their undeniable theoretical interest, such methods can be considered, from a practical viewpoint, as (costly) information procedures to be used in prior phases of the decision making process.

4.3 ITERATIVE OPTIMIZATIONS WITH PROGRESSIVE ARTICULATION OF PREFERENCES

This approach is usually referred to as the *interactive approach*. An interactive method is an iterative method which alternates two kinds of phases:

- calculation phases executed by a computer (or an analyst),
- dialogue phases involving the DM.

At each iteration, the procedure presents the DM with a proposal. Such a proposal usually consists of one or several alternatives which are potential candidates for being a best compromise solution. The DM is required to react against the current proposal and provide preference information (dialogue phase). This information is analyzed to adjust the preference parameter used by the scalarizing function. A new problem (3.1) is then solved (calculation phase), which allows the procedure to present a new proposal. After a certain number of iterations, determined either by the DM (if he is satisfied) or by the procedure (through a stopping rule), the exploration process is stopped with the current proposal from which a final prescription is derived.

Nowadays, this approach is certainly the most active in MOP. A very large number of interactive procedures have been proposed in the literature (see *e.g.* the surveys from Steuer (1986, § 13), Vanderpooten and Vincke (1989), White (1983)).

The recourse to an interactive approach may be justified according to various conceptions which widely depends on the role devoted to the DM and the assumptions regarding his preference structure (Vanderpooten (1989b)).

In a first conception, the DM's preference structure is assumed to pre-exist and to remain stable during the whole decision process. This assumption is sometimes stated explicitly by supposing the existence of an implicit value or utility function. Since the DM's preference structure remains stable, the preference information progressively collected can be considered as valid throughout the whole exploration process. The exploration of the set of alternatives is conducted by the procedure which makes use of accumulated preference information to construct a converging process which results in the determination of the prescription. It should be noticed that the convergence (which is sometimes presented as a mathematical convergence considering the underlying implicit value function) actually results from the very strong assumption concerning the stability of the DM's preference structure.

In a second conception, no prior assumption is made about the DM's preference structure. In this case, interaction is used to allow the DM to learn about his problem and to strengthen, or even create, his preference structure. The DM may change his mind during the exploration process, which means that pieces of preference information supplied at different iterations may be contradictory. For this reason, the procedure cannot construct a converging process. The exploration of the set of alternatives is conducted by the DM in a trial-and-error fashion. The determination of a prescription may then prove difficult.

In our opinion, the basic challenge concerning the interactive approach consists in conciliating both conceptions, *i.e.* supporting the DM's learning of preferences and changes of mind while guiding his exploration.

5. Conclusion

One of the major trends in MOP is the increasing recognition of the role devoted to the DM in the decision aid process. Conjointly, MOP methods may no longer be designed as algorithms which have recourse to the DM when they need additional information. Moreover, it is now widely recognized that certain types of information (such as trade-offs between criteria) are difficult to provide. A usual strategy consists then in requiring simple and possibly qualitative information (such as indicating the criteria to be improved). Such a strategy is clearly more adapted in the case of interactive procedures. Actually, we believe that MOP methods, and particularly interactive procedures, must be designed so as to allow the DM to provide information with a varying degree of precision according to the type of reaction he wishes or is able to express.

From a theoretical viewpoint, MOP methods can now rely on solid foundations (*e.g.* regarding the characterization of nondominated points). However, research in MOP must also address difficulties which often arise in real-world situation (multiple

DMs, uncertainty, partial information).

Beside theoretical aspects, it is our belief that a methodology for dealing with MCDA or MOP problems must primarily provide the DM with a convenient framework for improving his knowledge and reflection about his problem, confronting and strengthening his preference structure and, as a result, finding a satisfactory compromise solution.

References

Achilles, A., Elster, K.H., Nehse, R., (1979), "Bibliographie zur Vektor-optimierung (Theorie und Anwendungen)", *Optimization* **10**, 2, pp. 277-321.

Buchanan, J.T., (1986), "Multiple Objective Mathematical Programming", *New Zealand Operational Research* **14**, 1, pp. 1-27.

Chankong, V., Haimes, Y.Y., (1983), *Multiobjective Decision Making: Theory and Methodology*, North-Holland, New York.

Chankong, V., Haimes, Y.Y., Thadathil, J., Zionts, S., (1985), "Multiple Criteria Optimization: a State of the Art Review", in *Decision Making with Multiple Objectives*, Proceedings Cleveland, Ohio, (1984), Y.Y. Haimes and V. Chankong (eds.), LNEMS 242, Springer-Verlag, Berlin.

Charnes, A., Cooper, W.W., (1961), *Management Models and Industrial Applications of Linear Programming*, Wiley, New York.

Cohon, J.L., (1978), *Multiobjective Programming and Planning*, Academic Press, New York.

Ecker, J.G., Kouada, I.A., (1978), "Finding All Efficient Extreme Points for Multiple Objective Linear Programs", *Mathematical Programming* **14**, pp. 249-261.

Evans, G.W., (1984), "An Overview of Techniques for Solving Multiobjective Mathematical Programs", *Management Science* **30**, 11, pp. 1268-1282.

Evans, J.P., Steuer, R.E., (1973), "A Revised Simplex Method for Linear Multiple Objective Programs", *Mathematical Programming* **5**, 1, pp. 54-72.

Gal, T., (1986), "On Efficient Sets in Vector Maximum Problems - A Brief Survey", *European Journal of Operational Research* **24**, pp. 253-264.

Geoffrion, A.M., (1968), "Proper Efficiency and the Theory of Vector Maximi-zation", *Journal of Mathematical Analysis and Applications* **22**, pp. 618-630.

Goicoechea, A., Hansen, D.R., Duckstein, L., (1982), *Multiobjective Decision Analysis with Engineering and Business Applications*, Wiley, New York.

Henig, M., Ritz, Z., (1986), "Multiobjective Decision Rules for Multiobjective Decision Problems", *European Journal of Operational Research* **26**, 1, pp. 134-141.

Hwang, C.L., Masud, A.S.M., (1979), *Multiple Objective Decision Making - Methods and Applications*, LNEMS 164, Springer-Verlag, Berlin.

Ignizio, J.P., (1976), *Goal Programming and Extensions*, Lexington Books, Heath, Lexington.

Isermann, H., (1977), "The Enumeration of the Set of All Efficient Solutions for a Linear Multiple Objective Program", *Operational Research Quarterly* **28**, 3, pp. 711-725.

Isermann, H., Steuer, R.E., (1987), "Computational Experience Concerning Payoff Tables and Minimum Criterion Values Over the Efficient Set", *European Journal of Operational Research* **33**, pp. 91-97.

Jahn, J., (1984), "Scalarization in Vector Optimization", *Mathematical Programming* **29**, pp. 203-218.

Jahn, J., Krabs, W., (eds.) (1987), *Recent Advances and Historical Development of Vector Optimization*, Proceedings, Darmstadt, FRG, 1986, LNEMS 294, Springer-Verlag, Berlin.

Keeney, R.L., Raiffa, H., (1976), *Decisions with Multiple Objectives: Preferences and Value Tradeoffs*, Wiley, New York.

Lee, S.M., (1972), *Goal Programming for Decision Analysis*, Auerbach Publishers, Philadelphia.

MacCrimmon, K.R, (1973), "An overview of Multiple Objective Decision Making", in *Multiple Criteria Decision Making*, J.L. Cochrane and M. Zeleny (eds.), University of South Carolina Press, Columbia, pp. 18-44.

Narula, S.C., Weistroffer, H.R., (1989), "Algorithms for Multi-Objective Nonlinear Programming Problems: an Overview", in *Improving Decision Making in Organisations*, Proceedings Manchester, UK, 1988, A.G. Lockett and G. Islei (Eds), LNEMS 335, Springer-Verlag, Berlin, pp. 434-443.

Rasmussen, L.M., (1986), "Zero-one Programming with Multiple Criteria", *European Journal of Operational Research* **26**, 1, pp. 83-95.

Roubens, M., Vincke, Ph. (1985), *Preference Modelling*, LNEMS 250, Springer-Verlag, Berlin.

Roy, B., (1985), *Méthodologie Multicritère d'Aide à la Décision*, Economica, Paris.

Roy, B., (1989), "The Outranking Approach and the Foundations of ELECTRE Methods", Document du LAMSADE No. 53, Université de Paris Dauphine, France, to appear in *Readings in Multiple Criteria Decision Aid*, C.A. Bana e Costa (ed.), Springer-Verlag, Berlin.

Sawaragi, Y., Nakayama, H., Tanino, T., (1985), *Theory of Multiobjective Optimization*, Academic Press, New York.

Schärlig, A., (1985), *Décider sur plusieurs critères*, Presses polytechniques romandes, Lausanne.

Serafini, P., (ed.) (1985), *Mathematics of Multi-Objective Optimization*, CISM Courses and Lectures 289, Springer-Verlag, Berlin.

Stadler, W., (1984), "A Comprehensive Bibliography on MCDM", *MCDM: Past Decade and Future Trends*, M. Zeleny (Ed.), Jai Press Inc., Greenwich, Connecticut, pp. 223-328.

Steuer, R.E., (1986), *Multiple Criteria Optimization: Theory, Computation and Application*, Wiley, New York.

Steuer, R.E., Harris, F.W., (1980), "Intra-Set Point Generation and Filtering in Decision and Criterion Space", *Computers and Operations Research* **7**, pp. 41-53.

Teghem Jr, J., and Kunsch, P.L., (1986), "Interactive Methods for Multi-Objective Integer Linear Programming", in *Large-Scale Modelling and Interactive Decision Analysis*, Proceedings Eisenach GDR 1985, G. Fandel, M. Grauer, A. Kurzhanski and A.P. Wierzbicki (Eds), LNEMS 273, Springer-Verlag, Berlin, pp. 75-87.

Vanderpooten, D., (1989a), "The Construction of Prescriptions in Outranking Methods", Document du LAMSADE No. 58, Université de Paris Dauphine, France, to appear in *Readings in Multiple Criteria Decision Aid*, C.A. Bana e Costa (ed.), Springer-Verlag, Berlin.

Vanderpooten, D., (1989b), "The Interactive Approach in MCDA: a Technical Framework and some Basic Conceptions", *Mathematical and Computer Modelling* **12**, 10/11, pp. 1213-1220.

Vanderpooten, D., Vincke, Ph., (1989), "Description and Analysis of Some Representative Interactive Multicriteria Procedures", *Mathematical and Computer Modelling*, **12**, 10/11, pp. 1221-1238.

Vincke, Ph., (1989), *L'aide multicritère à la décision*, Ellipses, Paris.

White, D.J., (1983), "A selection of Multi-Objective Interactive Programming Methods", in *Multi-Objective Decision Making*, S. French, R. Hartley, L.C. Thomas and D.J. White (Eds.), Academic Press, London, pp. 99-126.

Wierzbicki, A.P., (1986), "On the Completeness and Constructiveness of Parametric Characterizations to Vector Optimization Problems", *OR Spektrum* **8**, 2, pp. 73-87.

Yu, P.L., (1985), *Multiple Criteria Decision Making*, Plenum Press, New York.

Yu, P.L., Zeleny, M., (1975), "The Set of All Non-Dominated Solutions in Linear Cases and a Multicriteria Simplex Method", *Journal of Mathematical Analysis and Applications* **49**, 2, pp. 430-468.

Zeleny, M., (1973), "Compromise Programming", in *Multiple Criteria Decision Making*, J.L. Cochrane and M. Zeleny (eds.), University of South Carolina Press, Columbia, pp. 262-301.

Zeleny, M., (1976), "The Theory of the Displaced Ideal", in *Multiple Criteria Decision Making*, Proceedings Kyoto 1975, M. Zeleny (ed.), LNEMS 123, Springer-Verlag, Berlin, pp. 153-206.

Zeleny, M., (1982), *Multiple Criteria Decision Making*, Mc Graw-Hill, New York.

Zionts, S., (1988), "Multiple Criteria Mathematical Programming: an Updated Overview and Several Approaches", in *Mathematical Models for Decision Support*, G. Mitra (ed.), NATO ASI Series, Vol. F48, Springer-Verlag, Berlin.

Part I
Chapter 3

STOCHASTIC PROGRAMMING: NUMERICAL SOLUTION TECHNIQUES BY SEMI-STOCHASTIC APPROXIMATION METHODS

K. MARTI
Universität der Bundeswehr München
Fakultät für Luft- und Raumfahrttechnik
Werner-Heisenberg-Weg 39
8014 Neubiberg / München, W-Germany

Abstract. Stochastic approximation procedures, e.g. stochastic gradient methods, for solving stochastic optimization problems, can be accelerated considerably by using deterministic descent directions or more exact gradient estimations at certain iteration points. Several methods for the computation of more exact gradient estimators and deterministic (feasible) descent directions are presented.

1. Optimization problems with random parameters

Given an ordinary optimization problem represented by

$$\min g_o(x) \tag{1}$$

s.t.

$$g_i(x) \le 0,\ i=1,\dots,m_1 \tag{1.1}$$

$$g_i(x) = 0,\ i=m_1+1,\dots,m \tag{1.2}$$

$$x \in D, \tag{1.3}$$

where $g_o(x), g_1(x), \dots, g_m(x)$ are certain real valued functions on $\mathbb{R}^n$, and D is a given convex subset of $\mathbb{R}^n$, as e.g. the positive orthant $\mathbb{R}^n_+$ of $\mathbb{R}^n$, in practice the functions $g_i(x)$, i=0,1,...,m, depend not only on the decision n-vector x, but in most cases it holds

$$g_i = g_i(\omega,x),\ i=0,1,\dots,m, \tag{2}$$

where ω is a certain vector of parameters representing certain physical, chemical, biological, technological, economical,... quantities of the underlying practical situation. However, these parameters have mostly a random nature or, if some are deterministic, then they can be obtained only by means of a measuring process having a certain random error. Thus, in the following the parameter(vector) ω is considered as the realization of a random variable $\tilde{\omega}$ taking values ω in a certain measurable space $(\Omega,\mathcal{A})$. It is assumed in the following that the probability distribution P of $\tilde{\omega}$ is known. By "E" we denote the expectation operator with respect to $\tilde{\omega}$.

Since the realization ω of $\tilde{\omega}$ is not revealed - if at all - before a decision vector x is chosen, the optimization problem (1) has no more a well defined meaning. Thus, (1) must be replaced by an appropriate **substituting** problem.

Among several possible approaches, we mention here the following

R. Slowinski and J. Teghem (eds.),
Stochastic versus Fuzzy Approaches to Multiobjective Mathematical Programming under Uncertainty, 23–43.

two basic methods:

A) Penalty-function approach. Here it is assumed that the violation of the constraints (1.1),(1.2), i.e.

$$g_i(\omega,x) \le 0, \ i=1,\dots,m_1 \tag{1.1)'$$

$$g_i(\omega,x) = 0, \ i=m_1+1,\dots,m \tag{1.2)'$$

causes costs

$$C = u(G(\omega,x)), \tag{2}$$

where

$$G(\omega,x) := \begin{pmatrix} g_1(\omega,x) \\ g_2(\omega,x) \\ \vdots \\ g_m(\omega,x) \end{pmatrix} \tag{2.1}$$

and $u: \mathbb{R}^m \longrightarrow \mathbb{R}$ is a certain cost- or loss function on $\mathbb{R}^m$. The goal is then to select the decision vector $x \in D$ such that the mean total costs are minimized. Hence, we have the following **mean value minimization** problem

$$\min F(x) \quad \text{s.t.} \quad x \in D \tag{3}$$

where

$$F(x) = Ef(\omega,x) \tag{3.1}$$

and

$$f(\omega,x) = g_o(\omega,x) + u(G(\omega,x)), \tag{3.2}$$

cf. [2],[6],[10],[14],[25].

Stochastic linear programs. In this case the basic problem (1) is a linear program

$$\min c'x \tag{4}$$

s.t.

$$Tx = h \ (Tx \le h, \text{ resp.}) \tag{4.1}$$

$$x \in D \tag{4.2}$$

having random data $\xi(\omega) = (T(\omega),h(\omega),c(\omega))$, where D denotes a fixed convex polyhedron, e.g. $D = \mathbb{R}^n_+$. Here, the random function $f(\omega,x)$ reads then

$$f(\omega,x) = \Psi(\xi(\omega),x) = c(\omega)'x + u(T(\omega)x-h(\omega)), \tag{5}$$

where again $u=u(z)$ designates a certain convex function evaluating the violation of the constraint (4.1). Especially, in **stochastic linear programming with recourse** [10],[14],[24] the loss $u(z)$ for a given deviation $z = T(\omega)x-h(\omega)$ of the equality constraint $T(\omega)x=h(\omega)$ is given by the optimal value of the so-called second stage linear program (after the realization of ω)

$$\min q'y \tag{6}$$
$$\text{s.t. } Wy = -z \tag{6.1}$$
$$y \geq 0, \tag{6.2}$$

where (q,W) are given data, see [10].

If in the basic linear program (4) we have the inequality constraint $T(\omega)x \leq h(\omega)$, as e.g. in **air quality planning problems** [3], then an appropriate convex loss function $u(z)$ should have the monotonicity property

$$z \leq w \Longrightarrow u(z) \leq u(w). \tag{7}$$

Besides stochastic linear programming there are many other concrete problems leading to a mean value problem (3) having a random function $f(\omega,x)$ of the type (5). Among many others we mention:

Error minimization problems [26]: Statistical linearization, fitting (linear in parameter) parametric models, e.g. linear regression analysis, learning problems.

Portfolio optimization problems [27]: Also in this case we have a problem of the type (3),(5), where $m=1$.

B) Chance-constrained programming method. Assume that in (1) there are only the constraints (1.1) and (1.3), hence $m_1=m$. Here, only decisions $x \in D$ are taken into consideration such that the constraints

$$g_1(\omega,x) \leq 0,\ g_2(\omega,x) \leq 0, \ldots, g_m(\omega,x) \leq 0$$

hold true - jointly or separately - at least with probability α, probabilities α_i, $i=1,\ldots,m$, resp., where α, α_i, $i=1,\ldots,m$, are given reliability levels. In many cases the objective function are then the mean primary costs $Eg_o(\omega,x)$. Hence, working e.g. with separate chance-constraints, we have to solve

$$\min Eg_o(\omega,x) \tag{7}$$
s.t.
$$P(g_i(\omega,x) \leq 0) \geq \alpha_i,\ i=1,\ldots,m_1; \tag{7.1}$$
$$x \in D; \tag{7.2}$$

in the case of joint chance-constraints, (7.1) is replaced by

$$P(G(\omega,x) \le 0) \ge \alpha, \tag{7.1)'$$

see [10],[22],[23].

Note that some relationships between the penalty-function and the chance-constrained programming approach are known, see [10].

For stochastic linear programs (4), the question of the convexity of the constraint (7.1), i.e.

$$P(T_i(\omega)x \le h_i(\omega)) \ge \alpha_i,$$

where (T_i,h_i) is the i-th row of (T,h), is well studied, see [10],[21].

2. Numerical methods for solving mean value minimization problems

Considering in the following the penalty-function approach (A) only, we have to solve (approximatively) a mean value minimization problem of the type

$$\min F(x) \quad \text{s.t.} \quad x \in D, \tag{8}$$

where D is a convex subset of $\mathbb{R}^n$ and

$$F(x) = Ef(\xi(\omega),x) \tag{8.1}$$

with a d-random vector $\xi=\xi(\omega)$ on a probability space $(\Omega,\mathfrak{A},P)$ and a function $f\colon \mathbb{R}^d \times \mathbb{R}^n \longrightarrow \mathbb{R}$. We assume that $F(x)$ is sufficiently smooth and the derivatives ∇F, $\nabla^2 F,\ldots$ can be obtained by interchanging differentiation and expectation operations. Hence it is

$$\nabla^k F(x) = E\nabla_x^k f(\omega,x), \quad k=1,2,\ldots . \tag{8.2}$$

Considering potential algorithms for solving (8), the main difficulty is that F and its derivatives $\nabla F,\nabla^2 F,\ldots$ are defined by means of certain multiple integrals which can be computed, in general, only with large computational expenses.

Presently there are the following main approaches to find solutions x^* of (8):

I. Discretization methods
II. Stochastic approximation methods
III. Methods based on structural properties of (8): Construction of descent direction without using derivatives

2.1. DISCRETIZATION METHODS

In this case the random vector $\xi(\omega)$ is replaced [1],[11],[14],[25] by a random vector $\xi^N(\omega)$ taking only a finite number of different values

$\xi^{N,j}, j=1,\dots,r_N$. If $\Omega^{N,j} = \{\omega\in\Omega: \xi^N(\omega) = \xi^{N,j}\}$, then F(x) is replaced by

$$F^N(x) = \sum_{j=1}^{r_N} f(\xi^{N,j},x)P(\Omega^{N,j}), \tag{9}$$

hence, we obtain then the approximating problem

$$\min F^N(x) \quad \text{s.t. } x\in D. \tag{10}$$

In stochastic linear programming with recourse, cf. (5),(6), the approximation (10) of (8) can be represented by a linear program of special type, see [10]. Error bounds and strategies for partitioning Ω can be found in [7],[12],[13].

2.2. STOCHASTIC APPROXIMATION METHODS

The time-consuming computation of derivatives of F(x) can be omitted if (8) is solved by means of a stochastic approximation algorithm, e.g. by the following stochastic gradient procedure [4],[5],[21]

$$X_{k+1} = p_D(X_k-\rho_k Y_k),\ k=1,2,\dots, \tag{11}$$

with the simple stochastic gradient

$$Y_k = \nabla_x f(\omega_k,X_k). \tag{11.1}$$

Here, $\omega_1,\omega_2,\dots,\omega_k,\dots$ is a sequence of independent realizations of the random parameter ω, $\nabla_x f(\omega,x)$ is the gradient of the stochastic function $f(\omega,x)$ with respect to x, called a stochastic gradient of F(x), $\rho_k>0$, k=1,2,..., designates a sequence of step sizes satisfying the conditions

$$\sum_{k=1}^{\infty} \rho_k = +\infty,\quad \sum_{k=1}^{\infty} \rho_k^2 < +\infty \tag{11.2}$$

and p_D denotes the projection operator from $\mathbb{R}^n$ onto the feasible domain D of (8). Unfortunately, due to its stochastic nature, algorithms of the type (11) have [8] only a very slow asymptotic convergence rate of the type

$$E||X_k-x^*||^2 = O(k^{-\lambda}) \text{ or } E(F(X_k) - F^*) = O(k^{-\gamma})$$

with some constants $\lambda,\gamma, 0<\lambda,\gamma<1$, where x^* is an optimal solution of (8) and F^* denotes the minimal value of (8). However, the main disadvantage is the nonmonotonicity of (11) which may be displayed in a highly oscillatory behavior. Hence, in many cases one does not know whether the procedure has reached already a certain neighborhood of a minimal point x^* of (8) or not. Consequently, the initial velocity of (11) is very small in general.

Improvements of stochastic gradient procedures can be obtained by

I) replacing the negative simple stochastic gradient $-Y_k$ at certain iteration points X_k, $k \in N_1$, by improved step directions h_k,

II) step size control.

Using method (I) we get the semi-stochastic approximation algorithm

$$X_{k+1} = \begin{cases} p_D(X_k + \rho_k h_k) & \text{for } k \in N_1 \\ p_D(X_k - \rho_k Y_k) & \text{for } k \in N_2, \end{cases} \tag{12}$$

where N_1, N_2 is a certain partition of the set $\mathbb{N}$ of integers which can be selected in advance or which is determined during the optimization process. The transition $X_k \longrightarrow X_{k+1}$ is called a deterministic, a stochastic step, resp., if $k \in N_1, k \in N_2$, respectively.

Mathematical results concerning the improvement of the convergence behavior of (12) in comparison with (11) - under certain standard assumptions - can be found in [16],[17],[18],[20]. Moreover, there are also many numerical results verifying these theoretical results.

In the following we show how improved step directions can be obtained.

3. Generators of improved step directions

In the following the construction of more exact gradient estimations as well as the construction of (feasible) descent directions is discussed.

3.1. MORE EXACT GRADIENT ESTIMATORS

An obvious improvement of the simple stochastic gradient $Y_k = \nabla_x f(\omega_k, X_k)$ is the arithmetic mean

$$\hat{\nabla F}(X_k) := \frac{1}{L} \sum_{j=1}^{L} \nabla_x f(\omega_{k,j}, X_k) \tag{13}$$

of a certain number L of simple stochastic gradients. Here, $\omega_{k,1}, \ldots, \omega_{k,L}$ are L independent realizations of ω to be generated at stage k, see e.g. [8]. Note that the number L may be dependent on (k, X_k).

A further interesting method is based on a regression analysis of F(x), called RESPONSE SURFACE METHODOLOGY (RSM), see [2],[9]:

Consider a subdomain of D

$$a \in S \subset D \tag{14}$$

containing the point a at which an estimation $\hat{\nabla F}(a)$ of ∇F must be computed, e.g. $a=X_k$. In order to estimate first $F(x)$ on S, estimates y_i of the function value $F(x^{(i)})$, $i=1,2,\dots,r$, are determined at certain points $x^{(i)} \in S$, $i=1,\dots,r$, to be selected by the optimizer. Simple estimators y_i of $F(x^{(i)})$ are given by

$$y^{(i)} = f(\omega^{(i)}, x^{(i)}),\ i=1,\dots,r, \tag{15}$$

where $\omega^{(1)},\dots,\omega^{(r)}$ are again independent realizations of ω. The objective function F is then approximated on S by a so-called polynomial response surface model, where the polynomial is usually of first and/or second order. In **Phase 1** of RSM, i.e. if the process (X_k) is far away from the optimum x^* of (8), then F is estimated on S, cf. (14), by the **linear** empirical model.

$$\hat{F}(x) = \beta_o + \beta_I'(x-a), \tag{16}$$

where

$$\beta' = (\beta_o, \beta_I') = (\beta_o, \beta_1, \dots, \beta_n) \tag{16.1}$$

are the unknown coefficients of the linear model (16). Using the estimates (16) of $F(x^{(i)})$, $i=1,\dots,r$, by least squares estimation (LSQ), an estimation $\hat{\beta}$ of β is given by

$$\hat{\beta} = (W'W)^{-1}W'y, \tag{17}$$

where W, y, resp., is the r×n matrix, the r-vector

$$W = \begin{pmatrix} \tilde{d}^{(1)'} \\ \tilde{d}^{(2)'} \\ \vdots \\ \tilde{d}^{(r)'} \end{pmatrix},\ d^{(i)} := x^{(i)}-a,\ \tilde{x} = \begin{pmatrix} 1 \\ x \end{pmatrix},\ \ y = \begin{pmatrix} y^{(1)} \\ y^{(2)} \\ \vdots \\ y^{(r)} \end{pmatrix}, \tag{17.1}$$

and the regularity of W'W is presupposed. Having $\hat{\beta}$, the gradient ∇F can be estimated by the gradient of the model (16), hence

$$\hat{\nabla F}(x) = \hat{\beta}_I \text{ for all } x \in S. \tag{18}$$

The accuracy of the estimation (18) for the gradient of F at a point $x \in S$, cf. (14), can be studied as follows: Denoting by $\epsilon^{(i)} = \epsilon^{(i)}(\omega, x^{(i)})$, $i=1,\dots,r$, the stochastic error in the estimation $y^{(i)}$ of $F(x^{(i)})$, see (15), we obtain by means of the first order Taylor expansion

$$y^{(i)} = F(x^{(i)}) + \epsilon^{(i)} = F(a) + \nabla F(a)'(x^{(i)}-a) + \rho_1^{(i)}(x^{(i)}-a) + \epsilon_i, \tag{19}$$

where $\rho_1^{(i)}(x^{(i)}-a)$ is the first order remainder of the Taylor expansion. Equation (19) yields, see (17.1), the data model

$$y = W\beta + \rho_1 + \epsilon, \tag{20}$$

where $\beta_o = F(a)$, $\beta_I = \nabla F(a)$ and

$$\rho_1 = \begin{pmatrix} \rho_1^{(1)}(x^{(1)}-a) \\ \rho_1^{(2)}(x^{(2)}-a) \\ \vdots \\ \rho_1^{(r)}(x^{(r)}-a) \end{pmatrix}, \quad \epsilon = \begin{pmatrix} \epsilon^{(1)} \\ \epsilon^{(2)} \\ \vdots \\ \epsilon^{(r)} \end{pmatrix}, \tag{20.1}$$

hence, ϵ is the (stochastic) noise term, and ρ_1 is the deterministic residual arising from the first order Taylor expansion of F on S. Putting (20) into (17), for β we get

$$\hat{\beta} = \begin{matrix} F(a) \\ \nabla F(a) \end{matrix} + (W'W)^{-1}W'(\rho_1+\epsilon), \tag{21}$$

hence

$$\hat{\beta}_I = \nabla F(a) + \Pi(W'W)^{-1}W'(\rho_1+\epsilon), \tag{21)'$$

where Π is the projection of $\mathbb{R}\times\mathbb{R}^n$ onto $\mathbb{R}^n$. Obviously, representation (21)' yields the accuracy of (18) as an estimator of $\nabla F(a)$. If $a=X_k$ is still far away from the minimal point x^* of (8), then in many cases the objective function F behaves almost linearly, hence,

$$e_{det} = (\approx)0 \text{ in the (almost) linearity range of F.}$$

If $e_{det} = 0$, then the empirical model (16) and the underlying mechanistic (linear) model coincide [2].

In **PHASE 2** of RSM, i.e. if the process (X_k) is approaching a certain neighborhood of a minimal point x^* of (8), then the linear model (16) of F on S is no more adequate and is replaced therefore by the more general **quadratic** empirical model

$$\hat{F}(x) = \beta_o + \beta_I'(x-a) + (x-a)'B(x-a), \tag{22}$$

where the parameters $\beta_o,\beta_i,i=1,\dots,n$, in the n-vector β_I, and β_{ij}, $i,j=1,\dots,n$, in the symmetric n×n matrix B, are estimated again by the LSQ method. The estimation $\hat{\beta}$ of the parameter $\frac{1}{2}(n^2+3n+2)$-vector

$$\beta' = (\beta_o,\beta_1,\dots,\beta_n,\beta_{12},\dots,\beta_{1n},\beta_{22},\dots,\beta_{2n},\dots,\beta_{nn}) \tag{23}$$

is given again by the above formulae (17),(17.1), provided that $\tilde{x}$ is defined for a given $x \in \mathbb{R}^n$ by

$$\tilde{x}=(1,x_1,\ldots,x_n,x_1^2,2x_1x_2,\ldots,2x_1x_n,x_2^2,2x_2x_3,\ldots,2x_2x_n,\ldots,x_n^2)'. \quad (24)$$

Having $\hat{\beta}=(\hat{\beta}_o,\hat{\beta}_I,\hat{B})$, the gradient of F on S is then estimated by the gradient of the quadratic model (22), hence

$$\hat{\nabla F}(x) = \hat{\beta}_I+2\hat{B}(x-a) \text{ for } x \in S \quad (24)$$

and therefore

$$\hat{\nabla F}(a) = \hat{\beta}_I.$$

Using second order Taylor expansion, corresponding to (19) for the observations we have

$$y^{(i)} = F(x^{(i)})+\epsilon^{(i)} = F(a)+\nabla F(a)'(x^{(i)}-a) + \frac{1}{2}(x^{(i)}-a)'\nabla^2F(a)(x^{(i)}-a) + \rho_2^{(i)}(x^{(i)}-a) + \epsilon^{(i)}, \quad (25)$$

where $\rho_2^{(i)}(x^{(i)}-a)$ is the remainder of this second order Taylor expansion. Thus, (25) yields the data model

$$y = W\beta + \rho_2 + \epsilon, \quad (26)$$

where $\beta_o=F(a)$, $\beta_I=\nabla F(a)$, $B=\frac{1}{2}\nabla^2F(a)$ and ρ_2, ϵ are defined corresponding to (20.1). Putting (26) into the estimation formula (17), in the present case we get, cf. (23),

$$\hat{\beta} = \begin{pmatrix} F(a) \\ \nabla F(a) \\ \frac{1}{2}\nabla^2F(a) \end{pmatrix} + (W'W)^{-1}W'(\rho_2+\epsilon), \quad (27)$$

hence,

$$\hat{\beta}_I = \nabla F(a) + e_{det} + e_{stoch}. \quad (27)'$$

Obviously,

$$e_{det} = (\approx)0 \quad \text{if F is a (nearly) quadratic function.}$$

For $e_{det} = 0$ the empirical model (22) and the underlying (quadratic) mechanistic model coincide. The covariance matrix V of the stochastic estimation error of β is given - for Phase 1 as well as Phase 2 - by

$$V = \mathrm{var}((W'W)^{-1}W'\epsilon) = (W'W)^{-1}W'QW(W'W)^{-1}, \quad (28)$$

where Q is the covariance matrix of the stochastic error term ϵ in

(26). Assuming that $y^{(1)},\ldots,y^{(r)}$ are stochastically independent observations of $F(x^{(i)})$, $i=1,\ldots,r$, we have that

$$Q = (\sigma_{ii}^2\ \delta_{ij})_{i,j=1,\ldots,r}, \tag{28.1}$$

where

$$\sigma_{ii}^2 = \operatorname{var}(\epsilon^{(i)}) = E(y^{(i)}-F(x^{(i)}))^2 \tag{28.2}$$

is the variance of the observation error. Considering the norm of V, for a vector $\eta \in \mathbb{R}^{n+1}$, $\eta \in \mathbb{R}^{\frac{1}{2}(n^2+3n+2)}$, resp., we find

$$\eta' V \eta = \lambda' Q \lambda = \sum_{j=1}^{r} \sigma_{jj}^2\ \lambda_j^2$$

and therefore

$$(\min_{1\le j\le r} \sigma_{jj}^2)||\lambda||^2 \le \eta' V \eta \le (\max_{1\le j\le r} \sigma_{jj}^2)||\lambda||^2, \tag{29.1}$$

where

$$\lambda = W(W'W)^{-1}\eta. \tag{29.2}$$

From (29.2) we get

$$||\lambda||^2 = \lambda'\lambda = \eta'(W'W)^{-1}W'(W(W'W)^{-1}\eta)$$
$$= \eta'(W'W)^{-1}(W'W)(W'W)^{-1}\eta = \eta'(W'W)^{-1}\eta,$$

hence, (29.1) yields

$$(\min_{1\le j\le r} \sigma_{jj}^2)\eta'(W'W)^{-1}\eta \le \eta' V \eta \le (\max_{1\le j\le r} \sigma_{jj}^2)\eta'(W'W)^{-1}\eta$$

and therefore

$$(\min_{1\le j\le r} \sigma_{jj}^2)||(W'W)^{-1}|| \le ||V|| \le (\max_{1\le j\le r} \sigma_{jj}^2)||(W'W)^{-1}||.$$

This yields

$$(\min_{1\le j\le r} \sigma_{jj}^2)\frac{1}{||W'W||} \le ||V|| \le (\max_{1\le j\le r} \sigma_{jj}^2)\frac{1}{\min\limits_{||\eta||=1} \eta'(W'W)\eta},$$

where

$$\frac{1}{||W'W||} \geq \frac{1}{||W||^2},$$

$$\min_{||\eta||=1} \eta'(W'W)\eta = \min_{||\eta||=1} ||W\eta||^2 = (\min_{||\eta||=1} ||W\eta||)^2.$$

$$= \min_{||\eta||=1} \sum_{j=1}^{r} ((\tilde{x}^{(j)}-\tilde{a})'\eta)^2$$

$$= \min_{||\eta||=1} \eta' \left(\sum_{j=1}^{r} (\tilde{x}^{(j)}-\tilde{a})(\tilde{x}^{(j)}-\tilde{a})' \right) \eta$$

Thus, in most practical cases it is

$$\min_{||\eta||=1} ||W\eta||^2 \longrightarrow +\infty \text{ for } r \longrightarrow +\infty, \tag{30}$$

hence $||V|| \longrightarrow 0$ for $r \longrightarrow +\infty$, provided that the error variances σ_{ii} are bounded from above.

3.2. CONSTRUCTION OF (FEASIBLE) DESCENT DIRECTIONS OF F

In the following we assume that the random function $f(\omega,x)$ is given by

$$f(\omega,x) = \bar{c}'x + u(T(\omega)x-h(\omega)), \tag{31}$$

where c is a given n-vector, $u: \mathbf{R}^m \longrightarrow \mathbf{R}$ is a convex loss function and $(T(\omega),h(\omega))$ is a random (m,n+1)-matrix with a given probability distribution. As was mentioned in section 1, there are many concrete situations in which a mean value minimization problem (8) with a random function of the type (31) must be solved!

The construction of descent directions for F at a given point x depends on the properties of the loss function u and the probability distribution of the random matrix $(T(\omega),h(\omega))$.

3.2.1 $(T(\omega),h(\omega))$ has a normal distribution. Let denote $(\bar{T},\bar{h})$ the mean and

$$Q = \begin{pmatrix} Q_{11} & Q_{12} & \cdots & Q_{1m} \\ Q_{21} & Q_{22} & \cdots & Q_{2m} \\ \vdots & \vdots & & \vdots \\ Q_{m1} & Q_{m2} & \cdots & Q_{mm} \end{pmatrix} \tag{32}$$

the $m(n+1)\times m(n+1)$ covariance matrix of $(T(\omega),h(\omega))$. Hence, Q_{ij} is the covariance matrix of the i-th and j-th row of $(T(\omega),h(\omega))$. Thus, it holds

$$Q_{ji} = Q_{ij}', \quad i,j=1,\ldots,m.$$

Consequently, $T(\omega)x-h(\omega)$ is a normal distributed random m-vector with mean $\bar{T}x-\bar{h}$ and covariance matrix Q_x given by

$$Q_x = \begin{pmatrix} \hat{x}'Q_{11}\hat{x} & \hat{x}'Q_{12}\hat{x} & \cdots & \hat{x}'Q_{1m}\hat{x} \\ \hat{x}'Q_{21}\hat{x} & \hat{x}'Q_{22}\hat{x} & \cdots & \hat{x}'Q_{2m}\hat{x} \\ \vdots & \vdots & & \vdots \\ \hat{x}'Q_{m1}\hat{x} & \hat{x}'Q_{m2}\hat{x} & \cdots & \hat{x}'Q_{mm}\hat{x} \end{pmatrix},$$

where $\hat{x} = \binom{x}{-1}$.

According to [15] we have this result.

<u>Theorem 3.1.</u> Suppose that the data $(\bar{T},\bar{h})$, Q and the loss function u are selected such that the function $x \longrightarrow Eu(T(\omega)x-h(\omega))$ is not constant on arbitrary line segments. If n-vectors x, $y \neq x$, are related such that

$$c'x \geq c'y \tag{33.1}$$
$$\bar{T}x = \bar{T}y \tag{33.2}$$
$$Q_x - Q_y \text{ is positive semidefinite,} \tag{33.3}$$

then $F(x) \geq F(y)$ and $h=y-x$ is a descent direction for F at x.

Whereas (33.1),(33.2) are simple linear relations, condition (33.3) has to be considered in more detail:

Using Gershgorin's circle theorem, we obtain the following general sufficient condition for (33.3)

$$\hat{x}'Q_{ii}\hat{x} - \hat{y}'Q_{ii}\hat{y} \geq \sum_{\substack{k=1 \\ k\neq i}}^{m} \left|\hat{x}'Q_{ik}\hat{x} - \hat{y}'Q_{ik}\hat{y}\right| \text{ for all } i=1,2,\ldots,m. \tag{34}$$

There are many other sufficient criterions for (33.3). Denoting by $A=(a_{ik})$ the symmetric matrix $A=Q_x-Q_y$, we mention still the following ones:

a) If the system of linear inequalities

$$\begin{pmatrix} a_{11}, & -|a_{12}|, & \ldots, & -|a_{1m}| \\ -|a_{21}|, & a_{22}, & \ldots, & -|a_{2m}| \\ \vdots & \vdots & \ddots & \vdots \\ -|a_{m1}|, & -|a_{m2}|, & \ldots, & a_{mm} \end{pmatrix} \begin{pmatrix} d_1 \\ d_2 \\ \vdots \\ d_m \end{pmatrix} \underset{(=)}{>} 0 \tag{35.1}$$

$$d_1>0, \ \ldots, \ d_m>0 \tag{35.2}$$

has a solution $d = (d_1,\ldots,d_m)'$, then A is positive (semi-)definite.

a1) If the system of linear inequalities

$$\begin{pmatrix} a_{11}, -|a_{12}|, & \dots, & -|a_{1m}| \\ -|a_{21}|, a_{22}, & \dots, & -|a_{2m}| \\ \vdots & \ddots & \vdots \\ -|a_{m1}|, -|a_{m2}|, & \dots, & a_{mm} \end{pmatrix} d \geq \begin{pmatrix} 1 \\ 1 \\ \vdots \\ 1 \end{pmatrix} \tag{36.1}$$

$$d \geq 0 \tag{36.2}$$

has a feasible solution d, then A is positive definite. Note that system (36) can be handeled by means of **linear programming**! A corresponding objective function is

$$Q(d): = \gamma' d \longrightarrow \min, \tag{36.3}$$

where γ is any fixed m-vector having positive components $\gamma_1>0,\dots,\gamma_m>0$.

b) Let

$$\chi_A(t) = \det(A-tI) = (-t)^m + S_1(-t)^{m-1} + S_2(-t)^{m-2} + \dots + S_m \tag{37}$$

denote the characteristic polynomial of the matrix A. Then the following criterion holds:

Lemma 3.1. A is positive (semi-) definite if and only if

$S_k > (=)\ 0$ for all $k=1,2,\dots,m$.

c) The coefficients S_k of (37) can be computed easily for matrices $A=Q_x-Q_y$ with a special structure which is implied by a special structure of the covariance matrix Q of the random matrix $(T(\omega),h(\omega))$.

An important example is represented by the following class of **tridiagonal matrices**

$$A = \begin{pmatrix} a_1 & b_2 & 0 & \dots & 0 & 0 \\ b_2 & a_2 & b_3 & \dots & 0 & 0 \\ 0 & b_3 & a_3 & \dots & 0 & 0 \\ \vdots & \vdots & \ddots & \ddots & \vdots & \vdots \\ & & & \ddots & & b_m \\ 0 & 0 & 0 & \dots & b_m & a_m \end{pmatrix}. \tag{38}$$

Obviously, this structure of $A=Q_x-Q_y$ occurs if the covariance matrix Q is **block-tridiagonal**, i.e. if

$$Q_{ij} = 0 \text{ for all } i,j \text{ such that } |j-i| \geq 2. \tag{38'}$$

In this case the coefficients S_k, $k=1,\dots,m$, of (37) can be obtained easily by the following two-stage recursion:

$$s_\ell^k = s_\ell^{k-1} + a_k\, s_{\ell-1}^{k-1} - b_k^2\, s_{\ell-2}^{k-2}, \qquad (39.1)$$

$$\ell=1,\ldots,k,\ k=2,\ldots,m,$$

where

$$s_o^o := s_o^k := 1,\ s_{-1}^k := s_{k+1}^k := 0,\ s_1^1 := a_1. \qquad (39.2)$$

d) By m-2 Householder-transformations of the type

$$A \longrightarrow S'AS \text{ with } S = I - \frac{2}{||a||2}\, aa',$$

and a certain vector $a \in \mathbb{R}^m$, each symmetric matrix A can be converted into a tridiagonal matrix. About $\frac{2}{3}m^3$ multiplications are needed for these transformations.

e) A further interesting case arise if $A=Q_x-Q_y$ is the **cyclic** symmetric matrix

$$A = \begin{pmatrix} a_o & a_1 & a_2 & \cdots & a_{m-1} \\ a_{m-1} & a_o & a_1 & \cdots & a_{m-1} \\ a_{m-2} & a_{m-1} & a_o & \cdots & a_{m-3} \\ \vdots & \vdots & \vdots & \ddots & \vdots \\ a_1 & a_2 & a_3 & \cdots & a_o \end{pmatrix}, \qquad (40)$$

where $a_1=a_{m-1}$, $a_2=a_{m-2}$, ... This situation occurs if the covariance matrix Q has the special structure

$$Q_{ii} = Q^{(o)} \text{ for all } i=1,\ldots,m \qquad (41.1)$$

$$Q_{ij} = Q^{(1)},\ \text{if } |j-i| = 1 \qquad (41.2)$$

$$Q_{ij} = Q^{(2)},\ \text{if } |j-i| = 2 \qquad (41.3)$$

$$\vdots$$

$$Q_{ij} = Q^{(m-1)},\ \text{if } |j-i| = m-1, \qquad (41.4)$$

where

$$Q^{(1)} = Q^{(m-1)},\ Q^{(2)} = Q^{(m-2)},\ldots,\ Q^{(r)} = Q^{(r+1)},\ r = [\tfrac{m-1}{2}]. \qquad (41.5)$$

In this case the (real, multiple) eigenvalues of A are given by

$$\lambda_j = \mathrm{Re}\{g(\epsilon^j)\},\ j=0,1,\ldots,\ [\tfrac{m}{2}], \qquad (42.1)$$

where

$$g(t) = a_o + 2a_1t + \ldots + 2a_rt^r + a_{r+1}t^{r+1} \qquad (42.2)$$

$$\text{with } a_{r+1} = \begin{cases} a_{[\frac{m}{2}]}, & m \text{ even} \\ 0, & m \text{ odd} \end{cases},$$

and

$$\epsilon = \exp(i\frac{2\pi}{m}), \quad i=\sqrt{-1}. \tag{42.3}$$

Thus, A is positive (semi-) definite if and only if

$$\mathrm{Re}(g(\epsilon^j)) >(\geq) 0 \text{ for all } j=1,\ldots,[\tfrac{m}{2}]. \tag{43}$$

3.2.2. Elliptically contoured distributions. Here we suppose that $(T(\omega),h(\omega))$ has an elliptically contoured distribution, cf. [19]. Among many other distributions, the multivariate normal, Cauchy, stable, Student t-, inverted Student t-distribution as also their truncations to elliptically contoured sets in $\mathbb{R}^{m(n+1)}$ and the uniform distributions on elliptically contoured sets in $\mathbb{R}^{m(n+1)}$ are members of the family of elliptically contoured distributions. The parameters of this class of distributions can be collected in an $m\times(n+1)$ location matrix

$$\Xi = (\Theta,\theta) \tag{44.1}$$

and a positive semidefinite symmetric $m(n+1)\times m(n+1)$ scale matrix

$$Q = (Q_{ij})_{i,j=1,\ldots,m} \tag{44.2}$$

having $(n+1)\times(n+1)$ submatrices Q_{ij}, $i,j=1,\ldots,m$, such that

$$Q_{ji} = Q_{ij}', \quad i,j=1,2,\ldots,m. \tag{44.3}$$

Using results of [19], we have this consequence:

<u>Theorem 3.2.</u> Suppose that the data Ξ,Q and the loss function u are selected such that the function $x \longrightarrow Eu(T(\omega)x-h(\omega))$ is not constant on arbitrary line segments. If the n-vectors $x,y\neq x$ are related such that

$$c'x \geq c'y \tag{45.1}$$
$$\Theta x = \Theta y \tag{45.2}$$
$$\hat{x}'\frac{Q_{ij}+Q_{ji}}{2}\hat{x} = \hat{y}'\frac{Q_{ij}+Q_{ji}}{2}\hat{y} \text{ for all } i\leq j, \tag{45.3}$$

then $F(x) \geq F(y)$ and $h=y-x$ is a descent direction for F at x.

Obviously, for a given vector x, (45) contains at most m+1, $\frac{1}{2}m(m+1)$, resp., linear, quadratic conditions, respectively.

3.2.3. Quadratic approximation of the loss function u. Suppose that there exists a closed convex subset $Z_o \subset \mathbb{R}^m$ such that the loss function u fulfills the relations

$$d'Vd \le d'\nabla^2 u(z)d \le d'Wd \text{ for all } z\in Z_o \text{ and } d\in \mathbb{R}^m, \tag{46.1}$$

where V,W are given fixed symmetric m×m matrices. Moreover, suppose that

$$T(\omega)x-h(\omega) \in Z_o \text{ for all } x\in D \text{ and } \omega\in\Omega. \tag{46.2}$$

Let $(\bar{T},\bar{h}) = E(T(\omega),h(\omega))$ be the mean of $(T(\omega),h(\omega))$, and let

$$\begin{aligned}\Sigma_{ij} &= \text{cov}((T_i(\cdot),h_i(\cdot)),(T_j(\cdot),h_j(\cdot))) \\ &= E((T_i(\omega),h_i(\omega))-(\bar{T}_i,\bar{h}_i))'((T_j(\omega),h_j(\omega))-(\bar{T}_j,\bar{h}_j))\end{aligned} \tag{47}$$

denote the covariance matrix of the i-th and j-th row $(T_i(\cdot),h_i(\cdot))$, $(T_j(\cdot),h_j(\cdot))$, resp., of $(T(\omega),h(\omega))$. For any symmetric m×m matrix $U = (u_{ij})$ we define then the symmetric (n+1)×(n+1) matrix $\hat{U}$ by

$$\begin{aligned}\hat{U} &= E(T(\omega)-\bar{T},h(\omega)-\bar{h})'U(T(\omega)-\bar{T},h(\omega)-\bar{h}) \\ &= \sum_{i,j=1}^{m} u_{ij}\,\Sigma_{ij}.\end{aligned} \tag{48}$$

For a given n-vector $x\in D$ we consider now the Taylor expansion of u at

$$\bar{z} = \bar{T}x - \bar{h} = E(T(\omega)x-h(\omega)), \tag{49}$$

hence

$$u(z) = u(\bar{z}) + \nabla u(\bar{z})'(z-\bar{z}) + \frac{1}{2}(z-\bar{z})'\nabla^2 u(w)(z-\bar{z}),$$

where $w=\bar{z}+\vartheta(z-\bar{z})$ for some $0<\vartheta<1$. Because of (46.2) it is $\bar{z} \in Z_o$ for all $x\in D$ and therefore $w \in Z_o$ for all $z \in Z_o$. Hence, (46.1) yields

$$\begin{aligned}u(\bar{z}) + \nabla u(\bar{z})'(z-\bar{z}) + \frac{1}{2}(z-\bar{z})'V(z-\bar{z}) &\le u(z) \le \\ \le u(\bar{z}) + \nabla u(\bar{z})'(z-\bar{z}) &+ \frac{1}{2}(z-\bar{z})'W(z-\bar{z}) \\ \text{for all } x \in D \text{ and } z \in Z_o.\end{aligned} \tag{50}$$

Putting $z:= T(\omega)x-h(\omega)$ into (50), because of (46.2) we find

$$u(\bar{z}) + \nabla u(\bar{z})'(T(\omega)x-h(\omega)-\bar{z}) + \frac{1}{2}(T(\omega)x-h(\omega)-\bar{z})'V(T(\omega)x-h(\omega)-\bar{z})$$
$$\le u(T(\omega)x-h(\omega)) \le \tag{51}$$
$$\le u(\bar{z}) + \nabla u(\bar{z})'(T(\omega)x-h(\omega)-\bar{z}) + \frac{1}{2}(T(\omega)x-h(\omega)-\bar{z})'W(T(\omega)x-h(\omega)-\bar{z})$$

for all $x\in D$ and all $\omega\in\Omega$.

Taking now expectations on both sides of (51), we get

$$u(\bar{z}) + \frac{1}{2}\hat{x}'\hat{V}\hat{x} \le Eu(T(\omega)x-h(\omega)) \le u(\bar{z}) + \frac{1}{2}\hat{x}'\hat{W}\hat{x} \tag{52.1}$$

for all $x\in D$.

Furthermore, putting $z := \bar{T}y-\bar{h}$, where $y\in D$, into (50), we find

$$u(\bar{T}y-\bar{h}) \le u(\bar{z}) + \nabla u(\bar{z})'\bar{T}(y-x) + \frac{1}{2}(y-x)'\bar{T}'W\bar{T}(y-x) \tag{52.2}$$
$$= u(\bar{z}) + (\bar{T}'\nabla u(\bar{z}))'(y-x) + \frac{1}{2}(y-x)'(\bar{T}'W\bar{T})(y-x)$$

for all $x\in D$ and $y\in D$.

Considering now $F(y)-F(x)$ for vectors $x,y\in D$, from (52.1) we get

$$F(y)-F(x) = (\bar{c}'y+Eu(T(\omega)y-h(\omega))) - (\bar{c}'x+Eu(T(\omega)x-h(\omega)))$$
$$\le \bar{c}'(y-x) + \frac{1}{2}\hat{y}'\hat{W}\hat{y} + u(\bar{T}y-\bar{h}) - (\frac{1}{2}\hat{x}'\hat{V}\hat{x}+u(\bar{T}x-\bar{h})),$$

hence we have the following result:

<u>Theorem 3.3.</u> If D, $(T(\omega),h(\omega))$, Z_o and V,W are selected such that (46.1),(46.2) hold, then for all $x,y\in D$ it is

$$F(y)-F(x) \le \bar{c}'(y-x) + \frac{1}{2}(\hat{y}'\hat{W}\hat{y}-\hat{x}'\hat{V}\hat{x}) + u(\bar{T}y-\bar{h}) - u(\bar{T}x-\bar{h}). \tag{53}$$

Using also inequality (52.2), we get this corollary:

<u>Corollary 3.1.</u> Under the assumption of Theorem 3.3 it holds

$$F(y) - F(x) \le \bar{c}'(y-x) + \frac{1}{2}(\hat{y}'\hat{W}\hat{y}-\hat{x}'\hat{V}\hat{x}) +$$
$$+ (\bar{T}'\nabla u(\bar{A}x-\bar{b}))'(y-x) + \frac{1}{2}(y-x)'\bar{T}'W\bar{T}(y-x). \tag{53.1}$$

Representing the $(n+1)\times(n+1)$ matrix $\hat{W}$ by

$$\hat{W} = \begin{pmatrix} \tilde{W} & \vdots & w \\ \cdots & \vdots & \cdots \\ w' & \vdots & \gamma \end{pmatrix}, \tag{53.2}$$

where $\tilde{W}$ is a symmetric $n\times n$ matrix, w is an n-vector and γ is a real number, then

$$\hat{y}'\hat{W}\hat{y} = \hat{x}'\hat{W}\hat{x} + (y-x)'\tilde{W}(y-x) + 2(\tilde{W}x-w)'(y-x)$$

and therefore

$$\begin{aligned} F(y)-F(x) &\le \bar{c}'(y-x) + \frac{1}{2}\hat{x}'(\hat{W}-\hat{V})\hat{x} + \frac{1}{2}(y-x)'(\tilde{W}+\bar{T}'W\bar{T})(y-x) + \\ &\quad + (\bar{T}'\nabla u(\bar{A}x-\bar{b}) + \tilde{W}x-w)'(y-x) \\ &= \frac{1}{2}\hat{x}'(\hat{W}-\hat{V})\hat{x} + (\bar{c}+\bar{T}'\nabla u(\bar{T}x-\bar{h}) + \tilde{W}x-w)'(y-x) + \\ &\quad + \frac{1}{2}(y-x)'(\tilde{W}+\bar{T}'W\bar{T})(y-x). \end{aligned} \tag{53.3}$$

Thus, $F(x+(y-x))-F(x)$ is estimated from above by a quadratic form in $h=y-x$.

Concerning the construction of descent directions of F at x we have this first immediate consequence from Theorem 3.3:

Theorem 3.4.1. Suppose that the assumptions of Theorem 3.3 hold. If vectors $x,y\in D, y\neq x$, are related such that

$$\bar{c}'x \ge (>) \bar{c}'y \tag{54.1}$$

$$\bar{T}x = \bar{T}y \tag{54.2}$$

$$\hat{x}'\hat{V}\hat{x} \ge (>) \hat{y}'\hat{W}\hat{y}, \tag{54.3}$$

then $F(x)\ge(>)F(y)$ and $h=y-x$ is a feasible descent direction of F at x.

Note. If the loss function u is **monotone** with respect to the partial order on $\mathbb{R}^m$ induced by $\mathbb{R}^m_+$, then (54.2) can be replaced by one of the weaker relations

$$\bar{T}x \ge \bar{T}y,\ \bar{T}x \le \bar{T}y, \text{ respectively.} \tag{54.2'}$$

The next result follows from (53.3).

Theorem 3.4.2. Suppose that the assumptions of Theorem 3.3 hold. If vectors $x,y\in D, y\neq x$, are related such that

$$(\tilde{W}+\bar{T}'W\bar{T})(y-x) = -(\bar{c}+\bar{T}'\nabla u(\bar{T}x-\bar{h}) + \tilde{W}x-w) \tag{55.1}$$

$$(\bar{c}+\bar{T}'\nabla u(\bar{T}x-\bar{h}) + \tilde{W}x-w)'(y-x) \le (<) -\hat{x}'(\hat{W}-\hat{V})\hat{x}, \tag{55.2}$$

then $F(y)\le(<)F(x)$ and $h=y-x$ is a feasible descent direction for F at x.

Proof. If (55) holds, then (53.3) yields

$$F(y)-F(x) \le \frac{1}{2}\hat{x}'(\hat{W}-\hat{V})\hat{x} + (\bar{c}+\bar{T}'\nabla u(\bar{T}x-\bar{h}) + \tilde{W}x-w)'(y-x) +$$
$$+ \frac{1}{2}(y-x)'(\tilde{W} + \bar{T}'W\bar{T})(y-x) \le$$
$$\le \frac{1}{2}\hat{x}'(\hat{W}-\hat{V})\hat{x} + (\bar{c}+\bar{T}'\nabla u(\bar{T}x-\bar{h}) + \tilde{W}x-w)'(y-x) +$$
$$+ (-1)\frac{1}{2}(y-x)'(\bar{c}+\bar{T}'\nabla u(\bar{T}x-\bar{h}) + \tilde{W}x-w)$$
$$= \frac{1}{2}(\hat{x}'(\hat{W}-\hat{V})\hat{x} + (\bar{c}+\bar{T}'\nabla u(\bar{T}x-\bar{h}) + \tilde{W}x-w)'(y-x)) \le (<) 0.$$

Discussion of (55): Suppose that

$$C = \tilde{W} + \bar{T}'W\bar{T} \text{ is positive definite,} \tag{56.1}$$

hence, there are positive numbers α,β such that

$$\alpha||x||^2 \le x'Cx \le \beta||x||^2. \tag{56.2}$$

According to (55.1) it is

$$y-x = -C^{-1}(\bar{c}+\bar{T}'\nabla u(\bar{T}x-\bar{h}) + \tilde{W}x-w),$$

and (55.2) reads

$$(\bar{c}+\bar{T}'\nabla u(\bar{T}x-\bar{h})+\tilde{W}x-w)'C^{-1}(\bar{c}+\bar{T}'\nabla u(\bar{T}x-\bar{h})+\tilde{W}x-w) \ge (>) \hat{x}'(\hat{W}-\hat{V})\hat{x}$$

or

$$a'C^{-1}a \ge \hat{x}'(\hat{W}-\hat{V})\hat{x},$$

where

$$a = \bar{c} + \bar{T}'\nabla u(\bar{T}x-\bar{h}) + \tilde{W}x - w.$$

Because of (56.2) it is

$$\frac{1}{\beta}||a||^2 \le a'C^{-1}a \le \frac{1}{\alpha}||a||^2.$$

Thus, we find the following criterion for (55.2):

i) If $\frac{1}{\beta}||\bar{c}+\bar{T}'\nabla u(\bar{T}x+\bar{h}) + \tilde{W}x-w||^2 \ge (>) \hat{x}'(\hat{W}-\hat{V})\hat{x}$, then (55.2) holds.

ii) If $\hat{x}'(\hat{W}-\hat{V})\hat{x} > \frac{1}{\alpha}||\bar{c}+\bar{T}'\nabla u(\bar{T}x-\bar{h}) + \tilde{W}x-w||^2$, then (55.2) is not valid.

References

[1] Birge, J.R., Wets. R.J-B.: Designing approximation schemes for stochastic optimization problems, in particular, for stochastic programs with recourse. Math. Programming Study 27, 54-102 (1986)

[2] Box, G.E., Draper, N.R.: Empirical Model-Building And Response Surfaces. New York-Chichester-Brisbane-Toronto-Singapore: J. Wiley 1987

[3] Ellis, J.H., Mc Bean, E.A., Farquhar, G.J.: Chance-constrained/ stochastic linear programming model for acid rain abatement - I. Complete colinearity and noncolinearity. Atmospheric Environment 19, 925-937 (1985)

[4] Ermoliev, Yu.: On the stochastic quasi-gradient method and stochastic quasi-Feyer sequences. Cybernetics transl. of Kibernetika 5, 208-220 (1969)

[5] Ermoliev, Yu.: Stochastic quasigradient methods and their applications to systems optimization. Stochastics 9, 1-36 (1983)

[6] Ermoliev, Yu., Wets, R.J-B. (eds.): Numerical Techniques for Stochastic Optimization. Berlin-Heidelberg-New York: Springer-Verlag 1988

[7] Frauendorfer, K., Kall, P.: A Solution Method For SLP Recourse Problems With Arbitrary Multivariate Distributions - The Independent Case. Probl. of Control and Information Theory 17(4), 177-205 (1988)

[8] Gaivoronski, A.: Implementation of Stochastic Quasigradient Methods. In: [6], pp. 313-351

[9] Jacobsen, Sh.H., Schruben, L.W.: Techniques For Simulation Response Optimization. Operations Research Letters 8, 1-9 (1989)

[10] Kall, P.: Stochastic Linear Programming. Econometrics and Operations Research, Vol. XXI. Berlin-Heidelberg-New York: Springer-Verlag 1976

[11] Kall, P.: Computational Methods for Solving Two-stage Stochastic Linear Programming Problems. ZAMP 30, 261-271 (1979)

[12] Kall, P., Stoyan, D.: Solving Stochastic Programming Problems with Recourse, Including Error Bounds. Math. Operationsforschung Statist. Ser. Optim. 13, 341-447 (1982)

[13] Kall, P.: Stochastic Programs with Recourse: An Upper Bound and the Related Moment Problem. ZOR 31, A119-A141 (1987)

[14] Marti, K.: Approximationen stochastischer Optimierungsprobleme. Math. Systems in Economics, Vol. 43. Königstein/Ts.: A Hain 1979

[15] Marti, K., Fuchs, E.: Computation Of Descent Directions And Efficient Points In Stochastic Optimization Problems Without Using Derivatives. Math. Programming Study 28, 132-156 (1986)

[16] Marti, K., Fuchs, E.: Rates of Convergence of Semi-Stochastic Approximation Procedures for Solving Stochastic Optimization Problems. Optimization 17, 243-265 (1986)

[17] Marti, K.: On accelerations of stochastic gradient methods by using more exact gradient estimations. Methods of Operations Research 53, 327-336 (1986)

[18] Marti, K.: Optimally Controlled Semi-Stochastic Approximation Procedures. In: Opitz/Rauhut (Hrsg.): Ökonomie und Mathematik, S. 216-230. Berlin-Heidelberg: Springer-Verlag 1987

[19] Marti, K.: Solving Stochastic Optimization Problems By Semi-Stochastic Approximation Procedures. In: Borne/Tzafestas (eds.): Applied Modelling and Simulation of Technological Systems, pp. 559-568. Amsterdam-New York-Oxford-Tokyo: Elsevier Science Publ. B.V. (North Holland) 1987

20] Marti, K., Plöchinger, E.: Optimal Step Sizes In Semi-Stochastic Approximation Procedures. To appear in Optimization 1990

[21] Pflug, G.Ch.: Stepsize Rules, Stopping Times and their Implementation in Stochastic Quasigradient Algorithms. In: [6], pp. 352-372

[22] Prekopa, A.: Logarithmic concave measures with applications to stochastic programming. Acta Sci. Math. 32, 301-316 (1971)

[23] Prekopa, A.: Programming under probabilistic constraints with a random technology matrix. Math. Operationsforsch. Statist. 5, 109-116 (1974)

[24] Wets, R.: Stochastic programs with fixed recourse: the equivalent deterministic program. SIAM Review 16, 309-339 (1974)

[25] Wets, R.J-B.: Stochastic Programming: Solution Techniques and Approximation Schemes. In: Bachem/Grötschel/Korte (eds.): Mathematical Programming - The State of the Art, pp. 566-603. Berlin-Heidelberg: Springer Verlag 1983

[26] Ziemba, W.T., Vickson, R.G. (eds.): Stochastic Optimization Models in Finance. New York-London: Academic Press 1975

[27] Zanakis, S.H., Rustagi, J.S. (eds.): Optimization in Statistics. TIMS Studies in the Management Sciences, Vol. 19. Amsterdam: North Holland 1982

Part I
Chapter 4

FUZZY PROGRAMMING: A SURVEY OF RECENT DEVELOPMENTS

M. INUIGUCHI, H. ICHIHASHI and H. TANAKA
Department of Industrial Engineering,
University of Osaka Prefecture,
804 Mozu-Umemachi 4-cho, Sakai, Osaka 591,
Japan

ABSTRACT. In this paper, a survey of the major types of fuzzy programming is provided classifying into the following three categories; mathematical programming with vagueness, mathematical programming with ambiguity and mathematical programming with vagueness and ambiguity.

1. INTRODUCTION

There are some difficulties when modeling a real world problem by a mathematical programming problem. One of such difficulties is in the fact that goals and constraints are often represented by the vague linguistic form, e.g., "we want the profit essentially larger than or equal to a dollars", "we can invest our money substantially smaller than b dollars" and so on. Another difficulty is in the fact that the parameters are not known exactly, e.g., "the production rate will be approximately c (pcs/min)", "the profit rate will be almost d (\$/min)" and so on.

Fuzzy programming is a tool dealing with these fuzziness which causes the difficulties in modeling. At the first stage of the research, the flexibility in constraints and the fuzziness in the objective functions were introduced into the conventional mathematical programming. The flexibility in constraints and the fuzziness in the objective functions are represented by fuzzy sets which are called "fuzzy constraints" and "fuzzy goals", respectively. This type of fuzzy programming deals with the first difficulty and is called the "flexible programming".

At the second stage, a mathematical programming with fuzzy coefficients have been independently initiated by some researchers. Since this mathematical programming problem is an ill-posed problem, various formulations can be considered depending on the interpretation of the problem. And indeed, many formulations have been proposed recently. The fuzzy coefficients represents the fuzzy regions where the coefficients possibly take and regarded as the possibility distributions. Hence, this type of fuzzy programming deals with the

R. Slowinski and J. Teghem (eds.),
Stochastic versus Fuzzy Approaches to Multiobjective Mathematical Programming under Uncertainty, 45–68.

second difficulty and is called the "possibilistic programming".

Here, note that the difference in the meanings of fuzziness between fuzzy goals (or fuzzy constraints) and fuzzy coefficients (i.e., possibility distributions). The former fuzzy goal represents the decision maker's satisfying set and its membership grade is corresponding to the degree of satisfaction. On the other hand, the latter fuzzy coefficient represents the set whose elements possibly occur and its membership grade is corresponding to the possibility degree of occurrence. Namely, the former means vagueness and the latter means ambiguity [1]. In this paper, a survey of the major types of fuzzy programming is provided classifying them into the following three categories:

1. mathematical programming with vagueness,
2. mathematical programming with ambiguity,
3. mathematical programming with vagueness and ambiguity.

2. MATHEMATICAL PROGRAMMING WITH VAGUENESS: FLEXIBLE PROGRAMMING

A flexible programming problem can be written as the following form [2,3,4]:

$$f_i(\mathbf{x}) \gtrsim g_i, \quad i=1,2,\ldots,m \,, \tag{2.1}$$

where $\mathbf{x}$ is an n-vector, each $f_i:\mathbf{R}^n \rightarrow \mathbf{R}$ is a real-valued function and each g_i is a real number. $\gtrsim$ is a fuzzified version of $\geqq$ and has the linguistic interpretation "essentially larger than or equal to". $\gtrsim g_i$ denotes "essentially larger than or equal to g_i" and is called a "fuzzy goal" or a "fuzzy constraint". It is assumed that the fuzzy goal or the fuzzy constraint is subjectively defined by the decision maker. Namely, fuzzy goals and fuzzy constraints reflect the vagueness of the decision maker. An example of fuzzy goals $\gtrsim g_i$ is shown in Figure 1. A flexible

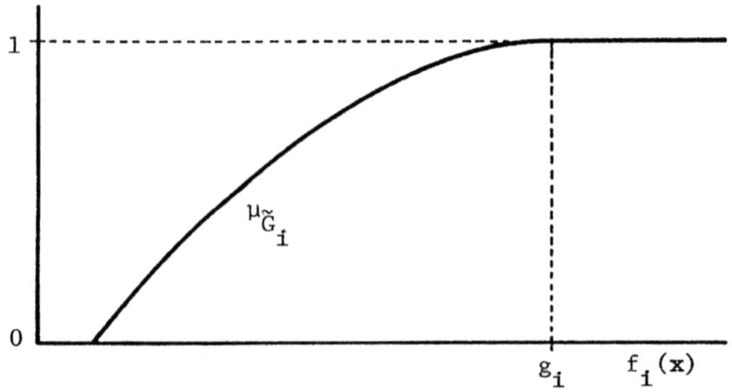

Figure 1. An example of fuzzy goal $\gtrsim g_i$.

programming problem can be interpreted as the problem to find a solution $\mathbf{x}$ such that $f_i(\mathbf{x})$ essentially larger than or equal to g_i for any $i \in \{1,2,\dots,m\}$.

Let us assume each fuzzy goal $\gtrsim g_i$ has the membership function $\mu_{\tilde{G}_i}: \mathbf{R} \to [0,1]$. In flexible programming, the solution of (2.1) is ones such that

$$\text{maximize } \mu_{\tilde{D}}(\mathbf{x}), \tag{2.2}$$

where $\mu_{\tilde{D}}: \mathbf{R}^n \to [0,1]$ is the membership function of a decision set $\tilde{D}$. Following Bellman and Zadeh's decision making method [5] in fuzzy environment, the decision set $\tilde{D}$ is defined by

$$\mu_{\tilde{D}} = \bigwedge_{i=1}^{m} \mu_{\tilde{G}_i}(f_i(\mathbf{x})), \tag{2.3}$$

where $a \wedge b = \min(a,b)$ and $\bigwedge_{i=1}^{m} a_i = a_1 \wedge a_2 \wedge \dots \wedge a_m$. Though the other definitions of the decision set have been proposed in the literature [4,6-10], the definition by (2.3) is usually adopted. Here, we assume the decision set $\tilde{D}$ is defined by (2.3).

Zimmermann [3,11] showed that (2.1) is reduced to a linear programming problem when each f_i is linear and each $\mu_{\tilde{G}_i}$ is a linear membership function. Let us consider the case when $f_i = \mathbf{a}_i \mathbf{x}$ and

$$\mu_{\tilde{G}_i}(y) = \begin{cases} 1, & y \geqq g_i, \\ 1 - \dfrac{g_i - y}{d_i}, & g_i - d_i \leqq y \leqq g_i, \\ 0, & y \leqq g_i - d_i, \end{cases} \tag{2.4}$$

where $\mathbf{a}_i$, $i=1,2,\dots,m$ are n-vectors and d_i is a positive number. g_i

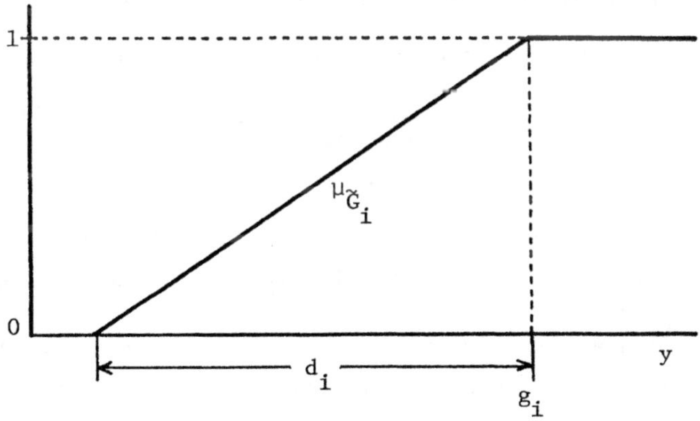

Figure 2. A linear membership function $\mu_{\tilde{G}_i}$.

represents the expectation level and g_i-d_i represents an admissibility level. The fuzzy goal having the membership function defined by (2.4) is depicted in Figure 2. Applying (2.2), (2.5) is reduced to a linear programming problem [3,4,11]:

$$\text{maximize } \lambda$$
$$\text{subject to } \mathbf{a}_i^* \mathbf{x} - \lambda \geqq g_i^*, \; i=1,2,\ldots,m \; , \qquad (2.5)$$

where $\mathbf{a}_i^*=\mathbf{a}_i/d_i$ and $g_i^*=1-g_i/d_i$.

Generally, when all $\mu_{\tilde{G}_i}$ are represented by

$$\mu_{\tilde{G}_i}(y) = \begin{cases} 1 \;, & y \geqq g_i, \\ \varphi\left(1 - \dfrac{g_i - y}{d_i}\right) \;, & g_i-d_i \leqq y \leqq g_i, \\ 0 \;, & y \leqq g_i-d_i, \end{cases} \qquad (2.6)$$

(2.1) with linear functions f_i is reduced to (2.5), where $\varphi:[0,1]\rightarrow[0,1]$ is a strictly increasing function. Tanaka et al. [2] showed that (2.1) can be solved by using a linear programming technique repetitionally even when $\mu_{\tilde{G}_i}$ are non-linear membership functions not necessarily represented by (2.6). Recently, Inuiguchi et al. [12] showed that when each $\mu_{\tilde{G}_i}$ is strictly quasi-concave piecewise linear in the range (0,1), (2.1) with linear functions f_i is reduced to a linear programming problem. Flexible programming is one of the most advanced applications of fuzzy systems theory. A comprehensive review of flexible programming is made by Zimmermann [4]. Sakawa and Yano [13-15] proposed an interactive decision making method using flexible programming.

3. MATHEMATICAL PROGRAMMING WITH AMBIGUITY

3.1. Possibilistic Linear Programming Proposed by Tanaka et al.

Tanaka and Asai [16] and Tanaka et al. [17] treated the following possibilistic linear programming problem:

$$\tilde{\mathbf{a}}_i'\mathbf{x}' \geqq 0, \; i=1,2,\ldots,m \; , \; \mathbf{x}' \geqq 0, \qquad (3.1)$$

where $\mathbf{x}'=(1,\mathbf{x}^t)^t=(1,x_1,x_2,\ldots,x_n)^t$ and $\tilde{\mathbf{a}}_i'$ is (n+1)-vector whose components are fuzzy numbers, i.e., $\tilde{\mathbf{a}}_i'=(\tilde{a}_{i0},\tilde{a}_{i1},\ldots,\tilde{a}_{in})$. Each $\tilde{a}_{ij}$ is a fuzzy number [18].

Tanaka and Asai [16] dealt with (3.1) when each $\tilde{a}_{ij}$ is a triangular fuzzy number and Tanaka et al. [17] dealt with (3.1) when each $\tilde{a}_{ij}$ is a trapezoid fuzzy number. Here, generalizing them, we assume each $\tilde{a}_{ij}$ is an L-R fuzzy number [18] $(a_{ij}^L, a_{ij}^R, \alpha_{ij}^L, \alpha_{ij}^R)_{L_{ij}R_{ij}}$. An L-R fuzzy number $M=(m^L, m^R, \alpha^L, \alpha^R)_{LR}$ is defined by

$$\mu_{\tilde{M}}(r) = \begin{cases} L((m^L-r)/\alpha^L), & r \leqq m^L, \\ 1, & m^L \leqq r \leqq m^R, \\ R((r-m^R)/\alpha^R), & m^R \leqq r, \end{cases} \tag{3.2}$$

where $\mu_{\tilde{M}}$ is the membership function of $\tilde{M}$ and L and R are reference functions, i.e., $L:[0,+\infty)\to[0,1]$ and $R:[0,+\infty)\to[0,1]$ are strictly decreasing in the range (0,1] and upper semi-continuous functions such that $L(0)=R(0)=1$. Usually, $L(1)=R(1)=0$ when $\mathrm{Supp}(\tilde{M})=\{r/\mu_{\tilde{M}}(r)>0\}$ is bounded. The L-R fuzzy number $\tilde{a}_{ij}$ is depicted in Figure 3. When $L_{ij}(r)=R_{ij}(r)=\max(1-r,0)$, each $\tilde{a}_{ij}$ is a trapezoid fuzzy number. And when $a_{ij}^L=a_{ij}^R$ and $L_{ij}(r)=R_{ij}(r)=\max(1-r,0)$, each $\tilde{a}_{ij}$ is a triangular fuzzy number.

Here, let us consider the calculation of $\tilde{\mathbf{a}}_i'\mathbf{x}'$. $\tilde{\mathbf{a}}_i'\mathbf{x}'$ is obtained via the extension principle [17], i.e.,

$$\mu_{\tilde{\mathbf{a}}_i'\mathbf{x}'}(y) = \sup_{\substack{\mathbf{a}' \\ \mathbf{a}'\mathbf{x}'=y}} \{\mu_{\tilde{a}_{i0}}(a_0)\wedge\cdots\wedge\mu_{\tilde{a}_{in}}(a_n)\}, \tag{3.3}$$

where $\mathbf{a}'=(a_0,a_1,\ldots,a_n)$. If $\mathbf{x}\geqq 0$, $L_{i0}=L_{i1}=\ldots=L_{in}=L_i$ and $R_{i0}=R_{i1}=\ldots=R_{in}=R_i$ then it is well known [17] that

$$\tilde{\mathbf{a}}_i'\mathbf{x}' = (\mathbf{a}_i'^L\mathbf{x}', \mathbf{a}_i'^R\mathbf{x}', \boldsymbol{\alpha}_i'^L\mathbf{x}', \boldsymbol{\alpha}_i'^R\mathbf{x}')_{L_iR_i}, \tag{3.4}$$

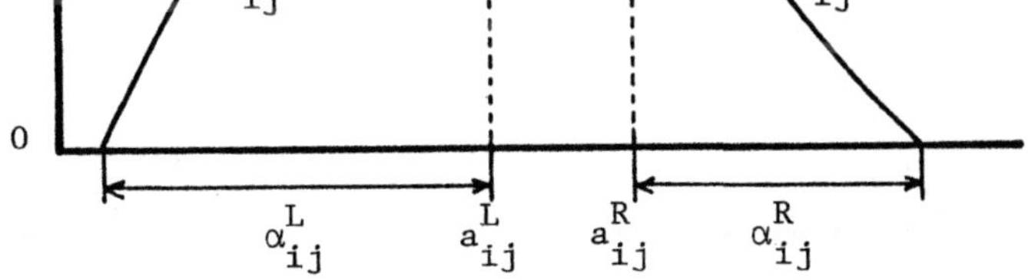

Figure 3. An example of L-R fuzzy number $\tilde{a}_{ij}$.

where $\mathbf{a}_i'^L=(a_{i0}^L,a_{i1}^L,\ldots,a_{in}^L)$, $\mathbf{a}_i'^R=(a_{i0}^R,a_{i1}^R,\ldots,a_{in}^R)$, $\boldsymbol{\alpha}_i^L=(\alpha_{i0}^L,\alpha_{i1}^L,\ldots,\alpha_{in}^L)$ and $\boldsymbol{\alpha}_i^R=(\alpha_{i0}^R,\alpha_{i1}^R,\ldots,\alpha_{in}^R)$. Even if $L_{i0}=L_{i1}=\ldots=L_{in}$ or $R_{i0}=R_{i1}=\ldots=R_{in}$ does not hold, it is known [18] that

$$[\tilde{\mathbf{a}}_i'\mathbf{x}']_h = [\tilde{a}_{i0}]_h + \sum_{j=1}^{n}[\tilde{a}_{ij}]_h x_j, \tag{3.5}$$

where $\mathbf{x}'\geqq 0$ and $[\tilde{M}]_h$ is a h-level set of a fuzzy set $\tilde{M}$, i.e.,

$$[\tilde{M}]_h = \left\{ r \,/\, \mu_{\tilde{M}}(r) \geqq h \right\}. \tag{3.6}$$

Tanaka and Asai [16] and Tanaka et al. [17] defined "a fuzzy number $\tilde{M}$ is almost positive" denoted by $\tilde{M}\geqq_h 0$ with the degree h of $\tilde{M}\geqq 0$ when $\tilde{M}$ is a triangular or trapezoid fuzzy number. Generalizing their idea, "an L-R fuzzy number $\tilde{M}=(m^L,m^R,\alpha^L,\alpha^R)_{LR}$ is almost positive" is defined by

$$\tilde{M}\geqq_h 0 \Leftrightarrow \mu_{\tilde{M}}(0)\leqq 1-h,\ \frac{m^L+m^R}{2}\geqq 0, \tag{3.7}$$

where h is the degree of $\tilde{M}\geqq 0$. And the larger h is, the stronger the meaning of "almost positive" is (see Figure 4).

As shown in Figure 4, since a reference function is strictly decreasing in the range (0,1] and upper semi-continuous, the following statement is valid;

$$\tilde{M}\geqq_h 0 \Leftrightarrow \inf\ [\tilde{M}]_{1-h}\geqq 0 \tag{3.8}$$

Using (3.7), Tanaka and Asai [16] and Tanaka et al. [17] formulated (3.1) as follows;

$$\begin{aligned}&\text{maximize} \quad h\\ &\text{subject to} \quad \tilde{\mathbf{a}}_i'\mathbf{x}'\geqq_h 0,\ i=1,2,\ldots,m\ ,\ \mathbf{x}'\geqq 0.\end{aligned} \tag{3.9}$$

Applying (3.5) and (3.8), we have

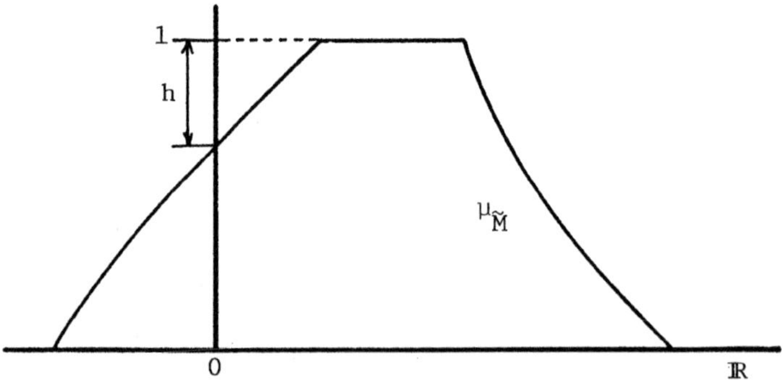

Figure 4. An illustration of $M\geqq_h 0$.

$$\tilde{a}_i'\mathbf{x}' \geqq_h 0 \Leftrightarrow \inf\,[\tilde{a}_{i0}]_{1-h} + \sum_{j=1}^{n} \inf\,[\tilde{a}_{ij}]_{1-h} x_j \geqq 0. \tag{3.10}$$

Since $\tilde{a}_{ij}=(a_{ij}^L, a_{ij}^R, \alpha_{ij}^L, \alpha_{ij}^R)_{L_{ij}R_{ij}}$, this lead to

$$\inf\,[\tilde{a}_{ij}]_{i-h} = a_{ij}^L - \alpha_{ij}^L L_{ij}^*(1-h),\ i=1,2,\dots,m,\ j=0,1,\dots,n, \tag{3.11}$$

where

$$L_{ij}^*(h) = \sup\left\{r \,/\, L_{ij}(r) \geqq h\right\},\ i=1,2,\dots,m\ ,\ j=0,1,\dots,n\ . \tag{3.12}$$

Hence, (3.9) is written as

$$\text{maximize}\ \ h$$

$$\text{subject to}\ \ \mathbf{a}_i'^{L}\mathbf{x}' - \alpha_i^{L*}(1-h)\mathbf{x}' \geqq 0,\ i=1,2,\dots,m\ ,\ \mathbf{x}' \geqq 0, \tag{3.13}$$

where

$$\alpha_i^{L*}(h) = (\alpha_{i0}^L L_{i0}^*(h), \alpha_{i1}^L L_{i1}^*(h), \dots, \alpha_{in}^L L_{in}^*(h)),\ i=1,2,\dots,m. \tag{3.14}$$

Namely, (3.9) is reduced to a usual non-linear programming problem (3.13). The constraints of (3.13) become linear when the parameter h is fixed. Hence, we can solve (3.13) by using simplex method repetitionally and changing the value of fixed parameter h [16,17].

3.2. Possibilistic Programming Proposed by Dubois

Dubois [19] treated a possibilistic programming in the setting of possibility theory. In [19], various treatments of constraints via the concept of possibility and necessity were proposed, and a possibilistic linear programming problem was formulated in much similar manner of chance constrained programming [20].

Here, we describe some treatments of the constraints and a formulation of a possibilistic programming problem.

(a) Equality constraints: Let us consider the constraints of the form

$$f(\mathbf{x},\tilde{\mathbf{a}}) = g(\mathbf{x},\tilde{\mathbf{b}}), \tag{3.15}$$

where $\tilde{\mathbf{a}}=(\tilde{a}_1,\tilde{a}_2,\dots,\tilde{a}_p)$ and $\tilde{\mathbf{b}}=(\tilde{b}_1,\tilde{b}_2,\dots,\tilde{b}_q)$. $\tilde{a}_i$, $i=1,2,\dots,p$ and $\tilde{b}_j$, $j=1,2,\dots,q$ are fuzzy numbers [18]. $f(\mathbf{x},\tilde{\mathbf{a}})$ and $g(\mathbf{x},\tilde{\mathbf{b}})$ are obtained via the extension principle [18].

Using possibility measure Π and necessity measure N [18,21], three indices of equality for each $\mathbf{x}$ are induced, i.e.,

$$WF(\mathbf{x}) = \Pi_{f(\mathbf{x},\tilde{\mathbf{a}})}(g(\mathbf{x},\tilde{\mathbf{b}})) = \Pi_{g(\mathbf{x},\tilde{\mathbf{b}})}(f(\mathbf{x},\tilde{\mathbf{a}})), \tag{3.16}$$

$$SF(\mathbf{x}) = N_{f(\mathbf{x},\tilde{\mathbf{a}})}(g(\mathbf{x},\tilde{\mathbf{b}})), \tag{3.17}$$

$$SF'(\mathbf{x}) = N_{g(\mathbf{x},\tilde{\mathbf{b}})}(f(\mathbf{x},\tilde{\mathbf{a}})), \tag{3.18}$$

where WF means "weak feasibility" and SF means "strong feasibility". The possibility measure Π and the necessity measure N are defined by

$$\Pi_{\tilde{A}}(\tilde{B}) = \sup_r \left\{\mu_{\tilde{A}}(r) \wedge \mu_{\tilde{B}}(r)\right\}, \tag{3.19}$$

$$N_{\tilde{A}}(\tilde{B}) = \inf_r \left\{(1-\mu_{\tilde{A}}(r)) \vee \mu_{\tilde{B}}(r)\right\}. \tag{3.20}$$

Here, $a \vee b = \max(a,b)$ and $\mu_{\tilde{A}}$ and $\mu_{\tilde{B}}$ are membership functions of fuzzy sets $\tilde{A}$ and $\tilde{B}$ respectively.

WF($\mathbf{x}$) represents the degree of possibility to which $f(\mathbf{x},\tilde{\mathbf{a}})$ and $g(\mathbf{x},\tilde{\mathbf{b}})$ are equal. SF($\mathbf{x}$) represents the degree of necessity to which $f(\mathbf{x},\tilde{\mathbf{a}})$ is equal to $g(\mathbf{x},\tilde{\mathbf{b}})$, and SF'($\mathbf{x}$) represents the degree of necessity to which $g(\mathbf{x},\tilde{\mathbf{b}})$ is equal to $f(\mathbf{x},\tilde{\mathbf{a}})$. These three indices are illustrated in Figure 5.

Let $f(\mathbf{x},\tilde{\mathbf{a}})=\tilde{\mathbf{a}}_i^{\prime}\mathbf{x}'$ and $g(\mathbf{x},\tilde{\mathbf{b}})=\tilde{\mathbf{b}}_i^{\prime}\mathbf{x}'$, where $\tilde{\mathbf{b}}_i^{\prime}=(\tilde{b}_{i0},\tilde{b}_{i1},\ldots,\tilde{b}_{in})$ and each $\tilde{b}_{ij}$ is an L-R fuzzy number $(b_{ij}^L, b_{ij}^R, \beta_{ij}^L, \beta_{ij}^R)_{L_{ij}^{\prime}R_{ij}^{\prime}}$. $\mathbf{x}'$ is assumed to be non-negative. Since each reference function is upper semi-continuous,

$$\mathrm{WF}(\mathbf{x}) \geqq h \iff \begin{cases} \mathbf{b}_i^{\prime L}\mathbf{x}' - \boldsymbol{\beta}_i^{L*}(h)\mathbf{x}' \leqq \mathbf{a}_i^{\prime R}\mathbf{x}' + \boldsymbol{\alpha}_i^{R*}(h)\mathbf{x}', \\ \mathbf{a}_i^{\prime L}\mathbf{x}' - \boldsymbol{\alpha}_i^{L*}(h)\mathbf{x}' \leqq \mathbf{b}_i^{\prime R}\mathbf{x}' + \boldsymbol{\beta}_i^{R*}(h)\mathbf{x}', \end{cases} \tag{3.21}$$

$$\mathrm{SF}(\mathbf{x}) \geqq h \iff \begin{cases} \mathbf{b}_i^{\prime L}\mathbf{x}' - \boldsymbol{\beta}_i^{L*}(h)\mathbf{x}' \leqq \mathbf{a}_i^{\prime L}\mathbf{x}' - \boldsymbol{\alpha}_i^{L*}(1-h)\mathbf{x}', \\ \mathbf{a}_i^{\prime R}\mathbf{x}' + \boldsymbol{\alpha}_i^{R*}(1-h)\mathbf{x}' \leqq \mathbf{b}_i^{\prime R}\mathbf{x}' + \boldsymbol{\beta}_i^{R*}(h)\mathbf{x}', \end{cases} \tag{3.22}$$

$$\mathrm{SF}'(\mathbf{x}) \geqq h \iff \begin{cases} \mathbf{b}_i^{\prime R}\mathbf{x}' + \boldsymbol{\beta}_i^{R*}(1-h)\mathbf{x}' \leqq \mathbf{a}_i^{\prime R}\mathbf{x}' + \boldsymbol{\alpha}_i^{R*}(h)\mathbf{x}', \\ \mathbf{a}_i^{\prime L}\mathbf{x}' - \boldsymbol{\alpha}_i^{L*}(h)\mathbf{x}' \leqq \mathbf{b}_i^{\prime L}\mathbf{x}' - \boldsymbol{\beta}_i^{L*}(1-h)\mathbf{x}', \end{cases} \tag{3.23}$$

hold [19], where $\mathbf{b}_i^{\prime L}=(b_{i0}^L, b_{i1}^L, \ldots, b_{in}^L)$, $\mathbf{b}_i^{\prime R}=(b_{i0}^R, b_{i1}^R, \ldots, b_{in}^R)$, $\boldsymbol{\alpha}_i^{R*}(h) =(\alpha_{i0}^R R_{i0}^*(h), \alpha_{i1}^R R_{i1}^*(h), \ldots, \alpha_{in}^R R_{in}^*(h))$, $\boldsymbol{\beta}_i^{L*}(h)=(\beta_{i0}^L L_{i0}^{\prime *}(h), \beta_{i1}^L L_{i1}^{\prime *}(h), \ldots, \beta_{in}^L L_{in}^{\prime *}(h))$ and $\boldsymbol{\beta}_i^{R*}(h)=(\beta_{i0}^R R_{i0}^{\prime *}(h), \beta_{i1}^R R_{i1}^{\prime *}(h), \ldots, \beta_{in}^R R_{in}^{\prime *}(h))$, $i=1,2,\ldots,m$. R_{ij}^*, $L_{ij}^{\prime *}$ and $R_{ij}^{\prime *}$ are pseudo-inverses of R_{ij}, $L_{ij}^{\prime}$ and $R_{ij}^{\prime}$ as defined in the same way of (3.12).

(b) Inequality constraints: Let us consider the constraints of the form

$$f(\mathbf{x},\tilde{\mathbf{a}}) \geqq g(\mathbf{x},\tilde{\mathbf{b}}). \tag{3.24}$$

Using ranking indices proposed by Dubois and Prade [21], four indices of inequality for each $\mathbf{x}$ are obtained, i.e.,

$$\mathrm{VSF}(\mathbf{x}) = N_{f(\mathbf{x},\tilde{\mathbf{a}})}(]g(\mathbf{x},\tilde{\mathbf{b}}),\infty)), \tag{3.25}$$

$$\mathrm{MSF}(\mathbf{x}) = N_{f(\mathbf{x},\tilde{\mathbf{a}})}([g(\mathbf{x},\tilde{\mathbf{b}}),\infty)), \tag{3.26}$$

$$\mathrm{MWF}(\mathbf{x}) = \Pi_{f(\mathbf{x},\tilde{\mathbf{a}})}(]g(\mathbf{x},\tilde{\mathbf{b}}),\infty)), \tag{3.27}$$

$$\mathrm{VWF}(\mathbf{x}) = \Pi_{f(\mathbf{x},\tilde{\mathbf{a}})}([g(\mathbf{x},\tilde{\mathbf{b}}),\infty)), \tag{3.28}$$

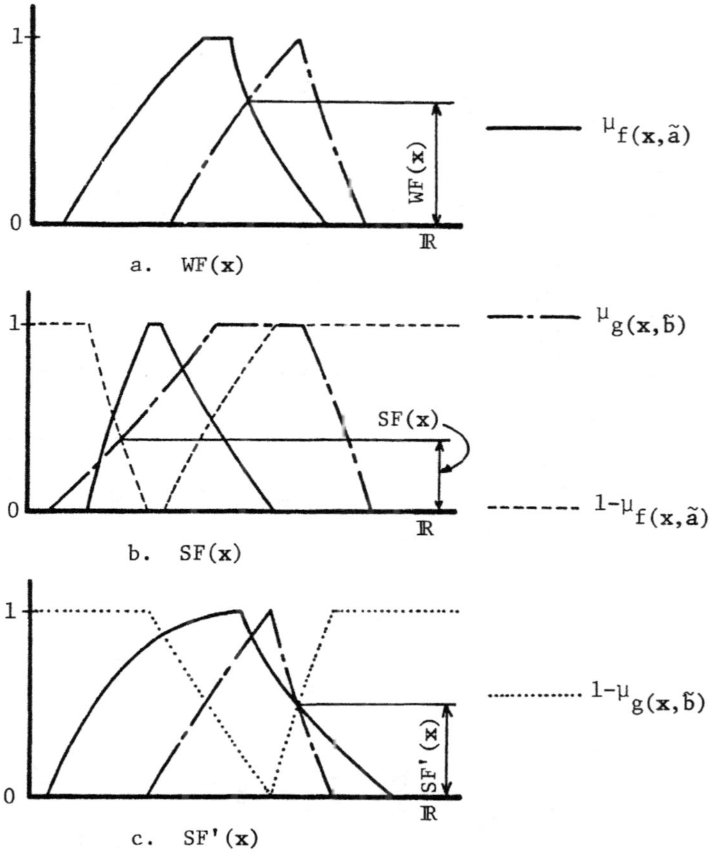

Figure 5. Illustrations of WF(**x**), SF(**x**) and SF'(**x**).

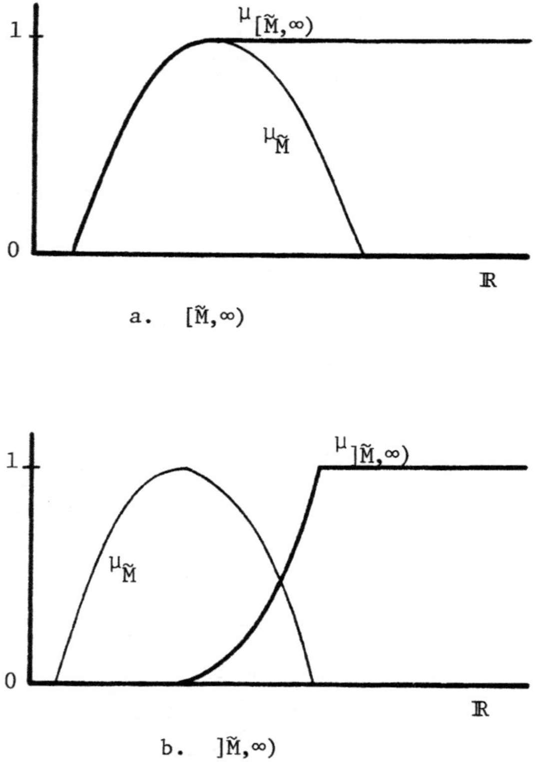

Figure 6. Illustrations of $[\tilde{M},\infty)$ and $]\tilde{M},\infty)$.

where VSF means "very strong feasibility", MSF means "medium strong feasibility", MWF means "medium weak feasibility" and VWF means "very weak feasibility". $]\tilde{M},\infty)$ and $[\tilde{M},\infty)$ are defined by

$$\mu_{]\tilde{M},\infty)}(r) = \inf_{s \geqq r} (1-\mu_{\tilde{M}}(s)), \tag{3.29}$$

$$\mu_{[\tilde{M},\infty)}(r) = \sup_{s \leqq r} \mu_{\tilde{M}}(s). \tag{3.30}$$

$\mu_{]\tilde{M},\infty)}$ and $\mu_{[\tilde{M},\infty)}$ are membership functions of $]\tilde{M},\infty)$ and $[\tilde{M},\infty)$ respectively. $]\tilde{M},\infty)$ and $[\tilde{M},\infty)$ are illustrated in Figure 6.

VSF($\mathbf{x}$) represents the degree of necessity to which $f(\mathbf{x},\tilde{\mathbf{a}})$ is necessarily greater than $g(\mathbf{x},\tilde{\mathbf{b}})$ and MSF($\mathbf{x}$) represents the degree of necessity to which $f(\mathbf{x},\tilde{\mathbf{a}})$ is possibly greater than or equal to $g(\mathbf{x},\tilde{\mathbf{b}})$. Likewise, MWF($\mathbf{x}$) represents the degree of possibility to which $f(\mathbf{x},\tilde{\mathbf{a}})$ is necessarily greater than $g(\mathbf{x},\tilde{\mathbf{b}})$ and VWF($\mathbf{x}$) represents the degree of possibility to which $f(\mathbf{x},\tilde{\mathbf{a}})$ is possibly greater than or equal to $g(\mathbf{x},\tilde{\mathbf{b}})$. They are depicted in Figure 7.

Let $f(\mathbf{x},\tilde{\mathbf{a}})=\tilde{\mathbf{a}}_i'\mathbf{x}'$ and $g(\mathbf{x},\tilde{\mathbf{b}})=\tilde{\mathbf{b}}_i'\mathbf{x}'$. Then we have

$$\mathrm{VSF}(\mathbf{x}) \geqq h \iff \mathbf{b}_i'^R\mathbf{x}' + \boldsymbol{\beta}_i^{R*}(h)\mathbf{x}' \leqq \mathbf{a}_i'^L\mathbf{x}' - \boldsymbol{\alpha}_i^{L*}(h)\mathbf{x}', \tag{3.31}$$

when R_{ij}', $j=0,1,\ldots,n$ or L_{ij}', $j=0,1,\ldots,n$ are continuous,

$$\mathrm{MSF}(\mathbf{x}) \geqq h \iff \mathbf{b}_i'^L\mathbf{x}' - \boldsymbol{\beta}_i^{L*}(h)\mathbf{x}' \leqq \mathbf{a}_i'^L\mathbf{x}' - \boldsymbol{\alpha}_i^{L*}(1-h)\mathbf{x}', \tag{3.32}$$

$$\mathrm{MWF}(\mathbf{x}) \geqq h \iff \mathbf{b}_i'^R\mathbf{x}' + \boldsymbol{\beta}_i^{R*}(1-h)\mathbf{x}' \leqq \mathbf{a}_i'^R\mathbf{x}' + \boldsymbol{\alpha}_i^{R*}(h)\mathbf{x}', \tag{3.33}$$

when R_{ij}', $j=0,1,\ldots,n$ or L_{ij}', $j=0,1,\ldots,n$ are continuous,

$$\mathrm{VWF}(\mathbf{x}) \geqq h \iff \mathbf{b}_i'^L\mathbf{x}' - \boldsymbol{\beta}_i^{L*}(h)\mathbf{x}' \leqq \mathbf{a}_i'^R\mathbf{x}' + \boldsymbol{\alpha}_i^{R*}(1-h)\mathbf{x}', \tag{3.34}$$

Since Dubois and Prade [21] use strict inequality relation > to define $]\tilde{M},\infty)$ [22], the conditions of (3.31) and (3.33) are needed. If we define $]\tilde{M},\infty)$ using inequality relation $\geqq$ by

$$\mu_{]\tilde{M},\infty)}(r) = \inf_{s>r} (1-\mu_{\tilde{M}}(s)), \tag{3.35}$$

then the conditions of (3.31) and (3.33) are eliminated.

(c) Formulation of a possibilistic programming problem: Let us formulate the following possibilistic programming problem;

$$\begin{aligned} &\text{maximize } \psi(\mathbf{x}), \\ &\text{subject to } \quad f_j(\mathbf{x},\tilde{\mathbf{a}}_j) \geqq g_j(\mathbf{x},\tilde{\mathbf{b}}_j),\ j=1,2,\ldots,w\ , \\ &\qquad\qquad\quad\ f_j(\mathbf{x},\tilde{\mathbf{a}}_j) = g_j(\mathbf{x},\tilde{\mathbf{b}}_j),\ j=w+1,w+2,\ldots,m\ , \end{aligned} \tag{3.36}$$

where ψ is an ordinary real-valued function. $f_j(\cdot,\tilde{\mathbf{a}}_j)$ and $g_j(\cdot,\tilde{\mathbf{b}}_j)$ are the functions with fuzzy parameters $\tilde{\mathbf{a}}_j$ and $\tilde{\mathbf{b}}_j$.

Dubois [19] proposed two ways of formulation:

(i) Similarly to flexible programming, defining a fuzzy goal $\tilde{G}$ with

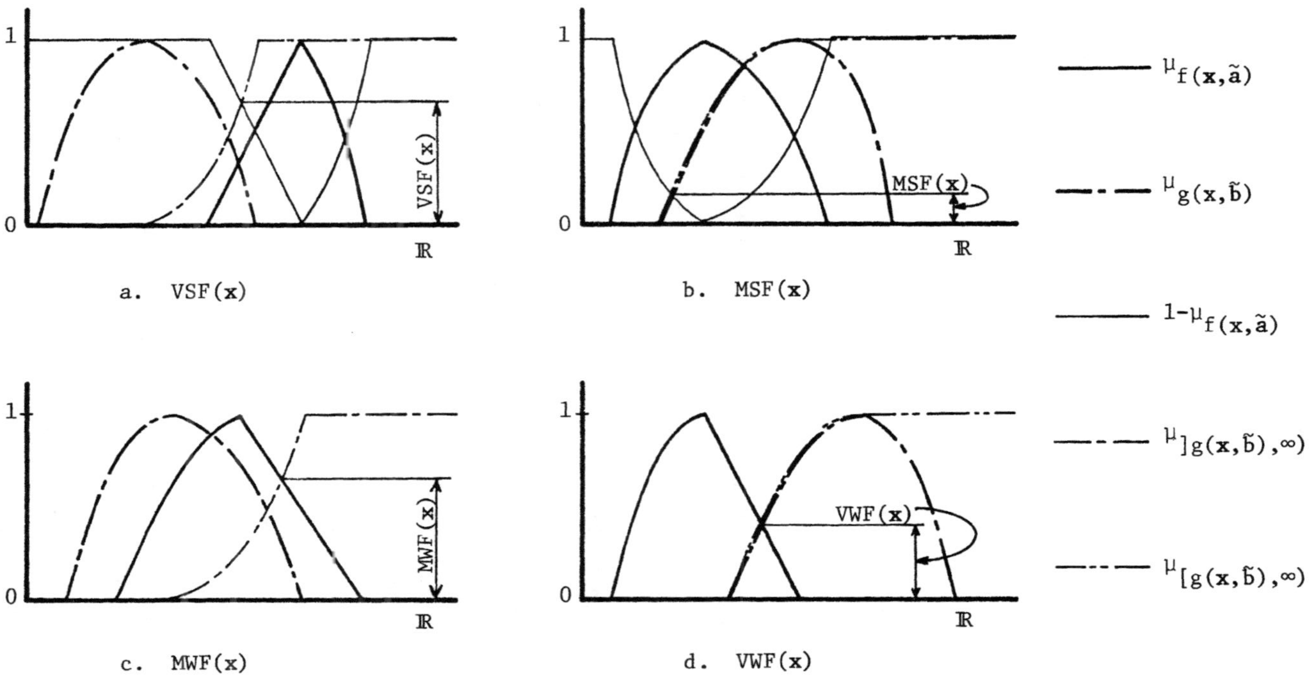

Figure 7. Illustrations of VSF(**x**), MSF(**x**), MWF(**x**) and VWF(**x**).

respect to $\psi(\mathbf{x})$, (3.36) is formulated as

$$\text{maximize} \quad \mu_{\tilde{G}}(\psi(\mathbf{x})) \wedge \bigwedge_{j=1}^{m} \mu_{\tilde{C}_j}(\mathbf{x}), \tag{3.37}$$

where $\mu_{\tilde{G}}$ is a membership function of $\tilde{G}$ and $\mu_{\tilde{C}_j}$ is defined by VSF($\mathbf{x}$), MSF($\mathbf{x}$), MWF($\mathbf{x}$) or VWF($\mathbf{x}$) for each $j \in \{1,2,\dots,w\}$ and defined by WF($\mathbf{x}$), SF($\mathbf{x}$) or SF'($\mathbf{x}$) for each $j \in \{w+1,w+2,\dots,m\}$. The selections from {VSF($\mathbf{x}$),MSF($\mathbf{x}$),MWF($\mathbf{x}$),VWF($\mathbf{x}$)} and {WF($\mathbf{x}$),SF($\mathbf{x}$),SF'($\mathbf{x}$)} are assumed to be made by the decision maker.

If ψ, $f_j(\cdot,\tilde{\mathbf{a}}_j)$, and $g_j(\cdot,\tilde{\mathbf{b}}_j)$, j=1,2,...,m are linear and $\mathbf{x} \geqq 0$ is added to the constraints, then the formulated programming problem can be solved by the same algorithm for possibilistic linear programming proposed by Tanaka et al. [16,17] (see Section 3.1)

(ii) Similarly to chance constrained programming, defining a satisfaction threshold t_j for each constraint, (3.36) is formulated as

$$\begin{aligned} &\text{maximize} \quad \psi(\mathbf{x}), \\ &\text{subject to} \quad \mu_{\tilde{C}_j}(\mathbf{x}) \geqq t_j, \quad j=1,2,\dots,m. \end{aligned} \tag{3.38}$$

This problem is called a "feasibility constrained programming problem" [19]. If ψ, $f_j(\cdot,\tilde{\mathbf{a}}_j)$ and $g_j(\cdot,\tilde{\mathbf{b}}_j)$, j=1,2,...,m are linear and $\mathbf{x} \geqq 0$ is added to the constraints, then (3.38) is reduced to a linear programming problem.

3.3. Possibilistic Linear Programming Using "Fuzzy Max"

Dubois and Prade [23] suggested "fuzzy max" can be applied to constraints with fuzzy parameters. Tanaka et al. [24] and Ramík and Římánek [25] formulated possibilistic linear programming problems using "fuzzy max". "Fuzzy max" is the extended maximum operation between real numbers to fuzzy numbers and defined by the extension principle as

$$\mu_{\tilde{M} \vee \tilde{N}}(r) = \sup_{r=s \vee t} \{\mu_{\tilde{M}}(s) \wedge \mu_{\tilde{N}}(t)\}, \tag{3.39}$$

where $\mu_{\tilde{M} \vee \tilde{N}}$ is a membership function of $\tilde{M} \vee \tilde{N}$. $\tilde{M} \vee \tilde{N}$ is depicted in Figure 8. Using fuzzy max, we can define an inequality relation $\tilde{M} \gtrsim \tilde{N}$ by

$$\tilde{M} \gtrsim \tilde{N} \iff \tilde{M} \vee \tilde{N} = \tilde{M}. \tag{3.40}$$

Applying this relation $\tilde{M} \gtrsim \tilde{N}$ to an inequality constraint, we can treat it as

$$f(\mathbf{x},\tilde{\mathbf{a}}) \gtrsim g(\mathbf{x},\tilde{\mathbf{b}}) \iff f(\mathbf{x},\tilde{\mathbf{a}}) \vee g(\mathbf{x},\tilde{\mathbf{b}}) = f(\mathbf{x},\tilde{\mathbf{a}}). \tag{3.41}$$

When $f(\mathbf{x},\tilde{\mathbf{a}})=\tilde{\mathbf{a}}_i^{!}\mathbf{x}'$, $g(\mathbf{x},\tilde{\mathbf{b}})=\tilde{\mathbf{b}}_i^{!}\mathbf{x}'$ and $\mathbf{x} \geqq 0$, (3.41) is equivalent to

$$\begin{aligned} &\mathbf{a}_i^{!L}\mathbf{x}' - \boldsymbol{\alpha}_i^{L*}(k)\mathbf{x}' \geqq \mathbf{b}_i^{!L}\mathbf{x}' - \boldsymbol{\beta}_i^{L*}(k)\mathbf{x}', \\ &\mathbf{a}_i^{!R}\mathbf{x}' + \boldsymbol{\alpha}_i^{R*}(k)\mathbf{x}' \geqq \mathbf{b}_i^{!R}\mathbf{x}' + \boldsymbol{\beta}_i^{R*}(k)\mathbf{x}', \quad \text{for any } k \in (0,1] \end{aligned} \tag{3.42}$$

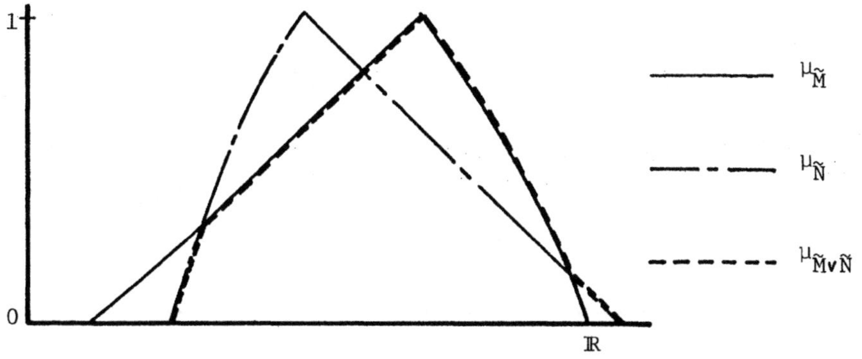

Figure 8. An illustration of "fuzzy max", $\tilde{M}\vee\tilde{N}$.

(3.42) has infinite inequalities. Hence, if we formulate a linear programming problem with inequality constraints $\tilde{\mathbf{a}}_i'\mathbf{x}'\gtrsim\tilde{\mathbf{b}}_i'\mathbf{x}'$, then the formulated problem become an infinitely constrained programming problem [26-28]. However, if all $\tilde{a}_{ij}$ and $\tilde{b}_{ij}$ have bounded supports, the left reference functions L_{ij} and L'_{ij} of $\tilde{a}_{ij}$ and $\tilde{b}_{ij}$, are all equal, i.e., $L_{i0}=L_{i1}=\ldots=L_{in}=L'_{i0}=L'_{i1}=\ldots=L'_{in}$, and the left reference functions R_{ij} and R'_{ij} of $\tilde{a}_{ij}$ and $\tilde{b}_{ij}$, are all equal, i.e., $R_{i0}=R_{i1}=\ldots=R_{in}=R'_{i0}=R'_{i1}=\ldots=R'_{in}$, then (3.42) is equivalent [25] to

$$\begin{aligned}&\mathbf{a}_i'^L\mathbf{x}'\geqq\mathbf{b}_i'^L\mathbf{x}',\quad \mathbf{a}_i'^R\mathbf{x}'\geqq\mathbf{b}_i'^R\mathbf{x}',\\&\mathbf{a}_i'^L\mathbf{x}'-\boldsymbol{\alpha}_i^L\mathbf{x}'\geqq\mathbf{b}_i'^L\mathbf{x}'-\boldsymbol{\beta}_i^L\mathbf{x}',\\&\mathbf{a}_i'^R\mathbf{x}'+\boldsymbol{\alpha}_i^R\mathbf{x}'\geqq\mathbf{b}_i'^R\mathbf{x}'+\boldsymbol{\beta}_i^R\mathbf{x}',\end{aligned}\tag{3.43}$$

where $\boldsymbol{\beta}_i^L=(\beta_{i0}^L,\beta_{i1}^L,\ldots,\beta_{in}^L)$ and $\boldsymbol{\beta}_i^R=(\beta_{i0}^R,\beta_{i1}^R,\ldots,\beta_{in}^R)$. While (3.42) has infinite inequalities, (3.43) has only four inequalities. Furthermore, if all $\tilde{a}_{ij}$ and $\tilde{b}_{ij}$ have bounded supports, $L_{i0}=L_{i1}=\ldots=L_{in}=L'_{i0}=L'_{i1}=\ldots=L'_{in}$, $R_{i0}=R_{i1}=\ldots=R_{in}=R'_{i0}=R'_{i1}=\ldots=R'_{in}$, $\alpha_{ij}^L=\alpha_{ij}^R$ and $\beta_{ij}^L=\beta_{ij}^R$, $j=1,2,\ldots,n$, then (3.42) is reduced to

$$\begin{aligned}&\mathbf{a}_i'^L\mathbf{x}'-\boldsymbol{\alpha}_i^L\mathbf{x}'\geqq\mathbf{b}_i'^L\mathbf{x}'-\boldsymbol{\beta}_i^L\mathbf{x}',\\&\mathbf{a}_i'^R\mathbf{x}'+\boldsymbol{\alpha}_i^R\mathbf{x}'\geqq\mathbf{b}_i'^R\mathbf{x}'+\boldsymbol{\beta}_i^R\mathbf{x}',\end{aligned}\tag{3.44}$$

which has only two inequalities. Tanaka et al. [24] showed (3.44) when all fuzzy numbers are isosceles triangular ones.

However, the inequality relation $\tilde{M}\gtrsim\tilde{N}$ defined by (3.40) does not

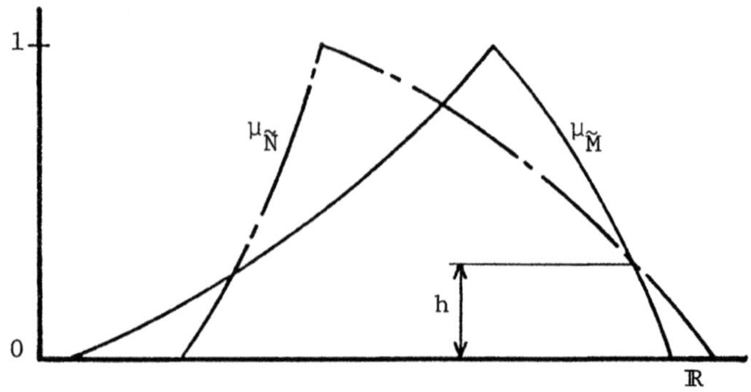

Figure 9. An illustration of $\tilde{M} \gtrsim^h \tilde{N}$.

yield a linear ordering but a partial ordering. This sometimes lead to the indecisive case neither $\tilde{M} \gtrsim \tilde{N}$ nor $\tilde{N} \gtrsim \tilde{M}$ holds. An improvement of (3.40) from this point of view was made by Tanaka et al. [24]. They introduced a level h corresponding to the degree of optimism of the decision maker and define $\tilde{M} \gtrsim^h \tilde{N}$ by

$$\tilde{M} \gtrsim^h \tilde{N} \Leftrightarrow \sup\ [\tilde{M}]_k \geqq \sup\ [\tilde{N}]_k \text{ and } \inf\ [\tilde{M}]_k \geqq \inf\ [\tilde{N}]_k \text{ for all } k \in [h,1]. \qquad (3.45)$$

Figure 9 shows an example of $\tilde{M} \gtrsim^h \tilde{N}$.

Applying (3.45) to a linear inequality, $\tilde{\mathbf{a}}_i'\mathbf{x}' \gtrsim^h \tilde{\mathbf{b}}_i'\mathbf{x}'$ is equivalent to (3.42) replacing $k \in (0,1]$ with $k \in [h,1]$. When the assumption of (3.43) is fulfilled, $\tilde{\mathbf{a}}_i'\mathbf{x}' \gtrsim^h \tilde{\mathbf{b}}_i'\mathbf{x}'$ is equivalent to

$$\begin{aligned} &\mathbf{a}_i'^L\mathbf{x}' \geqq \mathbf{b}_i'^L\mathbf{x}', \quad \mathbf{a}_i'^R\mathbf{x}' \geqq \mathbf{b}_i'^R\mathbf{x}', \\ &\mathbf{a}_i'^L\mathbf{x}' - \boldsymbol{\alpha}_i^{L*}(h)\mathbf{x}' \geqq \mathbf{b}_i'^L\mathbf{x}' - \boldsymbol{\beta}_i^{L*}(h)\mathbf{x}', \\ &\mathbf{a}_i'^R\mathbf{x}' + \boldsymbol{\alpha}_i^{R*}(h)\mathbf{x}' \geqq \mathbf{b}_i'^R\mathbf{x}' + \boldsymbol{\beta}_i^{R*}(h)\mathbf{x}'. \end{aligned} \qquad (3.46)$$

Since an equality r=s can be represented by two inequalities $r \geqq s$ and $s \geqq r$, we can define an equality relation $\tilde{M}=\tilde{N}$ using the inequality relation $\tilde{M} \gtrsim \tilde{N}$ defined by (3.40), i.e.,

$$\tilde{M}=\tilde{N} \Leftrightarrow \tilde{M} \gtrsim \tilde{N} \text{ and } \tilde{N} \gtrsim \tilde{M} \Leftrightarrow \tilde{M} \vee \tilde{N}=\tilde{M} \text{ and } \tilde{M} \vee \tilde{N}=\tilde{N}. \qquad (3.47)$$

Obviously, $\tilde{M}=\tilde{N}$ if and only if $\mu_{\tilde{M}}(r)=\mu_{\tilde{N}}(r)$ for any $r \in \mathbf{R}$, i.e., Zadeh's fuzzy set equality [29]. We can apply (3.47) to equality constraints, however, it is very rigid because membership functions are often only approximately estimated, as modeling subjective knowledge [19].

Now let us briefly describe the treatment of an objective function in the sense of (3.40). Tanaka et al. [24] treated an objective function, maximize $t(\mathbf{x},\tilde{\mathbf{c}})$, as

$$\text{maximize}\begin{cases}\inf\ [t(\mathbf{x},\tilde{\mathbf{c}})]_h \\ \sup\ [t(\mathbf{x},\tilde{\mathbf{c}})]_h, \quad h\in(0,1],\end{cases} \tag{3.48}$$

where $\tilde{\mathbf{c}}$ is a vector of fuzzy parameters. (3.48) has infinite objective functions. Since (3.48) is not so tractable, we can use

$$\text{maximize}\begin{cases}\inf\ [t(\mathbf{x},\tilde{\mathbf{c}})]_{h_i} \\ \sup\ [t(\mathbf{x},\tilde{\mathbf{c}})]_{h_i}, \quad i=1,2,\ldots,p\end{cases} \tag{3.49}$$

as an approximation of (3.48), where $0<h_1<h_2<\ldots<h_p\leqq 1$. Recently, Rommelfanger et al. [30] introduced fuzzy goals for treating multiple objective functions (3.49) in the similar manner of flexible programming [11].

3.4. Optimality and Efficiency for Possibilistic Programming Problems

Luhandjula [31,32] extended the optimality [32] and the efficiency [31] for usual mathematical programming problems to for possibilistic programming problems. Here, we describe the β-possibly efficient solution [31] proposed by him. Let us consider the following multiobjective possibilistic linear programming problem:

$$\begin{aligned}&\text{maximize}\quad (\tilde{\mathbf{c}}_1\mathbf{x},\tilde{\mathbf{c}}_2\mathbf{x},\ldots,\tilde{\mathbf{c}}_\ell\mathbf{x}),\\ &\text{subject to}\quad \mathbf{a}_i\mathbf{x}\leqq b_i,\ i=1,2,\ldots,m\ ,\ \mathbf{x}\geqq 0,\end{aligned} \tag{3.50}$$

where $\tilde{\mathbf{c}}_i$, $i=1,2,\ldots,\ell$ are n-vectors with fuzzy components whose membership functions are upper semi-continuous. $\mathbf{a}_i$, $i=1,2,\ldots,m$ are usual n-vectors and b_i, $i=1,2,\ldots,m$ are real numbers. For simplicity, F denotes the feasible set of (3.50). A β-possibly efficient solution to (3.50) is defined as follows: $\mathbf{x}^0\in F$ is β-possibly efficient for (3.50), if there is no $\mathbf{x}\in F$ such that

$$\Pi_{\tilde{C}}(\mathbf{x}D\mathbf{x}^0) = \sup_{C\in\mathbf{x}D\mathbf{x}^0}\mu_{\tilde{C}}(C)\geqq\beta\ , \tag{3.51}$$

where $\tilde{C}$ and $\mathbf{x}D\mathbf{x}^0$ is defined by

$$\mu_{\tilde{C}}(\mathbf{c}_1,\mathbf{c}_2,\ldots,\mathbf{c}_\ell) = \bigwedge_{i=1}^{\ell}\mu_{\tilde{\mathbf{c}}_i}(\mathbf{c}_i), \tag{3.52}$$

$$\begin{aligned}\mathbf{x}D\mathbf{x}^0 = \{&(\mathbf{c}_1,\mathbf{c}_2,\ldots,\mathbf{c}_p)/(\mathbf{c}_1\mathbf{x},\mathbf{c}_2\mathbf{x},\ldots,\mathbf{c}_p\mathbf{x})\geqq(\mathbf{c}_1\mathbf{x}^0,\mathbf{c}_2\mathbf{x}^0,\ldots,\mathbf{c}_p\mathbf{x}^0),\\ &(\mathbf{c}_1\mathbf{x},\mathbf{c}_2\mathbf{x},\ldots,\mathbf{c}_p\mathbf{x})\neq(\mathbf{c}_1\mathbf{x}^0,\mathbf{c}_2\mathbf{x}^0,\ldots,\mathbf{c}_p\mathbf{x}^0)\}.\end{aligned} \tag{3.53}$$

$\mu_{\tilde{C}}$ and $\mu_{\tilde{\mathbf{c}}_i}$ are membership functions of $\tilde{C}$ and $\tilde{\mathbf{c}}_i$. $C=(\mathbf{c}_1,\mathbf{c}_2,\ldots,\mathbf{c}_p)$ and each $\mathbf{c}_i$ is an n-vector.

Luhandjula [31] and Luhandjula et al. [28] showed that $\mathbf{x}^0 \in F$ is β-possibly efficient for (3.50) if and only if there exists $\varepsilon > 0$ such that $\mathbf{x}^0$ is efficient for a multiobjective interval linear programming problem,

$$\begin{aligned} &\text{maximize} \quad (\mathbf{c}_1\mathbf{x}, \mathbf{c}_2\mathbf{x}, \ldots, \mathbf{c}_\ell \mathbf{x}), \; \mathbf{c}_j \in [\tilde{\mathbf{c}}_j]_{\beta-\varepsilon}, \; j=1,2,\ldots,\ell, \\ &\text{subject to} \quad \mathbf{a}_i\mathbf{x} \leqq b_i, \; i=1,2,\ldots,m, \; \mathbf{x} \geqq 0, \end{aligned} \tag{3.54}$$

If β and ε are fixed, then (3.54) can be solved by implicit enumeration method proposed by Bitran [33].

Replacing the possibility measure $\Pi_{\tilde{C}}(\mathbf{x}D\mathbf{x}^0)$ in (3.51) by the necessity measure $N_{\tilde{C}}(\mathbf{x}D\mathbf{x}^0)$, we have another concept of efficient solution to (3.50), i.e., $\mathbf{x}^0 \in F$ is β-necessarily efficient for (3.50), if there is no $\mathbf{x} \in F$ such that

$$N_{\tilde{C}}(\mathbf{x}D\mathbf{x}^0) = \inf_{C \notin \mathbf{x}D\mathbf{x}^0} (1-\mu_{\tilde{C}}(C)) \geqq \beta. \tag{3.55}$$

A β-necessarily efficient solution is weaker than a β-possibly efficient solution [34]. Recently, Sakawa and Yano [35,36] and Sakawa et al. [37-39] proposed four kinds of efficiency for possibilistic programming problems based on Dubois and Prade's four ranking indices [21] (see also Section 3.2). Their definitions are different from β-possibly efficiency. However, the interaction [40] between two fuzzy objective values obtained from two different values of decision variables via the same objective functions with fuzzy numbers are discarded in their definitions. Inuiguchi and Kume [34] discussed the relationships among eight solution concepts for fuzzy multiobjective programming problems defined by using possibility measure and necessity measure.

4. MATHEMATICAL PROGRAMMING WITH VAGUENESS AND AMBIGUITY

4.1. Robust Programming

A mathematical programming problem of the form,

$$\begin{aligned} &\text{maximize} \quad \mathbf{c}\mathbf{x}, \\ &\text{subject to} \quad \hat{\mathbf{a}}_i'\mathbf{x}' \subseteq \tilde{b}_i, \; i=1,2,\ldots,m, \; \mathbf{x}'=(1,\mathbf{x}^t)^t \geqq 0, \end{aligned} \tag{4.1}$$

is called a "robust programming problem" [23,41], where $\tilde{\mathbf{c}}$ is a usual n-vector. Each $\tilde{b}_i$ is an L-R fuzzy number $(b_i^L, b_i^R, \beta_i^L, \beta_i^R)_{L_i R_i}$. The inclusion relation $\tilde{M} \subseteq \tilde{N}$ is defined by $\mu_{\tilde{M}}(r) \leqq \mu_{\tilde{N}}(r)$ for any $r \in \mathbf{R}$ [29] (see Figure 10).

Though the set-inclusive constraint $\tilde{a}_i'x' \subseteq \tilde{b}_i$ of (4.1) can be interpreted as an extension of a usual equality constraint as stated in [19], here, we interpret it as a constraint that the possible region of occurrence $\tilde{a}_i'x'$ should be contained in the satisfactory or tolerant region $\tilde{b}_i$ of the decision maker [23]. It is because if we take the former interpretation, the set-inclusive constraint $\tilde{b}_i \subseteq \tilde{a}_i'x'$ is also an extension of an inequality constraint. Besides, the latter interpretation is more understandable than the former. If we define

$$\tilde{b}_i = \{r \;/\; r \geqq b_i\}, \tag{4.2}$$

then the set-inclusive constraint $\tilde{a}_i'x' \subseteq \tilde{b}_i$ can be interpreted as an extension of an inequality constraint [41], where b_i is a real number. This interpretation is not violate the latter interpretation, since $\tilde{b}_i$ defined by (4.2) can be interpreted as the satisfactory region of the decision maker.

The constraint $\tilde{a}_i'x' \subseteq \tilde{b}_i$ is reduced to

$$\begin{aligned} &a_i'^{L}x' - \alpha_i^{L*}(h)x' \geqq b_i^{L} - \beta_i L_i^{*}(h), \\ &a_i'^{R}x' + \alpha_i^{R*}(h)x' \leqq b_i^{R} + \beta_i R_i^{*}(h), \text{ for any } h \in (0,1], \end{aligned} \tag{4.3}$$

where L_i^* and R_i^* are pseudo-inverses of L_i and R_i respectively, as defined in the same way of (3.12). Hence, (4.1) is reduced to an infinitely constrained linear programming problem. However, if all $\tilde{a}_{ij}$ and $\tilde{b}_i$ have bounded supports, $L_{i0}=L_{i1}=\ldots=L_{in}=L_i$ and $R_{i0}=R_{i1}=\ldots=R_{in}=R_i$, $i=1,2,\ldots,m$, then (4.3) is reduced to

$$\begin{aligned} &a_i'^{L}x' \geqq b_i^{L}, \quad a_i'^{R}x' \leqq b_i^{R}, \\ &a_i'^{L}x' - \alpha_i^{L}x' \geqq b_i^{L} - \beta_i^{L}, \\ &a_i'^{R}x' + \alpha_i^{R}x' \leqq b_i^{R} + \beta_i^{R}. \end{aligned} \tag{4.4}$$

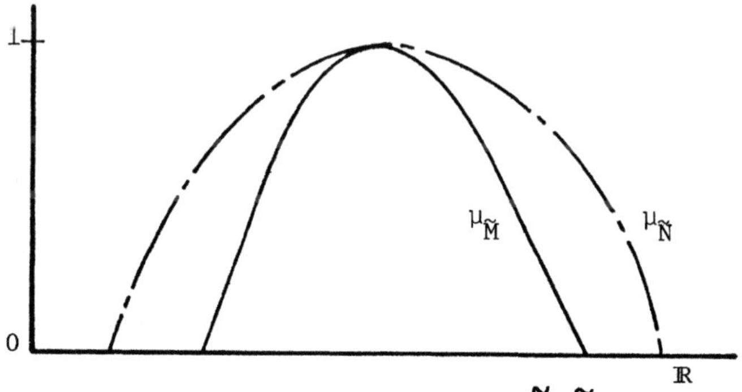

Figure 10. An example of $\tilde{M} \subseteq \tilde{N}$.

Hence, (4.1) is reduced to a linear programming problem with 4m inequality constraints [23].

Set-inclusive constraints with respect to usual (crisp) sets were considered by Soyster [42-44]. He treated it in the case of interactive [40] parameters and discussed the duality theorem with respect to the linear programming problem with set-inclusive constraints [43].

4.2. Possibilistic Programming with Fuzzy Preference Relations.

Orlovski [45] introduced fuzzy preference relations into possibilistic programming problems. He firstly extended a given fuzzy preference relations between elements to the relation between fuzzy sets.

Let $\tilde{\mathcal{R}}$ be a given fuzzy preference relation with the membership function $\mu_{\tilde{\mathcal{R}}}$. It is assumed that $\tilde{\mathcal{R}}$ is reflexive, i.e., $\mu_{\tilde{\mathcal{R}}}(r,r)=1$ for any element r. When two fuzzy sets $\tilde{A}$ and $\tilde{B}$ are compared by using this fuzzy preference relation $\tilde{\mathcal{R}}$, the preference relation $\tilde{\mathcal{R}}$ is extended by

$$\mu_{\tilde{\mathcal{R}}}(\tilde{A},\tilde{B}) = \sup_{r,s} \{\mu_{\tilde{A}}(r) \wedge \mu_{\tilde{B}}(s) \wedge \mu_{\tilde{\mathcal{R}}}(r,s)\}, \tag{4.5}$$

in the spirit of the extension principle. A strict fuzzy preference relation $\tilde{\mathcal{R}}^s$ between fuzzy sets is constructed from the extended fuzzy preference relation $\tilde{\mathcal{R}}$ defined by (4.5);

$$\mu_{\tilde{\mathcal{R}}^s}(\tilde{A},\tilde{B}) = \begin{cases} \mu_{\tilde{\mathcal{R}}}(\tilde{A},\tilde{B}) - \mu_{\tilde{\mathcal{R}}}(\tilde{B},\tilde{A}) \;, & \mu_{\tilde{\mathcal{R}}}(\tilde{A},\tilde{B}) \geqq \mu_{\tilde{\mathcal{R}}}(\tilde{B},\tilde{A}), \\ 0 \;, & \text{otherwise.} \end{cases} \tag{4.6}$$

Now, we describe how to apply these extended fuzzy preference relations to possibilistic programming. Let us consider the following possibilistic programming problem with usual (nonfuzzy) constraint;

$$\begin{aligned} &\text{maximize} \quad t(\mathbf{x},\tilde{\mathbf{c}}), \\ &\text{subject to} \quad \mathbf{x} \in F, \end{aligned} \tag{4.7}$$

where F is a feasible set.

Given a fuzzy preference relation $\tilde{\mathcal{R}}$ in the objective space, we can define a fuzzy preference relation $\tilde{R}$ and a strict fuzzy preference relation $\tilde{R}^s$ in the decision space by using (4.5) and (4.6), i.e.,

$$\mu_{\tilde{R}}(\mathbf{x}_1,\mathbf{x}_2) = \mu_{\tilde{\mathcal{R}}}(t(\mathbf{x}_1,\tilde{\mathbf{c}}),t(\mathbf{x}_2,\tilde{\mathbf{c}})), \tag{4.8}$$

$$\mu_{\tilde{R}^s}(\mathbf{x}_1,\mathbf{x}_2) = \mu_{\tilde{\mathcal{R}}^s}(t(\mathbf{x}_1,\tilde{\mathbf{c}}),t(\mathbf{x}_2,\tilde{\mathbf{c}})). \tag{4.9}$$

To find a solution to (4.7), we shall specify the fuzzy set of nondominated solutions, denoted $\widetilde{ND}$, in the form

$$\mu_{\widetilde{ND}}(\mathbf{x}^0) = 1 - \sup_{\mathbf{x}\in F} \mu_{\tilde{R}^s}(\mathbf{x},\mathbf{x}^0), \tag{4.10}$$

where $\mu_{\widetilde{ND}}$ is a membership function of $\widetilde{ND}$ and $\mu_{\widetilde{ND}}(\mathbf{x}^0)$ represents the degree of nondominance with respect to $\mathbf{x}^0$. Solutions having the maximal membership grade in $\widetilde{ND}$ may be naturally considered as rational choices for (4.7).

The determination of a complete explicit description $\mu_{\widetilde{ND}}$ of the fuzzy set $\widetilde{ND}$ may be difficult and/or not necessary. Orlovski [46] proposed that more realistic and practically important is determination of some solutions $\mathbf{x}^0$ satisfying

$$\mu_{\widetilde{ND}}(\mathbf{x}^0) \geqq h. \tag{4.11}$$

He showed that if $\widetilde{\mathcal{R}}$ is an inequality relation $\geqq$, i.e.,

$$\mu_{\widetilde{\mathcal{R}}}(r,s) = \begin{cases} 1, & r \geqq s, \\ 0, & r<s, \end{cases} \tag{4.12}$$

and the membership function of $\widetilde{\mathbf{c}}$, $\mu_{\widetilde{\mathbf{c}}}$ and t are continuous functions, then a solution satisfying (4.11) is a solution of the following usual mathematical programming problem;

$$\begin{aligned} &\underset{\mathbf{x},\ \mathbf{c}}{\text{maximize}} \quad t(\mathbf{x},\mathbf{c}), \\ &\text{subject to} \quad \mu_{\widetilde{\mathbf{c}}}(\mathbf{c}) \geqq h, \ \mathbf{x} \in F. \end{aligned} \tag{4.13}$$

Moreover, Orlovski discussed a possibilistic programming with fuzzy constraints [46] and a multiobjective possibilistic programming [47].

In this approach, the interaction [40] between two fuzzy objective values obtained from two different values of decision variables via same objective function $t(.,\widetilde{\mathbf{c}})$ is discarded in the definition of $\widetilde{\mathcal{R}}$ by (4.8) via (4.5).

4.3. Possibilistic Linear Programming with Fuzzy Goals

Inuiguchi et al. [48-51] formulated the following possibilistic linear programming regarded as an extension of flexible programming;

$$\widetilde{\mathbf{a}}_i'\mathbf{x}' \gtrsim g_i, \ i=1,2,\ldots,m, \quad \mathbf{x}' \geqq 0. \tag{4.14}$$

Each fuzzy goal $\gtrsim g_i$ has the membership function $\mu_{\widetilde{G}_i}:\mathbf{R}\rightarrow[0,1]$ and a decision set $\widetilde{D}$ is defined by

$$\mu_{\widetilde{D}}(\mathbf{y}) = \mathcal{T}\,(\mu_{\widetilde{G}_1}(y_1),\mu_{\widetilde{G}_2}(y_2),\ldots,\mu_{\widetilde{G}_m}(y_m)), \tag{4.15}$$

where $\mathbf{y}=(y_1,y_2,\ldots,y_m)$. $\mathcal{T}:[0,1]^m\rightarrow[0,1]$ is a monotonically increasing in each component. A decision set can be viewed as a satisfying set of the decision maker and its membership grade may be regarded as the degree of satisfaction.

Assuming each $\widetilde{\mathbf{a}}_i'$ is non-interactive [40] with $\widetilde{\mathbf{a}}_j'$, $i\neq j$, a possibility distribution $\widetilde{Y}$ in $\mathbf{R}^m$ can be defined by

$$\mu_{\widetilde{Y}}(\mathbf{y}) = \bigwedge_{i=1}^{m} \mu_{\widetilde{\mathbf{a}}_i'\mathbf{x}'}(y_i), \tag{4.16}$$

where $\mu_{\tilde{Y}}$ is a membership (possibility distribution) function of $\tilde{Y}$. The possibility distribution $\tilde{Y}$ represents the region of values which **y** possibly take with possibility degree.

Using possibility measure and necessity measure, two excessive decision procedures can be considered. One of them is maximizing possibility measure, i.e.,

$$\text{maximize} \quad \Pi_{\tilde{Y}}(\tilde{D}), \tag{4.17}$$

which means maximizing the possibility degree of satisfaction. Another is maximizing necessity measure;

$$\text{maximize} \quad N_{\tilde{Y}}(\tilde{D}), \tag{4.18}$$

which means maximizing the necessity degree of satisfaction.

Inuiguchi et al. [49] showed that (4.17) and (4.18) are reduced to the following non-linear programming problems respectively;

$$\begin{aligned} &\text{maximize} \quad \tau(h_1,h_2,\dots,h_m), \\ &\text{subject to} \\ &\qquad \mu^*_{\tilde{G}_i}(h_i) \leqq \mathbf{a}_i'^R\mathbf{x}' + \boldsymbol{\alpha}_i^{R*}(\tau(h_1,h_2,\dots,h_m))\mathbf{x}', \ i=1,2,\dots,m, \\ &\qquad \mathbf{x}' \geqq 0, \ 0 < h_i \leqq 1, \ i=1,2,\dots,m, \end{aligned} \tag{4.19}$$

$$\begin{aligned} &\text{maximize} \quad \tau(h_1,h_2,\dots,h_m), \\ &\text{subject to} \\ &\qquad \mu^*_{\tilde{G}_i}(h_i) \leqq \mathbf{a}_i'^L\mathbf{x}' - \boldsymbol{\alpha}_i^{L*}(1-\tau(h_1,h_2,\dots,h_m))\mathbf{x}', \ i=1,2,\dots,m, \\ &\qquad \mathbf{x}' \geqq 0, \ 0 < h_i \leqq 1, \ i=1,2,\dots,m, \end{aligned} \tag{4.20}$$

where $\mu^*_{\tilde{G}_i}$ is a pseudo-inverse of $\mu_{\tilde{G}_i}$ and defined by

$$\mu^*_{\tilde{G}_i}(h) = \inf\left\{r \mid \mu_{\tilde{G}_i}(r) \geqq 0\right\}, \ i=1,2,\dots,m. \tag{4.21}$$

They also propose the solution algorithm using simplex method repetitionally. When

$$\tau(r_1,r_2,\dots,r_m) = \bigwedge_{i=1}^{m} r_i, \tag{4.22}$$

then the algorithm [48,49] is almost equivalent to that proposed by Tanaka et al. [16,17] (see Section 3.1).

A decision procedure

$$\text{maximize} \quad \Pi_{\tilde{Y}}(\tilde{D}) + N_{\tilde{Y}}(\tilde{D}) \tag{4.23}$$

was proposed as an intermediate decision procedure between (4.17) and (4.18) by Inuiguchi et al. [50,51]. This decision procedure is induced via the concept of relative modalities. (4.23) is also reduced to a non-linear programming problem [50,51].

Buckley [52] formulates a possibilistic linear programming problem in the similar manner. He use an associate operation in stead of an minimum operation in (4.16) and instead of τ in (4.15).

5. CONCLUDING REMARKS

Various formulations of fuzzy programming problems have been surveyed classifying into three categories, i.e., mathematical programming with vagueness, mathematical programming with ambiguity and mathematical programming with vagueness and ambiguity. As demonstrated in the previous two sections, many formulations of fuzzy (possibilistic) programming problems can be derived depending on the interpretations of the problems. A unifying model for treating fuzzy programming problems, which encompasses as many formulations as possible, will be desired. A few attempts along this line are made by Ramík [53], Inuiguchi et al. [54] and Campos and Verdegay [55].

In most of formulations of possibilistic programming problems, the objective functions are treated as constraints using the information about the preferences or the aspirations of decision makers. Namely, they seem to intend to satisfy rather than to optimize. Hence, the new treatments of objective functions which are not reducing to constraints are of interest. Approaches proposed by Luhandjula [31,32], Sakawa et al. [35-39] and Inuiguchi and Kume [34] (see Section 3.4) will indicate this research directions.

An application of possibilistic linear programming to water supply planning was made by Słowiński [56,57]. An interactive method was proposed in [56,57] and many fuzzy linear programming methods were interestedly compared in [56]. Recently, Luhandjula [58] also reviewed many fuzzy programming methods.

References

[1] Klir, G.J. and Folger, T.A. (1988) Fuzzy Sets, Uncertainty, and Information, Prentice Hall, Englewood Cliffs, New Jersey.

[2] Tanaka, H., Okuda, T. and Asai, K. (1974) 'On fuzzy mathematical programming', **Journal of Cybernetics 3**, 37-46.

[3] Zimmermann, H.J. (1976) 'Description and optimization of fuzzy systems', **International Journal of General Systems 2**, 209-215.

[4] Zimmermann, H.J. (1985) 'Applications of fuzzy sets theory to mathematical programming', **Information Sciences 36**, 29-58.

[6] Luhandjula, M.K. (1982) 'Compensatory operations in fuzzy linear programming with multiple objectives', **Fuzzy Sets and Systems 8**, 245-252.

[7] Sommer, G. and Pollatschek, M.A. (1978) 'A fuzzy programming approach to an air pollution regulation problem', in G.J. Klir and L. Ricciardi (eds.), **Progress in Cybernetics and Systems Research**, Heimsphere Publ. Corp., Washington D.C., pp. 303-323.

[8] Zimmermann, H.J. and Zysno, P. (1980) 'Latent connectives in human decision making', **Fuzzy Sets and Systems 4**, 37-52.

[9] Dubois, D. and Prade, H. (1985) 'A review of fuzzy set aggregation

connectives', **Information Sciences 36,** 85-121.
[10] Ichihashi, H., Tanaka, H. and Asai, K. (1986) 'Fuzzy integrals and their applications to multiple attribute decision problems (in Japanese)', **Transactions of the Society of Instrument and Control Engineers 22,** 557-562.
[11] Zimmermann, H.J. (1978) 'Fuzzy programming and linear programming with several objective functions', **Fuzzy Sets and Systems 1,** 45-55.
[12] Inuiguchi, M., Ichihashi, H. and Kume, Y. (1989) 'A solution algorithm for fuzzy linear programming with piecewise linear membership functions', **Fuzzy Sets and Systems** (to appear).
[13] Sakawa, M. and Yano, H. (1985) 'An interactive fuzzy satisficing method using augmented minimax problems and its application to environmental systems', **IEEE Transactions on Systems, Man, and Cybernetics SMC-15,** 720-729.
[14] Sakawa, M. and Yano, H. (1986) 'An interactive fuzzy decision making method using constraint problems', **IEEE Transactions on Systems, Man, and Cybernetics SMC-16,** 179-182.
[15] Sakawa, M. and Yano, H. (1986) 'Interactive fuzzy decision making for multiobjective nonlinear programming Using augmented minimax problems', **Fuzzy Sets and Systems 20,** 31-43.
[16] Tanaka, H. and Asai, K. (1984) 'Fuzzy linear programming with fuzzy numbers', **Fuzzy Sets and Systems 13,** 1-10.
[17] Tanaka, H., Ichihashi, H. and Asai,K. (1985) 'Fuzzy decision in linear programming with trapezoid fuzzy parameters', in J. Kacprzyk and R.R. Yager (eds.), **Management Decision Support Systems Using Fuzzy Sets and Possibility Theory,** Verlag TÜV, Köln, pp. 146-154.
[18] Dubois, D. and Prade, H. (1987) 'Fuzzy numbers: An overview', in J.C. Bezek (ed.), **Analysis of Fuzzy Information, Volume 1:Mathematics and Logic,** CRC Press, Boca Raton, FL, pp. 3-39.
[19] Dubois, D. (1987) 'Linear programming with fuzzy data', in J.C. Bezek (ed.), **Analysis of Fuzzy Information, Volume 3: Applications in Engineering and Science,** CRC Press, Boca Raton, FL, pp. 241-263.
[20] Vajda, S. (1972) Probabilistic Programming, Academic Press, New York and London.
[21] Dubois, D. and Prade, H. (1983) 'Ranking fuzzy numbers in the setting of possibility theory', **Information Sciences 30,** 183-224.
[22] Inuiguchi, M., Ichihashi, H. and Tanaka, H. (1988) 'The extension of fuzzy preference relation based on the modal concept and its properties (in Japanese)', **Transactions of the Society of Instrument and Control Engineers 24,** 738-745.
[23] Dubois, D. and Prade, H. (1980) 'Systems of linear fuzzy constraints', **Fuzzy Sets and Systems 3,** 37-48.
[24] Tanaka, H., Ichihashi, H. and Asai, K. (1984) 'A formulation of fuzzy linear programming problem based on comparison of fuzzy numbers', **Control and Cybernetics 13,** 185-194.
[25] Ramík, J. and Římánek, J. (1985) 'Inequality relation between fuzzy numbers and its use in fuzzy optimization', **Fuzzy Sets and Systems 16,** 123-138.
[26] Blankenship, J.W. and Falk, J.E. (1976) 'Infinitely constrained optimization problems', **Journal of Optimization Theory and**

Applications 19, 261-281.
[27] Pence, J.A. and Soyster, A.L. (1989) 'The relationship between fuzzy set programming and semi-infinite linear programming', in G.W. Evans, W. Karwowski and M.R. Wilhelm (eds.), **Applications of Fuzzy Set Methodologies in Industrial Engineering**, Elesevir, Amsterdam, pp. 237-252.
[28] Luhandjula, M.K., Ichihashi, H. and Inuiguchi, M. (1989) 'Fuzzy and semi-infinite optimization', (submitted).
[29] Zadeh, L.A. (1965) 'Fuzzy Sets', **Information and Control 8**, 338-353.
[30] Rommelfanger, H., Hanuscheck, R. and Wolf, J. (1989) 'Linear programming with fuzzy objectives', **Fuzzy Sets and Systems 29**, 31-48.
[31] Luhandjula, M.K. (1987) 'Multiple objective programming problems with possibilistic coefficients', **Fuzzy Sets and Systems 21**, 135-145.
[32] Luhandjula, M.K. (1987) 'Linear programming with a possibilistic objective function', **European Journal of Operational Research 31**, 110-117.
[33] Bitran, G.R. (1980) 'Linear multiple objective problems with interval coefficients', **Management Science 26**, 694-706.
[34] Inuiguchi, M. and Kume, Y. (1989) 'Solution concepts for fuzzy multiple objective programming problems (in Japanese)', **Journal of Japan Society for Fuzzy Theory and Systems**, (to appear).
[35] Sakawa, M. and Yano, H. (1988) 'A unified approach for multiobjective linear programming problems with fuzzy parameters (in Japanese)', **The Transactions of the Institute of Electronics, Information and Communication Engineers A J71-A** 1569-1575.
[36] Sakawa, M. and Yano, H. (1989) 'Feasibility and Pareto optimality for multiobjective nonlinear programming problems with fuzzy parameters (in Japanese)', **The Transactions of the Institute of Electronics, Information and Communication Engineers A J72-A**, 1062-1068.
[37] Sakawa, M., Yano, H. and Takahashi, J. (1989) 'Extended Pareto optimality concept for multiobjective linear programming problems with fuzzy parameters and its properties (in Japanese)', **The Transactions of the Institute of Electronics, Information and Communication Engineers A J72-A**, 678-684.
[38] Sakawa, M., Yano, H. and Takahashi, J. (1989) 'Pareto optimality for multiobjective linear fractional programming problems with fuzzy parameters (in Japanese)', **The Transaction of the Institute of Electronics, Information and Communication Engineers A J72-A**, 931-937.
[39] Sakawa, M., Yano, H. and Takahashi, J. (1989) 'Pareto optimality for multiobjective linear fractional programming problems with fuzzy parameters', in J.C. Bezek (ed.), Proceedings of the 1989 IFSA Congress; The Coming of Age of Fuzzy Logic, Seattle (August 6-11), 303-306.
[40] Zadeh, L.A. (1978) 'Fuzzy sets as a basis for a theory of possibility, **Fuzzy Sets and Systems 1**, 3-28.

[41] Negoita, C.V. (1981) 'The current interest in fuzzy optimization', **Fuzzy Sets and Systems 6**, 261-269.
[42] Soyster, A.L. (1973) 'Convex programming with set-inclusive constraints and applications to inexact linear programming', **Operations Research 21**, 1154-1157.
[43] Soyster, A.L. (1973) 'A duality theory for convex programming with set-inclusive constraints, **Operations Research 22**, 892-898.
[44] Soyster, A.L. (1979) 'Inexact linear programming with generalized resource sets', **European Journal of Operations Research 3**, 361-321.
[45] Orlovsky, S.A. (1980) 'On formalization of a general fuzzy mathematical programming problem', **Fuzzy Sets and Systems 3**, 311-321.
[46] Orlovski, S.A. (1985) 'Mathematical programming problems with fuzzy parameters', in J. Kacprzyk and R.R. Yager (eds.), **Management Decision Support Systems using Fuzzy Sets and Possibility Theory**, Verlag TÜV, Köln, pp.136-145.
[47] Orlovski, S.A. (1984) 'Mulitobjective programming problems with fuzzy parameters', **Control and Cybernetics 13**, 175-183.
[48] Inuiguchi, M., Ichihashi, H., Tanaka, H. and Asai, K. (1986) 'Fuzzy multiobjective linear programming based on possibility and necessity (in Japanese)', Second Fuzzy System Symposium held by IFSA Japan Chapter, Tokyo (June 16-18), pp. 172-178.
[49] Inuiguchi, M., Ichihashi, H. and Tanaka, H. (1989) 'Possibilistic linear programming with measurable multiattribute value functions', **ORSA Journal on Computing 1**, 146-158.
[50] Inuiguchi, M., Ichihashi, H. and Tanaka, H. (1987) 'A formulation of linear programming problem with fuzzy coefficients based on the concept of relative modalities', Preprints of Second IFSA Congress, Tokyo (July 20-25), pp. 120-123.
[51] Inuiguchi, M. and Ichihashi, H. (1989) 'Relative modalities and their use in possibilistic linear programming', **Fuzzy Sets and Systems** (to appear).
[52] Buckley, J.J. (1988) 'Possibility and necessity in optimization', **Fuzzy Sets and Systems 25**, 1-13.
[53] Ramík, J. (1987) 'A unified approach to fuzzy optimization', Preprints of Second IFSA Congress, Tokyo (July 20-25), pp. 128-130.
[54] Inuiguchi, M., Ichihashi, H. and Kume, Y. (1989) 'A modalistic formalization of fuzzy mathematical programming problems (in Japanese)', **Transactions of the Institute of Systems Control and Information Engineers 2**, 69-79.
[55] Campos, L. and Verdegay, J.L. (1989) 'Linear programming problems and ranking of fuzzy numbers', **Fuzzy Sets and Systems 32**, 1-11.
[56] Słowiński, R. (1986) 'A multicriteria fuzzy linear programming method for water supply system development planning', **Fuzzy Sets and Systems 19**, 217-237.
[57] Słowiński, R. (1987) 'An interactive method for multiobjective linear programming with fuzzy parameters and its application to water supply planning', in J. Kacprzyk and S.A. Orlovski (eds.), **Optimization Models Using Fuzzy Sets and Possibility Theory**, D. Reidel Publishing Company, Dordrecht, pp. 396-414.
[58] Luhandjula, M.K. (1989) 'Fuzzy optimization : An appraisal', **Fuzzy Sets and Systems 30**, 257-282.

Part II.
THE STOCHASTIC APPROACH

Part II
Chapter 1

OVERVIEW OF DIFFERENT APPROACHES FOR SOLVING STOCHASTIC PROGRAMMING PROBLEMS WITH MULTIPLE OBJECTIVE FUNCTIONS

I.M.STANCU-MINASIAN
Laboratories of the Department of Economic Cybernetics
Calea Dorobanţi 15-17
71131 Bucharest, Romania

1. INTRODUCTION

Stochastic programming is one of the most exciting and challenging developments of mathematical programming. It aims to treat uncertainty within decision oriented models in a coherent and systematic way. Lack of such an approach is one of the objections raised to deterministic mathematical programming modelling. The requirement for a single objective or payoff functions is another objection; it can be argued that most decision makers usually have several decision criteria, and multi-objective programming aims to reflect this. Also simple examples show (similar to the Endorsed paradox and Arrow impossibility theorems) that there are, in general, no good ways of aggregating several criteria into one objective function. But maybe sometimes there are. Worse, even when there is a natural objective function, but stochastic elements come into play maximizing the expectation will often involve unacceptable large variances. In this way, a new interdisciplinary science is about to be born-stochastic programming with several objective functions.

The problem under consideration is encountered in several contexts. For instance, Lau /33/ considers the newsboy problem and three objective functions: maximizing the expected profit, b) maximizing expected utility, c) maximizing the probability of achieving a budget profit. Teghem, Jr. and Kunsch consider /32/, /66/ two energy planning problems. For more examples see also and Stancu-Minasian's book /54/ where applications are offered in areas commonly studied with single criteria deterministic models but which may be more appropriately modeled as stochastic multiple criteria decision maker problems: production planning, operation scheduling, assignment problems,

R. Slowinski and J. Teghem (eds.),
Stochastic versus Fuzzy Approaches to Multiobjective Mathematical Programming under Uncertainty, 71–101.

etc. Also, for applications of probabilistic goal programming see /11/ and /27/ and for stochastic approach in multicriteria structural design see /42/.

Consider the following problem of stochastic linear programming with multiple objective functions:

$$(1.1) \qquad \gamma_k(\omega) = \max_{x \in X(\omega)} \left\{ Z_k(x,\omega) = c_k'(\omega)x \right\} \quad k=1,\ldots,r$$

$$(1.2) \qquad X(\omega) = \left\{ x \mid A(\omega)x \leq b(\omega), x \geq 0 \right\}$$

where $A(\omega)=(a_{ij}(\omega))$, $1 \leq i \leq m$, $1 \leq j \leq n$, $b(\omega)=(b_i(\omega))$, $1 \leq i \leq m$ and $c_k(\omega)=(c_{kl}(\omega))$, $1 \leq k \leq r$, $1 \leq l \leq n$ are matrices of real-valued random variables defined on the probability space $\{\Omega, K, P\}$: Ω is a Borel set in R^s ($s = mn + m + rn$), K is the σ-algebra of all Borel subsets of Ω, P is a probability measure and x is an n-dimensional column vector.

Because the optimal values $\gamma_k(\omega)(k=1,\ldots,r)$ are random variables, they cannot be known beforehand. However, we wish to determine the probability distribution and/or certain moments (e.g. the mean value and the variance) of the random vector $\gamma(\omega)=(\gamma_1(\omega),\ldots,\gamma_r(\omega))$ subject to a certain a priori probability distribution of random elements contained in $\{A(\omega), b(\omega), (c_1(\omega),\ldots,c_r(\omega))\}$.

In this chapter we give an overview of different approaches for solving stochastic programming problems with multiple objective functions. The material presented is a synthesis and also a generalization of our papers /53/ and /54/.

Although in recent years there seems to be an increasing interest in stochastic programming problems and that of mathematical programming with multiple objective functions (for edification see the research bibliographies /45/ and /48/) however the field of stochastic programming with multiple objective functions was partly neglected in the European and American literature.

The methods discussed in this chapter are generalizations of the methods used for solving the deterministic and stochastic programming problems. Due to this fact, deterministic mathematical programming with multiple objectives is reviewed in § 2 and stochastic programming with one objective function is reviewed in § 3 with an emphasis on those results or methods which can be extended to the multiple criteria stochastic programming problem.

The solution of the stochastic programming problems with multiple objective functions can be obtained in the following manner:

1) either seek to replace the stochastic programming problem

with multiple objective functions by an equivalent deterministic problem with multiple objective functions (as done on some stochastic programming approaches for single criteria). Next, one uses the methods which will be presented in § 2.

2) or seek to reduce the problem to a stochastic problem with only one objective function (using synthesis functions, distance metrics, goal programming, etc.). Now solve the single criteria stochastic problem using the approaches which will be presented in § 3.

As we shall see, the solution of the mathematical programming problems with several deterministic objective functions raises several difficulties which stem from the fact that the solution vector is not uniquely defined. Further difficulties appear in solving the stochastic programming problems with one objective function. This explains why in solving the stochastic programming problems with multiple objective functions one is confronted with almost intractable difficulties and ambiguities.

2. POINTS OF VIEW IN SOLVING THE DETERMINISTIC MATHEMATICAL PROGRAMMING PROBLEM WITH SEVERAL OBJECTIVE FUNCTIONS

Consider the following mathematical programming problem with several objective functions

$$(2.1) \qquad \max_{x \in D} \{ F_k(x) = c_k' x \} \qquad k=1,\dots,r$$

where $D = \{ x \mid x \in R^n, Ax \leqslant b, x \geqslant 0 \}$, A, b, c_k and x are matrices of size $m \times n$, $m \times 1$, $1 \times n$, $n \times 1$, respectively.

The solving of this problem consists in finding the vector (or vectors) $x^* = (x_1^*, \dots, x_n^*) \in D$ which is the best for the assembly of objective functions F_k $(k=1,\dots,r)$.

Since the definition of x^* is somewhat ambiguous (the best is not quantifiable and insufficiently clear), several views have already been expressed with respect to the definition and solution of a multiple criteria problem. Thus, one has to introduce the notion of the best compromise solution, also know as nondominated solution, efficient solution, noninferior solution, Pareto's optimum solution, etc. The various ways of defining x^* can be grouped as follows:

1°. Let x^* be the vector which optimizes (maximizes or minimizes) a synthesis-function of the r efficiency functions: $h(F) = h(F_1,\dots,F_r)$

Several expressions are possible for h /54/:

a) $h(F_1,\dots,F_r) = \underset{i=1,\dots,r}{\text{optimum}} \{F_i(x)\}$

For instance, if F_i are maximum functions, one takes $h(F_1,\dots,F_r) = \min\limits_{i=1,\dots,r} \{F_i(x)\}$ and then maximizes it, i.e. $\max\limits_{x\in D} \min\limits_{1\leq i\leq r} \{F_i(x)\}$.

Thus, one obtains the so-called Chebyshev problem, the solution of which can be obtained through several algorithms available in the literature /7/.

b) For r=2, one consider that $h(F_1,F_2) = F_1(x)/F_2(x)$, which entails an ordinary fractional programming problem.

2°. Let x^* be the vector which minimizes the function

$$(2.2) \qquad h(x) = h(\Psi_1(x-X_1),\dots,\Psi_r(x-X_r)),$$

where $X_j=(x_{1j},\dots,x_{nj})'(j=1,\dots,r)$ is the optimal solution of the objective function F_j, and Ψ_j is the distance function between the vector $x\in D$ and X_j. Certain expressions for h and Ψ_j can lead to particular forms of (2.2).

3°. Let x^* be a point in the domain of feasible solutions which can be obtained through a search procedure based on various criteria. For instance, one can use the POP (Progressive Orientation Procedure) /2/ and STEM (Step Method) methods /3/.

4°. Let x^* be an optimum solution obtained by ordering the criteria in the following way /34/: solve r mathematical programming problems restraining each time the domain D by changing into constraints the optimum solutions obtained by solving a certain problem having only one objective function. That is, consider the following sequence of domains:

$$D_o^* = D,\ D_i^* = \{x \mid F_i(x) = \operatorname*{optimum}_{y\in D_{i-1}^*} F_i(y);\ x\in D_{i-1}^*\} (i=1,\dots,r)$$

Solving the mathematical programming problem with several objective functions means to find one or more points of the set D_r^*. Obviously, the set D_r^* depends on the way in which functions are ordered. Usually, to two different ordering schemes there correspond different sets of points.

5°. Let x^* belong to a set of efficient points defined as follows:

Definition 2.1. The point $x^*\in D$ is efficient point (solution) if and only if there exists no $x\in D$ such that $F_h(x)\geq F_h(x^*)$ for $h=1,\dots,r$ and such that for at least one h_o one has $F_{h_o}(x) > F_{h_o}(x^*)$ (assuming that maximum is sought for all objective functions). That is, x^* is an efficient point if it has the property that there exists no other point x to improve at least one objective function while the others remain unchanged.

The determination of efficient points is the most general method among the optimization methods of mathematical programming with multiple objective functions.

A procedure for obtaining efficient solutions by use parametric programming is given by the following.

Theorem 2.1. The feasible solution x^* is efficient for problem (2.1) if and only if it is optimal for the problem

$$P_\lambda \qquad \max_{x \in D} \sum_{i=1}^{r} \lambda_i F_i(x)$$

with a given $\lambda \in \Lambda = \left\{ \lambda = (\lambda_1, \ldots, \lambda_r) \in R^r : 0 < \lambda_i < 1, \sum_{i=1}^{n} \lambda_i = 1 \right\}$

The parameter space Λ can be decomposed into a finite number of polyhedra $\Lambda(x_i)$ (which correspond to different nondominated extreme points of D) and open half-spaces $\tilde{\Lambda}_j$. These polyhedra are connected by their boundary hyperplanes. The polyhedral set $\Lambda(x_i)$ consists of all λ such that x_i is an optimal solution of P_λ and the problem P_λ has no finite optimal solution for all $\lambda \in \tilde{\Lambda}_j$.

In connection with the structure of the space Λ of λ - values, in /40/, Sengupta, Podrebarac and Fernando put the following question "Given a basic feasible solution, what is the probability that it is also optimal? Or, equivalently, given that a solution is basic optimal, what is the probability it may be associated with an assigned region in the space Λ ?".

6°. The point $x^* \in D$ is obtained by using goal programming method introduced by Charnes and Cooper /10/.

Consider the goal vector $\bar{F} = (\bar{F}_1, \bar{F}_2, \ldots, \bar{F}_r)$ where the $\bar{F}_i$'s are the levels (goals, target values) to be reached by the objective functions $F_i (i \in I = \{1, \ldots, r\})$. Obviously, it is usually nearly imposible to find a point $x^* \in D$ such that the objective functions are of the required levels, $F_1(x^*) = \bar{F}_1, \ldots, F_r(x^*) = \bar{F}_r$.

For a given $x \in D$, certain differences (plus or minus) will exist between $F_i(x)$ and $\bar{F}_i$ and hence the problem consists in minimizing the distance between the vector $\bar{F}$ and the vector containing the possible values of the objective functions. Thus, one obtains the problem

(2.3) $\min g(x,y,z) = d(\bar{F},F(x))$ subject to

(2.4) $F_i(x)-y_i+z_i = \bar{F}_i,\ i \in I$

(2.5) $x \in D, y=(y_1,\ldots,y_r) \geqslant 0,\ z = (z_1,\ldots,z_r) \geqslant 0$

where y and z are unknown deviational vectors, whose components measure the deviation upwards and downwards, respectively, of $F_i(x)$ value from the corresponding goal values $\bar{F}_i$ $(i \in I)$, $g: R_+^{n+2r} \longrightarrow R_+$ represents a global deviation function depending on the deviational vectors y and z.

For any two points $x,\bar{x} \in R^n$, because $d(x,\bar{x}) = \|x-\bar{x}\|$ (the classical norm) we can use instead of the distance between x and $\bar{x}$, the norm of the difference of vectors x and $\bar{x}$, as a measure of approach between x and $\bar{x}$. The best known is norm L_p, or the Hölder's norm

$\|x\|_p = (\sum_{i=1}^{r} |x_i|^p)^{1/p}$, $p \geqslant 1$. In particular cases the following norms are obtained: $\|x\|_1 = \sum_i |x_i|$, $\|x\|_2 = (\sum_i |x_i|^2)^{1/2}$ (Euclidean distance), $\|x\|_\infty = \max_i\{|x_i|\}$.

Applying these norms to $F-\bar{F}$ leads to some goal programming models. If we consider for example $\|x\|_1$ and Euclidean distance we obtain

Model 1 : $\min_{x \in D} \|F-\bar{F}\|_1 = \sum_{i=1}^{r} |F_i-\bar{F}_i|$

Model 2: $\min_{x \in D} \|F-\bar{F}\|_2 = \left[\sum_{i=1}^{r} |F_i-\bar{F}_i|^2\right]^{1/2}$

Model 2': $\min_{x \in D} \|F-\bar{F}\|_2 = \sum_{i=1}^{r} \alpha_i (F_i-\bar{F}_i)^2$

where α_i are weighting coefficients for F_i.

3. STOCHASTIC PROGRAMMING WITH ONE OBJECTIVE FUNCTION

Consider the linear programming problem

(3.1) $\max\ (\min)\ Z = c'x$ subject to

(3.2) $x \in D = \{x \mid Ax \leqslant b, x \geqslant 0\}$

where $A = (a_{ij})$, $(i=1,\ldots,m;\ j=1,\ldots,n)$ is an $m \times n (m < n)$ matrix of rank m , c and x are n-vectors, and b is an m-vector.

Such a problem is called a stochastic programming problem if one or more of the coefficients from the set (A,b,c) are random variables on the probability space

$\{\Omega, K, P\}$.

Obviously, when the coefficients are random variables, the ordinary problem of linear programming applies no longer since (for a minimum problem, for example) one cannot find a feasible vector x_0 such that $c'(\omega)x_0 \leqslant c'(\omega)x$ for every $\omega \in \Omega$ and for every feasible x.

Usually, the stochastic problem is replaced by a suitable deterministic problem (called the Deterministic Equivalent) whose optimal solutions are extended to represent solutions to the stochastic problem.

The following sections may be distinguished in the field of stochastic programming : a) Distribution problems, b) Two-stage programming under uncertainty, c) Chance-constrained programming.

A more detailed classification of stochastic programming problems can be found in /48/.

The distribution problems consist in finding the distribution function and/or certain typical values (the mean and the variance) of $Z(\omega)$ subject to some a priori distributions of the triplet $(A(\omega), b(\omega), c(\omega))$.

In chance-constrained programming we do not require that the constraints (3.2) should always be satisfied, but rather that they be satisfied in a proportion of cases or, in other words, with certain given probabilities. Thus, the constraints (3.2) appear as $P(\sum_{j=1}^{n} a_{ij}x_j \leqslant b_i) \geqslant p_i \quad (i=1,\dots,m)$

where the p_i are given probabilities. The chance constraints can also appear as $P(\sum_{j=1}^{n} a_{ij}x_j \leqslant b_i,\ i=1,\dots,m) \geqslant p$, i.e. as a threshold imposed on the aggregated probability that the constraints are satisfied.

In two-stage programming under uncertainty a decision is first taken before the value of the random variables are know and then, after the random events have happened and their value is known, a second decision is used in order to minimize the "penalties" that may appear duo to any infeasibility.

Now consider the stochastic programming problem (3.1)-(3.2) under the form

$$(3.3) \qquad \max\left\{c'(\omega)x \mid Ax \leqslant b, x \geqslant 0\right\}$$

where c has a normal distribution with mean value $\bar{c}$ and covariance matrix V. Then, $c'(\omega)x$ is also a normal variable, with mean value $\bar{c}'x$ and variance $x'Vx$. The problem can be reformulated in deterministic terms by associating to it one of the following functions: a) The maximization of the mean value (called, E-model)

$$(3.4)\quad \max_{x \in D} f_1 = \max_{x \in D} E\,(c'(\omega)x) = \max_{x \in D} \bar{c}'x$$

b) The minimization of the variance (called, V-model)

$$(3.5)\quad \min_{x \in D} f_2 = \min_{x \in D} D^2(c'(\omega)x) = \min_{x \in D} x'Vx$$

c) The maximization of the mean value with a constraint on the variance. In this approach, the objective function is as for case a), but a constraint row is added to the formulation. An equivalent formulation is to minimize the variance subject to an equality constraint on the mean value and parametrize the mean value.

d) The minimization of the second-order moment.

$$(3.6)\quad \min_{x \in D} f_3 = \min_{x \in D} E\,(c'(\omega)x)^2$$

e) Maximization of the probability that the value of the function exceeds a given level k (one is satisfied if the objective function takes at least value k) (called, P-model)

$$(3.7)\quad \max_{x \in D} f_4 = \max_{x \in D} \left\{ P_k(c'x) = P\{\omega \mid c'(\omega)x \geq k\} \right\}$$

This is the minimum-risk model. Geoffrion /19/ calls it the "aspiration criterion". An extension of the minimum-risk criterion to the nonlinear mathematical programming case was given by Stancu-Minasian and Tigan /57/, /60/.

The solution of problem (3.7) is given by the solution of the following fractional programming problem

$$(3.7')\quad \max_{x \in D} (c'x-k)/(x'Vx)^{1/2}$$

f) Maximization of the α-fractile of the cumulative distribution of the objective function with the value of α preassigned by the decision-maker.

$$(3.8)\quad \max f_5 = \max\,(f = F_\alpha(c'x)) \text{ such that}$$

$$P\{\omega \mid c'(\omega)x \geq f\} \geq \alpha \qquad \text{and } x \in D.$$

This model was first introduced by Kataoka /30/ for the case of stochastic linear programming and was later described by Geoffrion /19/ as the "fractile criterion" model.

An interesting case of models e) and f) has been considered by Ishii and Nishida /28/ in which the lower allowable limit f and the specified probability level α are combined with the following type of objective function.

$$(3.9)\quad \min\,(\max)\ f - \lambda\alpha \qquad (\lambda \text{ positive scalar})$$

g) Gupta and Swarup /26/ considered also the minimization of the upper permisible limit "f" of $(c'x-\bar{c}'x)^2$ for a given probability level p.

$$(3.10) \quad \min f \text{ such that } P((c'x-\bar{c}'x)^2 \leqslant f) = p \text{ and } x \in D.$$

The optimal solution of this problem would tend to minimize the variance of $c'x$. This model might be prefered to model (3.5) (V-model) since an optimal solution x_0 of (3.5) may be more risky in the sense that the probability of obtaining a very high value of $\zeta = (c'x-\bar{c}'x)^2$ for x_0 is greater than for some other value of x because of the random variable ζ.

4. STOCHASTIC CHEBYSHEV PROBLEM

In /51/ Stancu-Minasian considered the following multiple criteria stochastic programming problem

$$(4.1) \quad \min_{x \in D} \left\{ Z_k(x,\omega) = (c^k(\omega))'x \right\} \qquad k=1,\ldots,r$$

where the elements of the vectors c^k are stochastic variables with known (joint) probability distribution.

Taking into account 1° - a (section 2) this problem can be reduced to one with a single objective function as follows: Instead of r objective functions $Z_k(x,\omega)$ we consider a single objective function $\eta(x,\omega) \triangleq \max_{1\leqslant k\leqslant r} \{Z_k(x,\omega) = (c^k(\omega))'x\}$ and minimize it. In this manner, we obtain the so-called Chebyshev problem

$$(4.2) \quad \zeta(\omega) = \min_{x \in D} \max_{1\leqslant k\leqslant r} \left\{ Z_k(x,\omega) = (c^k(\omega))'x \right\}$$

In /51/ is determined the probability distribution function of $\zeta(\omega)$, when the cost coefficients of $Z_k(x,\omega)$ are linear functions of the same random variable $t(\omega)$, that is they are under the form

$$(4.3) \quad Z_k(x,\omega) = (c_1^k + t(\omega)c_2^k)'x$$

where c_1^k and c_2^k are n-dimensional constant vectors and $t(\omega)$ is a random variable having a distribution function $T(z)$ continuous and strictly increasing.

For solving problem (4.2) we can consider the minimum-risk approach (see, Tigan and Stancu-Minasian /68/).

We shall consider an even more general case of (4.3) i.e.

$$(4.3') \quad Z_k(x,\omega) = (c_1^k + t(\omega)c_2^k)'x + \alpha_1^k + t(\omega)\alpha_2^k = (c^k(\omega))'x + \alpha^k(\omega)$$

where α_1^k, α_2^k $(k=1,\dots,r)$ are scalar constants.

We assume that

(4.4) $\quad (c_2^k)'x + \alpha_2^k > 0$ for all $x \in D$, and $k \in I = \{1,\dots,r\}$.

The minimum-risk approach to the Chebyshev problem (4.2) consists in finding the optimal solution of the following programming problem:

$$(4.5) \quad v(w) = \max_{x \in D} P\{\omega \mid \max_{k \in I} [(c^k(\omega))'x + \alpha^k(\omega)] \leqslant w\}$$

The following theorem is true.

Theorem 4.1. /68/ If assumption (4.4) holds and the distribution $T(z)$ of $t(\omega)$ is continuous and strictly increasing, then the minimum-risk solution of problem (4.5) does not depend on $T(z)$ and can be obtained by solving the deterministic piecewise linear fractional programming problem .

$$(4.6) \quad \max_x \min_{k \in I} \ (w-(c_1^k)'x-\alpha_1^k)/((c_2^k)'x+\alpha_2^k)$$

Remarks. 4.1). The piecewise linear fractional programming problem (4.6) can be solved by use of a parametric algorithm similar to Dinkelbach's algorithm /15/ for fractional programming;

4.2) If $c^k(\omega) = c_1^k + t^k(\omega)c_2^k$, $\alpha^k = 0, k \in I$ and $t^k(\omega)$ $(k \in I)$ are independent random variables whose distribution function T_k is continuous and strictly increasing, then the minimum-risk solution of problem (4.5) depends on T_k as follows:

$$\max_x F(x,w) = \max_x \prod_{k=1}^{r} T_k \ [(w-(c_1^k)'x)/((c_2^k)'x)]$$

4.3) When the functions $Z_k (k \in I)$ defining the objective function η are nonlinear (linear fractional, linear plus fractional or linear plus indefinite functions), the minimum-risk approach to the Chebyshev problem can be re-stated and solved under appropriate hypotheses in a manner similar to the linear case.

5. STOCHASTIC FRACTIONAL PROGRAMMING PROBLEM

In the particular case of two stochastic objective functions we can draw up a function of the type 1^0-b(section 2) which leads us to a stochastic fractional programming problem

(5.1) $$\gamma(\omega) = \underset{x \in D}{\text{optimum}}\ Z_1(x,\omega)/Z_2(x,\omega)$$

It is considered that Z_1 and Z_2 are linear functions of the same random variable $t(\omega)$ having a distribution function $T(z)$ continuous and strictly increasing i.e.

$$Z_1(x,\omega) = (c+t(\omega)c_1)'x \text{ and } Z_2(x,\omega) = (d+t(\omega)d_1)'x$$

For problem (5.1) we make the following assumptions:
a) The set D is nonempty and bounded,
b) The denominator of the objective function preserves the same sign (let us assume it to be positive) on D, consequently $P\{\omega \mid (d+t(\omega)d_1)'x > 0\} = 1$,
c) All basic solutions are nondegenerate.

In /50/ Stancu-Minasian gives a representation of the probability distribution of the optimal value $\gamma(\omega)$ by means of the characteristic values of the corresponding parametric programs.

Extensions of the fractile criterion model and aspiration criterion model to the case of stochastic linear fractional programming are given by Stancu-Minasian /46/, /49/, /54/ and Gupta and Swarup /24/, /25/ (see, also and Cambini /9/). Callahan and Bector /8/ consider Kataoka's problem for a product and for a quotient of general functions.

Theorem 5.1. /49/ The minimum-risk problem for the ratio of two linear functions whose numerator has random coefficients (normal non-degenerated, with mean value $\bar{c}$ and covariance matrix V), i.e. the problem

(5.2) $$\max_{x \in D} P\{\omega \mid c'(\omega)x / d'x \leq k\}$$

is equivalent to the following nonlinear fractional programming problem

(5.3) $$\max_{x \in D} (kd'x - \bar{c}'x)/(x'Vx)^{1/2}$$

Theorem 5.2. /49/ Kataoka's problem for the ratio of two linear functions whose numerator has random coefficients (normal non-degenerated, with mean value $\bar{c}$ and covariance matrix V), i.e. the problem

(5.4) $$\max\{k \mid P\{\omega \mid c'(\omega)x/d'x \leq k\} = \alpha,\ x \in D\}$$

is equivalent to the following nonlinear programming problem

(5.5) $$\max_{x \in D} (\bar{c}'x - q(x'Vx)^{1/2})/d'x$$

where $\Phi^{-1}(\alpha) = -q$ and $\Phi(.)$ is Laplace's function.

It should be noted that in some conditions, problem (5.5) has a finite optimal solution and that the objective function is explicitely quasi-concave.

However, the approach in these models has been to assume the random variables involved to be normally distributed and to assume the denominator of the fractional objective functional to be fixed. Gupta et al. /23/ studied a stochastic linear fractional programming problem, in which the parameters of both the numerator and the denominator are assumed to be mutually independent Cauchy variates

$$(5.6)\qquad \max\{f \mid P\{(C'x-\bar{c}'x)/(D'x-\bar{d}'x) \leqslant f\} = p,\ x \in D\}$$

where the vectors C and D are assumed to be random with their components C_j and D_j $(j=1,\dots,n)$ all being independent Cauchy variates with parameters $(\bar{c}_j, q_j)$ and $(\bar{d}_j, r_j)$, respectively.

Gupta and Jain /22/ consider Kataoka's problem for the linear fractional objective function under chance constraints with both the right hand and the left hand coefficients being random.

Extensions of the minimum-risk approach to the linear fractional max-min problem

$$(5.7)\qquad \max_x \min_y \{(cx + dy + f)/(ex + gy + h) \mid Ax + By \leqslant b, x \geqslant 0, y \geqslant 0\}$$

are given by Stancu-Minasian and Tigan /58/, /60/.

Similarly, the results presented above for a quotient of functions can be generalized for the case $F(x) = (c'x)(d'x)$ (see Swarup et al. /61/ and Stancu-Minasian /46/, /49/, /54/) where c is a random vector with components having a multinormal distribution and d is vector of constants. Stancu-Minasian /46/, /49/, /54/ considers also the case when c and d are independent random vectors.

6. STOCHASTIC GOAL PROGRAMMING

The aim of this section is to outline certain variants of goal programming in the stochastic case.

6.1. Contini /14/ considered that the relation between the objectives $F=(F_1,\dots,F_r)'$ and x contains a random factor $u = (u_1,\dots,u_r)$ i.e.

$$(6.1)\qquad F_k(x) = c_{k1}x_1+\dots+c_{kn}x_n+u_k = C_k'x+u_k, \quad k=1,\dots,r$$

where the random vector u follows a normal distribution N(0,V) (V is a non-singular covariance matrix). We assume that the known matrix $C = (c_{ij})_{\substack{i=1,\dots,r\\ j=1,\dots,n}}$ is of rank at least r.

Due to the existence of a perturbation factor u the equality $\bar{F}_i = F_i$ ($\bar{F}_i$ is the required level for F_i) cannot hold. Then we choose Y^* in the space E^r such that $\bar{F} = (\bar{F}_1, \ldots, \bar{F}_r) \in Y^*$ and determine the vector x such that this maximize the probability that $(F_1(x), \ldots, F_r(x))$ belonging to Y^*. One gets the following model:

$$(6.2) \quad \max_{x \in D} P(F(x) \in Y^*)$$

Since u is normal the domain Y^* is usually, an ellipsoid in E^r centered in $(\bar{F}_1, \ldots, \bar{F}_r)$ and expressed as $Y^* = \{ y = (y_1, \ldots, y_r) \mid Q^* = (y-\bar{F})' V^{-1} (y-\bar{F}) \leqslant c^2 \}$, where c is a convenient scalar.

Theorem 6.1. /14/ The optimal solution of problem (6.2) is obtained by solving the quadratic programming problem

$$(6.3) \quad \min_{x \in D} (\bar{F}' V^{-1} \bar{F} + x' (C' V^{-1} C) x + 2 (C' V^{-1} \bar{F})' x)$$

If we assume that u_i are independent random variables, then the covariance matrix V is diagonal and the model (6.2) becomes

$$(6.4) \quad \min_{x \in D} R(x) = \sum_{i=1}^{r} (1/v_{ii}^2)(\bar{F} - C_i' x)^2$$

One notes that this model coincides with the deterministic model 2′ (section 2) if it is assumed that $\alpha_i = 1/v_{ii}^2$. This means that if u_i are indepedent random variables, then the solution of the stochastic model (6.2) is similar to that of the deterministic model 2′ (section 2) provided that the weighting factors are $\alpha_i = 1/v_{ii}^2$.

6.2. Let us reconsider now the previous problem in a different version.

Consider that the elements of matrix C are independent random variables with mean values $\bar{c}_{ij}$ and variances σ_{ij}^2. Choose domain Y^* such that

$$Y^* = \{ (y_1, \ldots, y_r) \in R^r \mid \varepsilon_{1i} \leqslant y_i \leqslant \varepsilon_{2i}, \; i = 1, \ldots, r \}$$

and set two limiting thresholds $\alpha_1 = (\alpha_{1i})$, $\alpha_2 = (\alpha_{2i})$ $(i = 1, \ldots, r)$ for the probability that $F(x) \in Y^*$.

We have to determine the extent to which Y^* can be decreased provided that probability that $F(x) \in Y^*$ ranges within the given limits. Thus, the problem is

(6.5) $\min \|\varepsilon_2 - \varepsilon_1\|$ subject to

(6.6) $P\{C_i x \geqslant \bar{F}_i - \varepsilon_{1i}\} \geqslant \alpha_{1i}, \; P\{C_i x \leqslant \bar{F}_i + \varepsilon_{2i}\} \geqslant \alpha_{2i}, i=1,\ldots,r$
$\varepsilon_{1i}; \; \varepsilon_{2i} \geqslant 0$

6.3. Assume that in $F(x)=Cx$ the elements of C are random variables. Consider that the optimization criteria chosen is that of minimizing the mean value of the norm of vector $\bar{F}-Cx$. If the norm $\|\cdot\|_p$ is considered, then we obtain the following model /12/.

(6.7) $\min_{x \in D} E(\|\bar{F}-Cx\|_p)$

For $p=2$ one obtains the model

(6.8) $\min_{x \in D} E\left(\sum_{i=1}^{r} |\bar{F}_i - C_i x|^2\right)$

We assume that the elements of matrix C are independent random variables with mean values $\bar{c}_{ij}$ and variances σ^2_{ij}.

Theorem 6.2. /12/ The optimal solution of model (6.8) is obtained by solving the quadratic programming problem

$$\min_{x \in D} \left[\sum_{i=1}^{r} (\bar{F}_i - \sum_{j=1}^{n} \bar{c}_{ij} x_j)^2 + \sum_{j=1}^{n} (\sum_{i=1}^{r} \sigma_{ij})^2 x_j^2 \right]$$

6.4. In what follows we shall assume that the functions F_i are well determined, but the thresholds $\bar{F}$ are not specified at the moment of decision. Assume that the elements of $\bar{F}$ are independent random variables with known distribution $T_i(.)(i=1,\ldots,r)$. We can consider the problem /4/:

(6.9) $\min \|\varepsilon\|$ subject to

(6.10) $P\{C_i x \geqslant \bar{F}_i - \varepsilon_i\} \geqslant \alpha_{1i}, \; P\{C_i x < \bar{F}_i + \varepsilon_i\} \geqslant \alpha_{2i} \; i=1,\ldots,r$
$\varepsilon = (\varepsilon_1, \ldots, \varepsilon_r) \geqslant 0$

6.5. Stancu-Minasian and Tigan /59/ considered an extension of the goal programming problem (2.3)-(2.5) (section 2) which is replaced by the following equivalent Problem GP1:

Minimize $g(x,y,z)$ subject to

(6.11) $F_i(x) - y_i \leqslant \bar{F}_i; \; F_i(x) + z_i \geqslant \bar{F}_i, \; i \in I = \{1,\ldots,r\}$

(6.12) $x \in D, y_i \geqslant 0, \; z_i \geqslant 0, \; i \in I$

It can be easily shown that problems (2.3)-(2.5) and GP1 have the same optimal solution set when the function g is of the form

$$(6.13)\qquad g(x,y,z) = \sum_{i\in I} (\alpha_i y_i + \beta_i z_i)$$

where α_i and β_i are intragoal positive weights and specify the relative penalties to be applied for over-or underachievements in $\bar{F}_i$.

In /59/ it is considered that the target values $\bar{F}_i$ are random variables, the constraints (6.11) are satisfied with certain given probabilities, the objective functions are linear fractional, and the global deviation function g is of Chebyshev type, i.e.

$$g(x,y,z) = \max_{i\in I} (\alpha_i y_i + \beta_i z_i)(\alpha_i, \beta_i > 0,\ i\in I)$$

Then, the problem GP1 is reduced to a linear fractional min-max problem with linear constraints. If the objective functions are random normal, then the problem is equivalent to a deterministic linear fractional min-max problem with nonlinear constraints.

In /59/ it is considered the case when the fractional objective functions F_i are simply randomized.

7. GROUP DECISION-MAKING IN STOCHASTIC PROGRAMMING

Consider the r-objective function stochastic programming problem

$$(7.1)\qquad \max_{x\in D}\ Z_k(x,\omega) = c_k'(\omega)x, \qquad (k=1,\dots,r)$$

where $c_k'(\omega) = (c_{k1}(\omega),\dots,c_{kn}(\omega)),\ (k=1,\dots,r)$.

For a given $\omega\in\Omega$ the above problem becomes a r-objective function deterministic problem.

One can construct a synthesis function by the weighted sum of the r functions using weights $\alpha_k (k=1,\dots,r)$.

$$F = \sum_{k=1}^{r}\alpha_k Z_k = \sum_{k=1}^{r}\alpha_k \sum_{j=1}^{n} c_{kj}x_j = \sum_{j=1}^{n}\Big(\sum_{k=1}^{r}\alpha_k c_{kj}\Big)x_j =$$

$$= \sum_{j=1}^{n} c_j(\alpha)x_j = c'(\alpha)x.$$

The weights α_k are usually normalized such that $0\leqslant\alpha_k\leqslant 1$, $\sum_{k=1}^{r}\alpha_k = 1$. Hence, the problem becomes

$$(7.2) \quad \max_{x \in D} \sum_{k=1}^{r} \alpha_k Z_k(x,\omega)$$

In practice, the above summation is often meaningless, since the functions Z_k represent entirely different magnitudes. To overcome this difficulty one may transform the functions Z_k into utility function $Z_k'(x,\omega)$ through the von Neumann-Morgenstern method /38/.

The r utility functions Z_k' can be algebrically summated and hence one can maximize a synthesis function, i.e.

$$(7.3) \quad \max_{x \in D} Z^* (x,\omega) = \sum_{k=1}^{r} \alpha_k Z_k'(x,\omega)$$

Thus the r-objective function stochastic programming problem has been replaced by a mathematical programming problem with one stochastic objective function.

Another aspect to the problem is the following: In a group decision-making process the weights α_k are random variables since the members of the group will may assign various weights to individual objective functions. Without loss of generality, in the following argument we consider the coefficients c_k deterministic whereas the weights α_k are random variables. Note that these weights may be the result of consultations with one or more individuals in a decision-making group.

If the group is sufficiently small, then α_k can be mean values of the weights given by each individual. But if the group is rather large, it becomes impossible to consult with each member in turn. If the decision-making collectivity is large enough one has to deal with the so-called welfare functions. Thus, one can assume that α_k are random variables.

Let $c(\alpha) = \sum_{k=1}^{r} \alpha_k c_k$ and let $F_\alpha(.)$ be the distribution function of the random vector $\alpha=(\alpha_1,\dots,\alpha_r)$. If the weights α_k are known then, according to Theorem 2.1 (section 2) the solutions of problem (7.3) are efficient solutions.

The knowledge of the distribution function $F_\alpha(.)$ can be achieved in two ways:

1) by use of the statistic analysis,
2) or by assuming that $F_\alpha(.)$ is an a priori distribution of the weights α_k.

The weights α_k can be also established through the Simulation Method /13/.

For the problem (7.3) one can consider the following two problems presented in section 3.

1) Finding the minimum-risk solution corresponding to level u i.e. the solution which maximizes the function

(7.4) $P\{\alpha \mid c'(\alpha)x \geqslant u, x \in D\}$

2) Determine the distribution function of the random variable

(7.5) $\zeta(\alpha) = \max \{c'(\alpha)x \mid x \in D\}$

given the distribution $F_\alpha(.)$ and the values of the probabilities that various possible bases are optimal.

Bereanu shows /6/ that under certain conditions the optimal solution of problem (7.4) is independent of the distribution of α or of the value of u.

8. EFFICIENT SOLUTIONS IN STOCHASTIC PROGRAMMING

Let us now consider the stochastic programming problem with one objective function

(8.1) $$\gamma(\omega) = \max_{x \in D} c'(\omega)x$$

where $D = \{x \mid Ax \leqslant b;\ x \geqslant 0\}$, the constraints coefficients are deterministic and the c vector is random.

As shown in section 3 the expression max $c'(\omega)x$ becomes meaningless if c becomes random. Thus, we shall introduce the concept of efficiency which is analogous to the concept of optimality in deterministic mathematical programming multiple objective functions.

Definition 8.1. A point $x^o \in D$ is said to be efficient with probability 1 if there is no other vector $x^1 \in D$ almost surely as good as x^o (with respect to the objective function $c'(\omega)x$) and even better than x^o with a positive probability.

That is, $x^o \in D$ is efficient with probability 1 if there exists no $x^1 \in D$ such that $P\{\omega \mid c'(\omega)x^o \leqslant c'(\omega)x^1\}=1$ and $P\{\omega \mid c'(\omega)x^o < c'(\omega)x^1\} > 0$.

The problem confronting us is that of finding the set of points $D_1 \subset D$ which are efficient with probability 1 or at least another reasonable subset of D_1.

On the other hand, Tammer /62/ assumed that the objective functions are random $f(x,y(\omega))=f_o(x) + \sum_{i=1}^{m} y_i(\omega) f_i(x)$ where $y(\omega)=(y_1(\omega),\dots,y_m(\omega))$ is a random vector, and defined the notion of "optimality" as follows:

Definition 8.2. A point $x^0 \in D$ is said to be ε - efficient (efficient with the probability $1-\varepsilon$) for the problem

(8.2) $\quad \max \{f(x,y(\omega)) \mid x \in D\}$,

if there exists no $x^1 \in D$ such that

$$P\ \{ f(x^1,y(\omega)) \geq f(x^0,y(\omega))\} \geq 1-\varepsilon$$

and

$$P\ \{ f(x^1,y(\omega)) > f(x^0,y(\omega))\} > 0$$

Note that the notion of ε - efficiency is a generalization of the notion of efficiency with probability 1. In order to determine the ε - efficient points one can use a result similar to Theorem 2.1 (section 2). In /62/ it is shown that for certain parameters $\lambda \in R^m$, the optimal solutions of the parametric program $\max\{ f(x,\lambda) \mid x \in D\}$ are ε - efficient for the stochastic program (8.2). (For other variants of the notion of efficiency with probability 1 see also /63/ and /64/). Thus, one can calculate all ε - efficient solutions of the stochastic program (8.2) by solving the corresponding parametric programs.

Consider now the following stochastic programming problem with several objective functions:

(8.3) $\quad \max\limits_{x \in D} c_i'(\omega)x \qquad i=1,\dots,r$

Consider that $F_i(z)$ is the cumulative distribution function of $c_i'(\omega)x$ and $z_i=z_i(x)$ ranges over $c_i'(\omega)x$. (In particular, $z_i(x)$ can be $E(c_i'(\omega)x)$). Let $0<\alpha<1$ such that $P\left[c_i'(\omega)x \geq z_i(x)\right] = 1-F_i(z_i(x))=1-\alpha$.

The set of stochastic, efficient solutions at level α (denoted S_α) is defined as follows /21/: $S_\alpha = \{x \mid x \in D,\ F_i(z_i(x))=\alpha,\ i=1,\dots,r$, there exists no other $x' \in D$ such that $F_i(z_i(x'))=\alpha$ for all $i=1,\dots,r$, $z_i(x') \geq z_i(x)$ for all $i=1,\dots,r$ and at least for one i_0, $z_i(x') > z_{i_0}(x)\}$.

It is interesting to note that this definition is essentially that given in section 2, provided α remains constant.

As we did in section 3 for an objective function, the problem (8.3) can be replaced by a suitable deterministic problem. Several solution variants exist in this respect.

1. For each function, one maximizes its mean value

(8.4) $\quad \max\limits_{x \in D} E\,(c_i'(\omega)x) \qquad i=1,\dots,r$

2. For each $i=1,\dots,r$, one maximizes the variance

$$(8.5)\qquad \min_{x\in D} \mathrm{Var}\,(c_i'(\omega)x) \qquad i=1,\dots,r$$

If c_k are assumed to be random vectors having a multinormal distribution the problem (8.5) becomes

$$(8.6)\qquad \min_{x\in D} x'V_k x \qquad k=1,\dots,r$$

Hence, problem (8.3) becomes a deterministic problem with r quadratic objective functions.

The cases analogous for e) and f) ones from section 3 will be considered in sections 9 and 11.

Dubov /16/ considers problem (8.4) for a n given criteria $f_1(x,\omega)$, $f_2(x,\omega),\dots,f_n(x,\omega)$ depending continuously on a random variable ω with continuous distribution density $p(\omega)$. The set of possible solution D is a finite set. Denote by $F(\alpha)$ the image of D in the space of vector estimate with $\omega=\alpha$. Consider the vector $\varphi(x)=(\varphi_1(x),\dots,\varphi_n(x))$ $x\in D$ where $\varphi_i(x) = = \int_\omega f_i(x,\omega)p(\omega)d\omega$. The elements of $\varphi(x)$ constitute a set of averaged vector estimates Φ. The point which optimize $\varphi(x)$ is called Pareto optimal in mean.

It is shown that for a solution x to be Pareto optimal in mean, it is necessary and sufficient that there exists nonnegative weight coefficients α_i, for $i=1,\dots,n$ such that the function $\alpha_1\varphi_1(.)+\dots+\alpha_n\varphi_n(.)$ attains an extremum at the point x. This is a result similar to Theorem 2.1. (section 2). If the set D is not finite, the necessary conditions for optimality holds but the sufficient conditions take, in the general case, a more complicated form.

9. MULTIPLE MINIMUM-RISK SOLUTIONS IN STOCHASTIC PROGRAMMING

Several papers /44/, /46/, /47/, /52/, /54/ by Stancu-Minasian considered a more general case of the minimum-risk problem, i.e. the case when the probabilities that the values of r linear objective functions exceed some levels of performance u_k are maximized.

$$(9.1)\qquad \max_{x\in D} P\{\omega \mid c_i'(\omega)x \geq u_i\} = z_i \qquad (i\in I=\{1,\dots,r\})$$

where $u_1,\dots,u_r$ are given positive values and $D=\{x \mid Ax \leq b, x\geq 0\}$.

Such a problem is called /44/ a "multiple minimum-risk problem with levels $u_1,\dots,u_r$".

Suppose that the random vectors $c_k(\omega)(k=1,\dots,r)$ follow a non-degenerate n-dimensional normal distribution with the mean values $\bar{c}_k$ and the covariance matrices V_k.

For problem (9.1) a "compromise" can be reached by using a synthesis function for the r functions.

(9.2) $F=F(P\{\omega \mid c_1'(\omega)x \geq u_1\},\dots,P\{\omega \mid c_r'(\omega)x \geq u_r\})$

A possible expression for such a function is /11/:

$$\max_{x \in D} \sum_{k=1}^{r} w_j P\{\omega \mid c_k'(\omega)x \geq u_k\}$$

where w_j are weights of function Z_j, or to maximize the joint probability /29/.

$$P\{\omega \mid c_1'(\omega)x \geq u_1,\dots,c_r'(\omega)x \geq u_r\}.$$

In order to solve problems of type (9 .1) Stancu-Minasian suggested a method (analog to the case 4°, section 3) consisting in the sequential solving of some minimum-risk problems which include also quadratic constraints. The following hypotheses are necessary:

i) the programs of the mean values are bounded, i.e. $T_i=\max \bar{c}_i'x<\infty$ $(i\in I)$, ii) $u_i<T_i$, iii) $0\notin D$.

Then one proceeds as follows: First order the criteria according to their importance, i.e. if Z_1 is more important than Z_2, Z_2 than Z_3 a.s.o. then one gets the sequence $Z_1,Z_2,\dots,Z_r$.

Solve the minimum-risk problem:

(9.3) $$\max_{x\in D} P\{\omega \mid c_1'(\omega)x \geq u_1\},$$

which correspond to function Z_1 and level u_1.

Let this solution be $x^1_{u_1}\in D$ and let p_1 be the optimum value of (9.3). Next, define the domain $U_1=\{x \mid x\in D, P\{\omega \mid c_1'(\omega)x \geq u_1\}\geq p_1-\varepsilon_1\}$, ($\varepsilon_1>0$, known) on which we solve the minimum-risk problem

$$\max P\{\omega \mid c_2'(\omega)x \geq u_2\}$$

Let now the maximum value be p_2 and let $x^2_{u_2}\in U_1$ be the corresponding optimum solution. Proceed similarly on U_2, where

$U_2=\{x \mid x\in U_1,\ P\{\omega \mid c_2'(\omega)x \geqslant u_2\} \geqslant p_2-\varepsilon_2\}$ ($\varepsilon_2>0$, known), and solve the minimum-risk problem max $P\{\omega \mid c_3'(\omega)x \geqslant u_3\}$, etc.

Assume now that for the first r-1 optimum values $p_1, p_2, \ldots, p_{r-1}$ the correspon ding domains $U_1, U_2, \ldots, U_{r-2}$ have been identified. Then define

$U_{r-1}=\{x \mid x\in U_{r-2}, P\{\omega \mid c_{r-1}'(\omega)x \geqslant u_{r-1}\} \geqslant p_{r-1}-\varepsilon_{r-1}\}$

($\varepsilon_{r-1}>0$, known)

and solve the minimum-risk problem max $P\{\omega \mid c_r'(\omega)x \geqslant u_r\}$ whose solution, $x^r_{u_r}$, is called /44/, /46/, /47/ "the multiple minimum-risk solution corresponding to levels $u_1, u_2, \ldots, u_r$ and to the relaxation values $\varepsilon_1, \varepsilon_2, \ldots, \varepsilon_{r-1}$.

The solution $x^r_{u_r}$ is "the best compromise solution" for (9.1) and the ordering scheme described above. Obviously, for a different ordering scheme, one may get other solutions. Hence, on can formulate the following

<u>Definition 9.1.</u> /44/ A multiple minimum-risk solution corresponding to levels $u_1, \ldots, u_r$ and relaxation values $\varepsilon_1, \varepsilon_2, \ldots, \varepsilon_{r-1}$ is the vector $x=(x_1, \ldots, x_n)'$ which maximizes $P\{\omega \mid c_r'(\omega)x \geqslant u_r\}$, subject to $x\in D$,

$P\{\omega \mid c_i'(\omega)x \geqslant u_i\} \geqslant p_i-\varepsilon_i$, $(i=1, \ldots, r-1)$

where $p_i = \max_{x\in U_{i-1}} P\{\omega \mid c_i'(\omega)x \geqslant u_i\}$, $(i=1, \ldots, r-1)$

$U_0=D, U_{i-1}=\{x \mid x\in U_{i-2}, P\{\omega \mid c_{i-1}'(\omega)x \geqslant u_{i-1}\} \geqslant p_{i-1}-\varepsilon_{i-1}\}$, $(i=1, \ldots, r-1)$

On the other hand, Stancu-Minasian and Tigan /56/ take the concept of efficient points according to Definition 2.1 (section 2) as optimal solutions for problem (9.1). They consider that the feasible set D is an arbitrary, nonempty and compact set in R^n.

<u>Definition 9.2.</u> A point $x^0\in D$ is said to be a <u>vectorial minimum-risk solution</u> for problem (9.1) if it is an <u>efficient solution</u>, according to the Definition 2.1.

Concerning to problem (9.1), we adopt the following assumptions:

(H1) The objective functions are simply randomized, i.e. $c_k(\omega)=c_k^1+t_k(\omega)c_k^2$ where $c_k^i\in R^n$ $(i=1,2)$ are given constant vectors and $t_k(\omega)$ is a random variable,

(H2) $c_k^2x>0$ for all $x\in D$ and for all $k\in I$,

(H3) The distribution functions T_k of the $t_k(\omega)$ are continuous and strictly increasing.

Theorem 9.1. Let assumptions (H1)-(H3) be satisfied. Then a point $x' \in D$ is a vectorial minimum-risk solution for problem (9.1) if and only if x' is an efficient solution for the following linear fractional programming problem with several objective functions

$$(9.4) \qquad \max_{x \in D} \; (c_k^1 x - u_k)/c_k^2 x \qquad k \in I$$

By theorem 9.1, we can conclude that the vectorial minimum-risk problem can be solved with a method used for solving the multiobjective linear fractional programming problem (9.4). For instance, an interactive procedure to solve (9.4) was given by Weber /69/.

In /56/ similar results was obtained for the vectorial minimum-risk problem with linear fractional objective functions

$$(9.5) \qquad \max_{x \in D} \; P\left\{\omega \,|\, (c_k' x + \alpha_k)/(d_k' x + \beta_k) \geqslant u_k\right\} \; k \in I$$

10. THE INTERACTIVE METHODS

To solve stochastic programming problem with multiple objective functions, Goicoechea, Duckstein and Bulfin /21/ introduce the PROTRADE (Probabilistic Tradeoff Development) method. This method can be considered to be the stochastic analog to the deterministic POP and STEM methods. It is an interactive method providing a so-called "dialogue" between the user and the model, the decision-maker having the option of conducting the search process for the efficient solutions by modifying the initial conditions according to the partial results obtained. During the iterations the decision-maker can improve upon the value already attained by one objective function as well as upon the probability of reaching the corresponding objective (or both).

The great interest of this method is to be interactive and to concern a very general problem, having nonlinear constraints and general distribution for the random coefficients of the linear objectives.

In /65/ Teghem, Jr. pointed some disadvandages of the Protrade method and propose another interactive approach (STRANGE), generalizing the method STEM, for the problem: min $z^k = c^k x$ $k=1,\dots,K$ $x \in D(x)=\{x \mid Tx=d, x \geqslant 0\}$ where c^k and d are "discrete random variables". The STRANGE method (see also /67/) is described in details and illustrated with a numerical example in this volume.

Leclercq /35/, /36/, /37/, obtained a model somewhat similar to (9.1) starting from a problem with one objective *function (which does not exclude the case of more functions), where the coefficients are random variables and some of the constraints contain random variables. Consider the problem*

(10.1) $\max\{ c'x \mid D_j'x \geq d_j \quad j=1,\ldots,q;$

$E_k'x \geq e_k \quad k=1,\ldots,r; \quad x \geq 0 \}$

where the elements of vectors $c(n \times 1)$, $D_j(n \times 1)$, $d=(d_1,\ldots,d_q)'$ *are random variables and those of vectors* $E_k(n \times 1)$ *and* $e=(e_1,\ldots,e_r)'$ *are constants.*

One sets out to determine the vector x such that to maximize the probability that constraints $D_j'x \geq d_j, j=1,\ldots,q$ *are true. The following problem is obtained:*

$\max P\{\omega \mid c'(\omega)x \geq s\}$

$\max P\{\omega \mid D_j'x \geq d_j\} \qquad j=1,\ldots,q$

subject to $x \in D^* = \{ x \in R^n \mid E_k'x \geq e_k,\ k=1,\ldots,r;\ x \geq 0\}$

In this light, with random elements appearing in constraints as well, the distinction between the objective function and the constraints fades away. Leclercq consider that the coefficients are random multinormal or Cauchy variables independent or not.

For the normal distribution case, the following problem is obtained

(10.2) $\max f_i(z) = (A_i z)/(z'V^i z) \qquad i=1,\ldots,q+1$

on a suitable domain $\overline{D}$.

In order to solve problem (10.2) Leclercq /37/ propose an interactive algorithm similar to the STEM method. The decision maker has an active role in the process of searching for the optimum solutions.

First, for each functions f_i *one considers a threshold* B_i *and the weights* p_i *expressing its importance.*

A solution z is satisfactory in terms of the j-th criterion if $f_j(z) \geq B_i$. *Leclercq defines the notion of non-improvable solution referring to* $B_1,\ldots,B_{q+1}$. *This concept is a generalization of that of efficient solution (Definition 2.1).*

Definition 10.1. The point $x \in \overline{D}$ *is said to be non-improvable with respect to* $B_1,\ldots,B_{q+1}$ *if and only if there is no* $y \in \overline{D}$ *such that*

a) $f_i(y) \geq f_i(x)$ *if* $i \in I = \{ i \mid f_i(x) < B_i \}$

b) $\sum_{i \in I} f_i(y) > \sum_{i \in I} f_i(x)$

c) $f_j(y) \geqslant B_j$ if $j \in \{ j \mid f_j(x) \geqslant B_j \}$

This definition implies that x is efficient if and only if it is non-improvable with respect to $\infty,\dots,\infty$.

The algorithm consists of a series alternation of the computational and decisional stages. In the decision phase, the decision maker uses the available information to determine the thresholds $B_1,\dots,B_{q+1}$ and then, in the computational phase, a non-improvable solution with respect to $B_1,\dots,B_{q+1}$. The thresholds $B_1,\dots,B_{q+1}$ differ from one iteration to the next. The decison-maker can choose to decrease a certain B_i in order to improve the values of other functions.

11. MULTIPLE FRACTILE CRITERION MODEL

In a recent paper /55/ Stancu-Minasian deals with a more general case of the Kataoka's problem. i.e. the determination of the largest value of r fractional objective functions such that the probability of obtaining such result exceed (or is equal) with some given thresholds. Probabilistic constraints are introduced instead of linear conditions. Problem P_I

(11.1) "maximize" $F=(f_1,f_2,\dots,f_r)$ subject to

(11.2) $P(g^i x/(h^i x+h_o^i) \geqslant f_i) = p_i,\ i=1,\dots,r,$

(11.3) $P(\sum_j a_{kj}x_j \leqslant b_k) \geqslant q_k,\quad k=1,\dots,m,$

(11.4) $x_j \geqslant 0,\quad j=1,\dots,n$

where x_j are decision variables, $g^i=(g_1^i,\dots,g_n^i), h^i = (h_1^i,\dots,h_n^i)$ $(i=1,\dots,r)$, $a_j=(a_{j1},\dots,a_{jn})$ $(j=1,\dots,m)$ are random vectors, $b_1,\dots,b_m$ are random variables, $p_1,\dots,p_r$, $q_1,\dots,q_m$ are preassigned probability levels of the chance constraints and $h_o^1,\dots,h_o^r$ are real constants. Here the term "maximize" is used for finding all efficient solutions in a maximization sense.

Suppose that

(i) $1 \geqslant p_i > 1/2;\ 1 \geqslant q_j > 1/2\quad (i=1,\dots,r;\ j=1,\dots,m)$

(ii) For each $i=1,\dots,r$ the random variables $g_j^i (j=1,\dots,n)$ have a known multi-normal distribution with mean vector $G^i=(G_1^i,G_2^i,\dots,G_n^i)$ and dispersion matrix $\bar{\bar{D}}^i=(d_{jk}^i)$

(iii) $h^1, \ldots, h^r$ are constant vectors.

Also, we assume that one of the following statements is true

(iv) For each $i=1,\ldots,m$, the random variables $(a_{i1}, a_{i2}, \ldots, a_{in}, -b_i)$ have a known multi-normal distribution with mean vector $(A_{i1}, A_{i2}, \ldots, A_{in}, -B_i)$ and known $(n+1)$ order dispersion matrix $\bar{\bar{E}}_i$.

(v) For each $i=1,\ldots,m$, the random variables $(a_{i1}, a_{i2}, \ldots, a_{in}, -b_i)$ have unknown distributions but known mean vector $(A_{i1}, A_{i2}, \ldots, A_{in}, -B_i)$ and known $(n+1)$ order dispersion matrix $\bar{\bar{E}}_i$.

If the above assumptions holds, the Problem P_I becomes a deterministic multiple non-linear fractional programming problem with non-linear inequality constraints

Problem P_{II}

(11.5) $$\text{"Maximize"}\ F = \left(\frac{\bar{G}^1 X - s_1 (X'\bar{D}^1 X)^{1/2}}{H^1 X + h_o^1}, \ldots, \frac{\bar{G}^r X - s_q (X'\bar{D}^r X)^{1/2}}{H^r X + h_o^r}\right)$$

(11.6) $$X \in \{ A_i X + (X'\bar{E}_i X)^{1/2} \leqslant B_i;\ i=1,\ldots,m;\ A_{m+1}X=1, X \geqslant 0\}$$

where $\bar{G}^i = (G_1^i, \ldots, G_n^i, 0), H^i = (h_1^i, h_2^i, \ldots, h_n^i, 0) (i=1,\ldots,r)$, $A_{m+1} = (0,0,\ldots,0,1)$ are $(n+1)$- vectors, $\bar{E}_i = t_i^2 \bar{\bar{E}}_i$ and $D^i = \begin{pmatrix} \bar{\bar{D}}^i & 0 \\ 0 & 0 \end{pmatrix}$ are $(n+1)x(n+1)$ symmetric positive semi-definite matrices and $-s_i = \Phi^{-1}(1-p_i)$.

Use is made of the transformation $y=tX$, where $t \geqslant 0$ is so chosen that, $H^i y + h_o^i t = \eta_i \quad \eta_i \in R$, $i=1,\ldots,r$.

The following convex vectorial mathematical programming is obtained.

Problem P_{III} $$\text{"Maximize"}F^*(y,t) = (\bar{G}^1 y - s_1 (y'\bar{D}^1 y)^{1/2}, \ldots, \bar{G}^r y - s_r (y'\bar{D}^r y)^{1/2})$$

$A_i y + (y'\bar{E}_i y)^{1/2} \leqslant B_i t$; $i=1,\ldots,m$

$H^i y + h_o^i t = \eta_i \quad i=1,\ldots,k$

$A_{m+1} y = t; \quad y, t \geqslant 0$.

The relation between the optimal solutions of the problems P_{II} and P_{III} is given by

Theorem 11.1. /55/. If

a) $0 < \operatorname{sgn} \gamma_i = \operatorname{sgn} (H^i X + h_o^i)$ (for all $i=1,\ldots,k$) for X^+ an optimal solution of Problem P_{II} and

b) (y_1, t_1) is an efficient solution for Problem P_{III}, then y_1/t_1 is efficient for Problem P_{II}.

REFERENCES

/1/ Armstrong,R.D. and Balintfy,J.L. (1975) 'A chance-constrained multiple choice programming algorithm'. Operations Res. 23, 494-510.

/2/ Benayoun,R. and Tergny,J. (1969) 'Critères multiples en programmation mathématique: une solution dans le cas linéaire', Rev.Française Informat. Recherche Operátionnelle, 3(5-2) 31-56.

/3/ Benayoun,R., J. de Montgolfier, Tergny,J. and Larichev, O.I.(1971) 'Linear programming with multiple objective functions: STEP Method (STEM)', Math.Programming,1(3), 366-375 .

/4/ Ben-Israel,A.,Charnes,A. and Kirby,M.J.L.(1970) 'On stochastic linear approximation problems'. Oper.Res.18(3), 555-558.

/5/ Bereanu,B. (1964) 'Programme de risque minimal en programmation linéaire stochastique', C.R.Acad.Sci.Paris, 259,5,981-983.

/6/ Bereanu,B.(1976) 'Large group decision making with multiple criteria', Multiple Criteria Decision Making, Proceedings of a Conference held at Jouy-en-Josas 1975, Thiriez H. and Zionts S.(eds.).Lecture Notes in Econom. and Math.Systems.Vol.130,Springer, Berlin-Heidelberg-New York, 87-101.

/7/ Blau,R. (1973) 'Decomposition techniques for the Chebyshev problem', Oper.Res.21 (5), 1157-1163.

/8/ Callahan,J.R. and Bector,C.R.(1973) 'Optimization with general stochastic objective functions', Proceedings of the Third Manitoba Conference on Numerical Mathematics, 127-137; also in: Z.Angew.Math.Mech.55,(1975), 528-530.

/9/ Cambini,A. 'Sulla programmazione lineare frazionaria stocastica', Publication No.50, series A,Dept.di Ricerca Operativa e Scienze Statistiche, Universitá di Pisa, Italy.

/10/Charnes,A. and Cooper,W.W.(1961) Management Models and Industrial Applications of Linear Programming.Vol.1, John Wiley, New York.

/11/ *Charnes, A. and Stedry,A.C.(1966) 'Search-Theoretic models of organization control by budgeted multiple goals', Management Sci. 12(5), 457-481.*

/12/ *Chobot,M. (1973) 'Stochastiky pristup k cilovému programovaniu', Ekon.Mat.Obzor,3,305-320.*

/13/ *Ciobanu,Gh. (1976) 'Applying digital simulation in solving problems of linear programming with several objective functions.Econom.Comp.Econom.Cybernet.Stud. Res.1, 63-68.*

/14/ *Contini,B. (1968) 'A stochastic approach to goal programming', Oper.Res.16(3), 576-586.*

/15/ *Dinkelbach,W.(1967) 'On nonlinear fractional programming', Management Sci.,13(7), 492-498.*

/16/ *Dubov,Ju.A.(1979) 'Necessary and sufficient conditions for Pareto optimality in mean', Izv.Akad.Nauk SSSR, Tehn.Kibernet.6,137-141,200 (in russian); translated as Engrg.Cybernetics,17(1979),6,(1980),109-114.*

/17/ *Dürr,W. (1972)'Stochastische programmierungsmodelle als wektormaximumprobleme.' In: Proceedings Operations Research, Würzburg-Wien 1, 189-199.*

/18/ *Fandel,G. and Jochen,W. (1974) 'Das Vektormaximumproblem bei Sicherheit und Unsichercheit', Tech.Report, Institut für Geselbschafts und Wirtschaftswissenschaften der Universität Bonn.*

/19/ *Geoffrion,A.M. (1967) 'Stochastic programming with aspiration or fractile criteria', Management Sci., 13, 672-679.*

/20/ *Goicoechea,A. (1977) 'The PROTRADE-method: a multiobjective approach to decision-making', in Star M.K. and Zeleny M.(eds.), Multiple Criteria Decision Making,TIMS studies in the management sciences.Vol. 6, North-Holland Publ.Comp.,Amsterdam.*

/21/ *Goicoechea,A.,Hansen,D.R. and Duckstein,L. (1982) Multiobjective Decision Analysis with Engineering and Bussiness Applications.John Wiley and Sons.*

/22/ *Gupta, S.N. and Jain,R.K. (1986) 'Stochastic fractional programming under chance constraints with random technology matrix', Acta Ciencia Indica 12(3),191-198.*

/23/ *Gupta,S.N., Jain,A.K. and Swarup,K.(1987) 'Stochastic linear fractional programming with the ratio of independent Chauchy variates', Nav.Res.Logist.34(2), 293-305.*

/24/ *Gupta,S.N. and Swarup,K. (1979) 'Chance constrained programming with stochastic fractional functional ob-*

jective', Preprint volume of the AIRO Conference, Italy (Sept.1979,pp.M163-M172, Airo-Associazione Italiana Di Ricerca Operativa).

/25/ Gupta,S.N. and Swarup,K.(1980) 'Duality in stochastic fractional programming under chance constraints', Ricerca Operativa (Italy) N13, 53-56.

/26/ Gupta,S.N. and Swarup,K.(1980) 'Note on stochastic programming for minimization of variance', New Zealand Oper.Res.8(2), 185-187.

/27/ Hendrix,G.C. and Stedry,A.(1974) 'The elementary redundancy-optimization problem: a case study in probabilistic multiple-goal programming', Oper.Res.22,639-653.

/28/ Ishii,H. and Nishida,T.(1985) 'The stochastic linear continuous type knapsack problem: A generalized P model', European J.Oper.Res.19(1),118-124.

/29/ Kaplan,R. and Soden,J. (1971) 'On the objective function for the sequential P-model of chance-constrained programming', Oper.Res.19(1), 105-114.

/30/ Kataoka,S.(1962) 'On stochastic programming II.A preliminary study of a stochastic programming model', Hitotsubashi J.Arts Sci.2, 36-44.

/31/ Kunsch,P.L.(1983) 'STRANGE-Multicriteria decision making', Belgonucléaire, Internal report.

/32/ Kunsch.P.L. and J.Teghem,Jr.(1987) 'Nuclear Fuel Cycle Optimization using Multi-Objective Stochastic Linear Programming', European J.Oper.Res.31, 240-249.

/33/ Lau,H.S.(1980) 'The newsboy problem under alternative optimization objectives', J.Oper.Res.Soc.31,525-535.

/34/ Lebedev,B.D.,Podinovskii,V.V. and Styrikovic,R.S.(1971) 'An optimization problem with respect to the order totality of criteria', (in russian), Ekon.Mat.Metody, SSSR,7, 612-616.

/35/ Leclerq,J.P.(1979) Résolution de programmes linéaires stochastiques par des techniques multicritères. Thèse de doctorat. Faculté des Sciences de l'Université de Namur.

/36/ Leclerq,J.P.(1981) 'La programmation linéaire stochastique: une approache multicritère. Partie I-ère. Formulation. Partie II-ème: Un algorithme interactif adapté aux distributions multinormales,' Cahiers Centre Études Rech.Opér.23(1),31-41;ibidem 23(2),121-132.

/37/ Leclercq,J.P.(1982) 'Stochastic programming: an interactive multicriteria approach', European J.Oper.Res. 10(1), 33-41.

/38/ Neumann von J.,Morgenstern,O.(1953) Theory of Games and Economic Behavior.Princeton University Press.

/39/ Roubens,M. and J.Teghem,Jr.(1988) 'Comparison of methodologies for multicriteria feasibility-constrained fuzzy and multiple-objective stochastic linear programming. Combining fuzzy imprecision with probabilistic uncertainty in decision making', Lect.Notes Econ.Math. Syst.310, 240-265.
/40/ Sengupta,S.S., Podrebarac,M.L. and Fernando,T.D.H.(1974) 'Probabilities of optima in multiobjective linear programmes', in J.L.Cochrane and M.Zeleny (eds.) Multiple Criteria Decision Making, University of South Carolina Press,Columbia,S.C., 217-235.
/41/ Slowinski,R. and J.Teghem,Jr.(1988) 'Fuzzy versus stochastic approaches to multicriteria linear programming under uncertainty', Naval Res.Logist.35(6), 673-695.
/42/ Stenvers,Karl-Heinz(1985) 'Stochastische Vectoroptimierung zür Auslegung von Tragwerksstrukturen-Lösungsalgorithmen und Versuchsergebnisse', (Stochastic vector optimization for construction of load-bearing structures-Solution algorithms und experimental results) (Dissertation). Fachbereich Maschinentechnik I der Universität-Gesamthochschule Siegen.
/43/ Stancu-Minasian,I.M.(1974) 'Stochastic programming with multiple objective functions', Econom.Comp.Econom. Cybernet.Stud.Res.1, 49-67.
/44/ Stancu-Minasian,I.M.(1975) 'Notă asupra programării stohastice cu mai multe funcţii obiectiv, avînd vectorul c aleator', Stud.Cerc.Mat.,27(4), 453-459.
/45/ Stancu-Minasian,I.M.(1975) 'A selected bibliography of works related to the multiple criteria decision making,' Preprint nr.2,Catedra de Cibernetică Economică A.S.E., Bucharest,Roumania.
/46/ Stancu-Minasian,I.M.(1976) 'Criterii multiple în programarea stohastică', Ph.D.Dissertation, Centrul de Statistică Matematică, Bucureşti.
/47/ Stancu-Minasian,I.M.(1976) 'Asupra problemei de risk minim multiplu I: cazul a două funcţii obiectiv. II: cazul a r (r > 2) funcţii obiectiv', Stud.Cerc.Mat., 28(5), 617-623, ibidem 28(6), 723-734.
/48/ Stancu-Minasian,I.M. and Wets,M.J.(1976) 'A research bibliography in stochastic programming', Operations Res.24(6), 1078-1119.
/49/ Stancu-Minasian,I.M.(1976) 'Asupra problemei lui Kataoka,' Stud.Cerc.Mat.28(1), 95-111.
/50/ Stancu-Minasian,I.M.(1977) 'On stochastic programming with multiple objective functions', Proceedings of the Fifth Conference on Probability Theory. September 1-6, 1974, Braşov, Romania, Editura Academiei R.S.R., 428-436.
/51/ Stancu-Minasian,I.M.(1978) 'Problema Cebîşev stocastică. Funcţia de repartiţie a optimului', Stud.Cerc.Mat. 30(5), 567-577.

/52/ Stancu-Minasian,I.M.(1979) 'On the multiple-Minimum risk problem', Bull.Math.Soc.Sci.Math.R.S.Roumanie (N.S.) Tome 23(71), nr.4, 427-437.
/53/ Stancu-Minasian,I.M. (1982) 'Recent results in stochastic programming with multiple objective functions', in M.Grauer,A.Lewandowski,A.P.Wierzbicki(eds.),Multiobjective and Stochastic Optimization. IIASA Collaborative Proceedings Series CP-S12, 79-96.
/54/ Stancu-Minasian,I.M.(1984) Stochastic Programming with Multiple Objective Functions, Ed.Academiei,Bucureşti and D.Reidel Publishing Company, Dordrecht/Boston/ Lancester/Tokio.
/55/ Stancu-Minasian,I.M.(1989) 'O generalizare a problemei lui Kataoka la cazul criteriilor multiple'.Simpozionul Probleme actuale ale teoriei probabilităţilor şi statisticii matematice. Academia R.S.R.,Centrul de Statistică Matematică, Bucureşti.
/56/ Stancu-Minasian,I.M.and St.Tigan(1984) 'The vectorial minimum-risk problem', Proceedings of the Colloquium on Approximation and Optimization, Cluj-Napoca,October 25-27, 321-328.
/57/ Stancu-Minasian,I.M. and St.Tigan(1987)'Criteriul riscului minim în programarea stohastică', Lucrările Sesiunii ştiinţifice a Centrului de Calcul al Universităţii Bucureşti, 20-21 februarie, 392-397.
/58/ Stancu-Minasian,I.M. and St.Tigan(1987) 'The stochastic linear-fractional max-min problem', Itinerant Seminar on Functional Equations, Approximation and Convexity, Cluj-Napoca, 275-280.
/59/ Stancu-Minasian,I.M. and St.Tigan(1988) 'A stochastic approach to some linear fractional goal programming problems', Kybernetika(Prague)24(2), 139-149.
/60/ Stancu-Minasian,I.M. and St.Tigan(1988)'On some fractional programming models occuring in minimum-risk problem'. International Workshop on Generalized Concavity. Fractional Programming and Economic Applications. Mai 30-31 and June 1, Pisa.
/61/ Swarup,K.,Aggarwal,S.P. and Gupta,R.K.(1972) 'Stochastic indefinite quadratic programming', Z.Angew.Math. Mech.52, 371-373.
/62/ Tammer,K.(1978) 'Relations between stochastic and parametric programming for decision problems with a random objective function', Math.Operationsforsch.Statist. Ser.Optim.9(4), 523-535.
/63/ Tammer,K.(1979) 'Beiträge zur theorie der parametrischen optimierung,zu den mathematischen grundlagen ihren anwendung und zu lösungsverfahren', Diss.(B),Humboldt-Universität, Berlin.

/64/ Tammer,K.(1980) 'Behandlung stochastischer optimierungsprobleme unter dem geichtspunkt des strategie der vektoroptimierung', Wiss.Z.TH Leipzig, 4, 295-302.
/65/ Teghem,J.Jr.,(1983) 'Multiobjective and stochastic linear programming,' Found.Control Engrg.8(3-4),225-232 (1984).
/66/ Teghem,J.Jr. and Kunsch,P.(1985) 'Multi-objective Decision Making under Uncertainty: an Example for Power Systems', in Y.Y.Haimer and V.Chankong(eds.), Decision Making with Multiple Objectives,Springer-Verlag, Berlin.
/67/ Teghem,J.Jr.,Dufrane,D.,Thauvoye,M. and Kunsch,P.L. (1986) 'Strange: an interactive method for multi-objective linear programming under uncertainty', European J.Oper.Res.26(1), 65-82.
/68/ Tigan,St. and Stancu-Minasian,I.M.(1983) 'Criteriul riscului minim pentru problema Cebîşev', Lucrările celui de al IV-lea Simpozion"Modelarea cibernetică a proceselor de producţie", 26-28 mai, A.S.E.-Bucureşti, Vol.I, 338-342.
/69/ Weber,R.(1983) 'Pseudomonotonic multiobjective programming', Cahiers Centre Études Rech.Opér.25,115-128.
/70/ Wolf,H.(1985) 'Die ermittlung effizienter lösungen zur stochastischen linearen optimierungsaufgabe', OR. Spectrum 7(2), 81-90.

Part II
Chapter 2

"STRANGE" : AN INTERACTIVE METHOD FOR MULTIOBJECTIVE STOCHASTIC LINEAR PROGRAMMING, AND "STRANGE-MOMIX" ITS EXTENSION TO INTEGER VARIABLES

J. TEGHEM
Faculté Polytechnique de Mons
Rue de Houdain, 9,
7000 MONS
Belgium

ABSTRACT. Many practical problems, such as investment planning within a long time horizon, can be formulated as multi-objective linear programming in which discrete random coefficients are present in the objective functions and in some constraints. This uncertainty is modelized with the comprehensive language of scenarios, well known in the fields of planning and forecasting.

The STRANGE method is designed to obtain a best compromise for such problem : it involves in particular some concepts of stochastic programming and uses a parametric analysis at each interactive phase to provide the decision maker with detailed information on a large set of efficient solutions. STRANGE-MOMIX is the extension of the method to problems including some integer variables.

1. INTRODUCTION.

The method STRANGE solves the following Multi-Objective Stochastic Linear Programming problem (MOSLP) :

$$\begin{aligned} &\text{"min"}\ z_k = c_k \cdot X \qquad\qquad k = 1, \ldots, K \\ &\text{s.t. } X \in D(X) = \{ X \mid TX \leqq d,\ X \geqq 0 \} \end{aligned} \qquad (1)$$

where c_k and (T,d) are "discrete random variables"; more precisely :

- each linear objective function depends on different scenarios, each of them being affected by a subjective probability; let c_{ks_k} (s_k = $1,\ldots,S_k$) be the possible values of c_k and p_{ks_k} the corresponding subjective probabilities with

$$\sum_{k=1}^{S_k} p_{ks_k} = 1$$

R. Slowinski and J. Teghem (eds.),
Stochastic versus Fuzzy Approaches to Multiobjective Mathematical Programming under Uncertainty, 103–115.

- some elements of the constraints are uncertain : let (T_r, d_r), $(r = 1,\ldots,R)$, be the possible outcomes of matrix (T,d) and q_r the corresponding probabilities with

$$\sum_{r=1}^{R} q_r = 1$$

This particular structure of MOSLP is often obtained for the modelization of investment planning problems within a time horizon. It is the case for the energy systems planning problems (see [10], [5] and chapter II-3 of this book) for which STRANGE (STRAtegy for Nuclear Generation of Electricity) has been designed. The initial presentation of this method has been done in [11] and preceded by an analysis of the literature [9]. At this time, very few papers exist concerning MOSLP. Stancu Minasian (see [8] and chapter II-1 of this book) has presented extensive surveys of this subject. Before STRANGE, only the "Protrade method" of Goicoecha [4] described an interactive approach for MOSLP.

The advantage of the Protrade method is that it treats a more general problem, with non linear constraints and general distributions for the random coefficients of the linear objectives. However, Protrade does not seem to be suitable to solve problem (1) and practical applications like in [5] and [10]; effectively :

- a utility function is introduced, the practical construction of which is often a difficult problem; moreover, the use of such a function is based on certain assumptions which are often not satisfied in practice and thus limit its area of application;
- at each step, the information is reduced to the mean value of the objectives, hence not taking enough the dispersion into account. This can be insufficient for the decision maker to correctly appreciate a compromise.

In the initial version of STRANGE [11], uncertain coefficients in the constraints are only present in the right-hand-side; the extension to the left-hand-side coefficients has been proposed in [7] in the framework of the comparison with the FLIP method of Slowinski, using a fuzzy approach. An other improvement has been brought when a demo software on micro-computer has been developed (see [13]) : the user has now the choice between two options for the interactive phases.

Some extensions of STRANGE has been recently presented :

- RB-STRANGE is an extension of STRANGE in the case of non-linear objective functions (see [2]);
- STRANGE-MOMIX is developed for the case where problem (1) contains some integer variables (see [12]);
- Urli and Nadeau used some ideas of STRANGE to analyse the situation where the information on the stochastic coefficients is incomplete (see chapter II-4 of this book);
- Recently, we have found an interesting use of the parametric analysis of STRANGE to easily solve fractional linear programming problems after the transformation of this problem in a bi-criterion LP problem (see [18]).

In the next section, we present the STRANGE method. In section 3, the extension STRANGE-MOMIX is described. Some general comments and conclusions are presented in section 4.

2. THE STRANGE METHOD.

Three steps can be distinguished :

- the definition of the associate deterministic problem
- the search of a first compromise
- the description of the interactive phases.

2.1. The associate deterministic problem.

The uncertainties are managed in two steps :
(a) Each situation (k,s_k) is defined as a criteria to take into account the different scenarios affecting the K objectives. This large set of $\sum_{k=1}^{K} s_k$ new objectives has the advantage of not reducing the complexity of the problem; so, the DM will always have a complete view on the consequences of a solution.
(b) The vectors of slack variables V_r, by default, and W_r, by excess, are introduced for each possible outcome r, so that the set of constraints becomes :

$$D_1 = \{(X,V_r,W_r),\ (r=1,\dots,R)\ |T_r X+V_r-W_r=d_r,\ r=1,\dots,R,\ X\geq 0, V_r\geq 0, W_r\geq 0\}$$

The slack variables W_r measure the violation of the constraints in case of outcome r; so, an additional objective, denoted z_{K+1} and called risk objective, expresses a global measure of violations of the constraints due to uncertainty :

$$z_{K+1} = \sum_{r=1}^{R} q_r\ (\beta_r . W_r)$$

where β_r is a vector of possible penalties.

Remarks.

(i) In the real case studies ([5] and [10]) analysed with STRANGE, the variables W_r are measured in the same unit and there is no reason to introduce penalties because

$$z_{K+1} = \sum_{r=1}^{R} q_r\ (\sum_i (W_r)_i)$$

has a precise meaning (in fact the safety of supply in these applications).

(ii) Criterion z_{K+1} is scenario independant; nevertheless to unify the notations, we note :

$$z_{K+1} = z_{K+1,\ s_{K+1}} \qquad \text{with} \qquad S_{K+1} = 1$$

So, the following deterministic MOLP problem is obtained :

$$\text{"min"}\ z_{ks_k}(X) = c_{ks_k} \cdot X \qquad k=1,\dots,K+1;\ s_k=1,\dots,S_k$$
$$\text{s.t.}\ (X, V_r, W_r) \in D_1 \tag{2}$$

Problem (2) is considered as "equivalent" to problem (1).

2.2. The search of the first compromise.

The single objective problem defined by objective $z_{k\,s_k}$ and outcome r is called (k, s_k, r) problem. Each of these (k, s_k, r) problem is solved and an optimal solution $X^{(r)}_{k\,s_k}$ is determined. The best solution $\tilde{X}_{ks_k}$ is selected :

$$z_{ks_k}(\tilde{X}_{ks_k}) = \min_{r\in\{1,\dots R\}} z_{ks_k}(X^{(r)}_{k\ s_k})$$

This provides the ideal point in the criterion space

$$M_{ks_k} = z_{ks_k}(\tilde{X}_{ks_k})$$

- If solution $\tilde{X}_{k\ s_k}$ is unique, the pay-off table can be defined without ambiguity by the entries :

$$z_{(lt_l)(ks_k)} = z_{lt_l}(\tilde{X}_{ks_k})$$

- If $\tilde{X}_{ks_k}$ is not a unique solution, a similar technique as this proposed by Despontin [3] is applied to define the table. The single criterion problem (3) is solved.

$$\min z_{lt_l}(X)$$
$$\text{s.t.}\ (X, V_r, W_r) \in D_1 \tag{3}$$
$$z_{ks_k}(X) = M_{ks_k}$$

and $z_{(lt_l)(ks_k)}$ is defined as the optimal value of (3).

Like in the STEM method [1], this pay-off table is used to define some normalizing weights Π_{ks_k}, giving for each objective z_{ks_k} the relative importance of the distances from the ideal point : the worst value of z_{ks_k} in the pay-off table is first determined :

$$m_{ks_k} = \max_{(lt_l)} z_{(ks_k)(lt_l)}$$

and the normalizing weights are defined by :

$$\Pi_{ks_k} = \frac{\alpha_{k\, s_k}}{\sum_{k=1}^{K+1} \sum_{s_k=1}^{S_k} \alpha_{k\, s_k}}$$

where $$\alpha_{ks_k} = \frac{m_{ks_k} - M_{ks_k}}{m_{ks_k}} \cdot \frac{1}{\|c_{ks_k}\|}$$

The first compromise is then obtained by the minmax optimization :

$$\begin{aligned} & \min \delta \\ \text{s.t.} \quad & \sum_{s_k=1}^{S_k} p_{ks_k} (c_{ks_k} X - M_{ks_k}) \Pi_{ks_k} \leqq \delta \quad k = 1, \ldots, K+1 \\ & (X, V_r, W_r) \in D_1 \end{aligned} \tag{4}$$

However in case the optimal solution of problem (4) is not unique, the obtained solution may not be efficient; to ensure the efficiency of the first compromise $\tilde{X}_1$, the Despontin modification [3] might be used, giving problem (5).

$$\begin{aligned} & \min M.\delta - \sum_{k=1}^{K+1} \varepsilon_k \\ \text{s.t.} \quad & \sum_{s_k=1}^{S_k} p_{ks_k} (c_{ks_k}.X - M_{ks_k}) \Pi_{ks_k} \leqq \delta - \varepsilon_k \quad k = 1, \ldots, K+1 \\ & (X, V_r, W_r) \in D_1, \quad \varepsilon_k \geqq 0 \quad k = 1, \ldots, K+1 \end{aligned} \tag{5}$$

where M stands for a very large number.

2.3. The interactive phases.

For each compromise $\tilde{X}_m$, the DM receives three pieces of information :

(a) The set of values $z_{ks_k}^{(m)} = z_{ks_k}(\tilde{X}_m)$ together with interval of variations $[M_{ks_k}, m_{ks_k}]$; by this information, the DM has a complete view on the consequences of a compromise.

(b) It can be useful to also provide the mean value of each objective

$$\bar{z}_k^{(m)} = \sum_{s_k=1}^{S_k} p_{ks_k} z_{ks_k}^{(m)}$$

(c) Possibly, if the number of scenarios islarge enough, it can be interesting to determine the confidence levels of the compromise for each objective

$$\alpha_k^{(m)} = P\,(c_k \cdot \tilde{X}_m > \bar{z}_k^{(m)}) = \sum_{s_k \mid z_{ks_k}^{(m)} > \bar{z}_k^{(m)}} p_{ks_k}$$

If the DM is not satified with the compromise $\tilde{X}_m$, one asks him to indicate :

- a criterion $(ks_k)^*$ to be relaxed
- possibly an upper limit $\Lambda_{(ks_k)^*}$ for the value of $z_{(ks_k)^*}^{(m+1)}$.

STRANGE explore the consequences of the relaxation indicated by the DM : in order to completely analyse the results of this relaxation, a parametric LP problem is considered :

$$\min M.\delta - \sum_{k=1}^{K+1} \varepsilon_k$$

$$\text{s.t.} \sum_{s_k=1}^{S_k} p_{ks_k} (c_{ks_k}.X - M_{ks_k})\, \Pi_{ks_k} \leq \delta - \varepsilon_k, \quad k = 1,\dots,K+1 \quad (6)$$

$$c_{(ks_k)^*} \cdot X = M_{(ks_k)^*} + \lambda(m_{(ks_k)^*} - M_{(ks_k)^*}), \qquad \underline{\lambda} \leq \lambda \leq \bar{\lambda}$$

$$(X, V_r, W_r) \in D_{m+1}, \ \varepsilon_k \geq 0 \qquad k = 1,\dots,K+1$$

where $\underline{\lambda}$ and $\bar{\lambda}$ are given respectively by :

$$M_{(ks_k)^*} + \underline{\lambda}(m_{(ks_k)^*} - M_{(ks_k)^*}) = z_{(ks_k)^*}^{(m)}$$

$$M_{(ks_k)^*} + \bar{\lambda}(m_{(ks_k)^*} - M_{(ks_k)^*}) = \Lambda_{(ks_k)^*}$$

For the definition of D_{m+1}, two versions are proposed to the DM.

1°) D_{m+1} is defined, like in STEM method [1], by

- $D_2 = D_1$
- $D_{m+2} = D_{m+1} \cap \{X \mid z_{(ks_k)^*}(X) \leq z_{(ks_k)^*}(\tilde{X}^{(m+1)})\}$

  for $m \geq 1$, after compromise $\tilde{X}^{(m+1)}$ has been determined

. the weight $\Pi_{(ks_k)*}$ is set equal to zero when $z_{(ks_k)*}$ is relaxed and in the course of all the procedure.

Thus, in this option, the objectives which have been relaxed will not be tightened again during the next interactive phases. So, a kind of "convergence" is imposed : at each interactive phase, the DM must choose a different criterion that in thepreceding phases; the procedure thus comes to an end, after a maximum of $\sum_{k=1}^{K+1} S_k$ interactive phases.

However, this option presents the drawback that the DM has no possibility of retraction ... except to come back at an preceding phase. This is the reason why another option has been considered.

2°) . $D_m = D_1$ for all m.

. the weight $\Pi_{(ks_k)*}$ is set to zero only during the relaxation of criterion $z_{(ks_k)*}$, but re-established to its initial value for the next interactive phases.

So at each iteration, the DM can obtain any possible value of a criterion relaxed before.

Using a dual simplex technique, the optimal solution $X^{(m+1)}(\lambda)$ of problem (6) is found and the sequence of bounds $\underline{\lambda} = \lambda_o < \lambda_1 < ... <\lambda_n = \overline{\lambda}$ corresponding to the stable intervals of the optimal bases are determined. For each objective, a piecewise linear function

$z_{ks_k}(X^{(m+1)}(\lambda))$ is displayed graphically so that the DM can assess the complete range of variation of all objectives when $z_{(ks_k)*}(X)$ is relaxed. To make easier the comparison for the DM, the same information is also given in relative values belonging to the interval [0,1] :

$$\frac{z_{ks_k}(X^{(m+1)}(\lambda)) - M_{ks_k}}{m_{ks_k} - M_{ks_k}}$$

Basing on this information, the DM is able to define an acceptable level $\tilde{\lambda}$ of relaxation corresponding to the new compromise :

$$\tilde{X}^{(m+1)} = X^{(m+1)}(\tilde{\lambda})$$

In the first option, $\tilde{\lambda}$ corresponds to the maximum level of relaxation accepted for the relaxed criterion.

An illustrative example of STRANGE is described in chapter IV-2 of this book.

3. THE STRANGE-MOMIX METHOD.

The investment problems, for which STRANGE has been designed, often include some integer variables. For this reason, it is interesting to consider the problem (1'), similar to the problem (1), but with some integer variables. The presence of integer variables in a MOLP structure introduces supplementary difficulties to the characterization of the efficient solutions [14] and to the interactive work [15]. Nevertheless, the minmax optimization in problem (5') corresponds to the use of a generalized augmented Tchebytchef norm and thus assures the efficiency of the first compromise $\tilde{X}^{(1)}$. So, the sections 2.1 and 2.2 can be applied without any change to problem (1').

However, it is not the case for section 2.3 : the parametric analysis becomes a quite difficult problem when there are some integer variables and no adequate methods to solve this problem are available in the literature. It is the reason why we proposed a new method (see [16]), called MOMIX, to solve a multiobjective mixed integer linear programming (MOILP). Notice that MOMIX is of course also suitable for the continuous case as shown recently (see [17]). The interactive phases of MOMIX are based on a different approach, using the concept of interactive Branch and Bound (see [6]). If in addition, uncertainties are present in this problem, the treatment of integer variables by MOMIX and the treatment of uncertainties by STRANGE can be combined to solve the MOSILP problem (see [12]).

In STRANGE-MOMIX, the way to manage the uncertainties and to determine a first compromise are thus identical. We now describe the interactives phases of STRANGE-MOMIX.

The interactive Branch and Bound includes two steps :

- the first step (a "depth first" progression in the tree) leads to the determination of a good compromise;
- the second step (a backtracking procedure) confirms the degree of satisfaction achieved by the decision-maker or it finds a better compromise if necessary.

3.1. For $m \geq 1$, let $\tilde{X}^{(m)}$ the mth compromise, $z^{(m)}_{ks_k}$ the values of criteria and $[M^{(m)}_{ks_k}, m^{(m)}_{ks_k}]$, $\Pi^{(m)}_{ks_k}$ repectively the variation interval and the weight of criterion ks_k at this m^{th} iteration (The variation intervals and the weights defined in section 2.2 correspond to $m = 1$).

The decision-maker has to choose at each iteration, the criterion $l_m(1) \in \{ks_k, k = 1,\ldots, K+1, s_k = 1, \ldots, S_k\}$ he is willing to improve in priority; a new constraint is then introduced :

$$D_{m+1} = D_m \cap \{ X \mid z_{l_m(1)} < z^{(m)}_{l_m(1)} \} ;$$

further the variation intervals $[M^{(m+1)}_{ks_k}, m^{(m+1)}_{ks_k}]$ and the weights $\Pi^{(m+1)}_{ks_k}$ are updated on the new feasible set D_{m+1}.

The new compromise $\tilde{X}^{(m+1)}$ is obtained by the resolution of problem (7) :

$$\min M\delta - \sum_{k=1}^{K+1} \varepsilon_k$$

$$\text{s.t.} \sum_{s_k=1}^{S_k} p_{ks_k}(c_{ks_k} . X - M_{ks_k}^{(m+1)})\, \Pi_{ks_k}^{(m+1)} \leq \delta - \varepsilon_k \quad k=1,\ldots,K+1 \tag{7}$$

$$(X, V_r, W_r) \in D_{m+1}, \; \varepsilon_k \geq 0, \; k = 1,\ldots,K+1$$

The first branch of this interactive tree is described in figure 1.

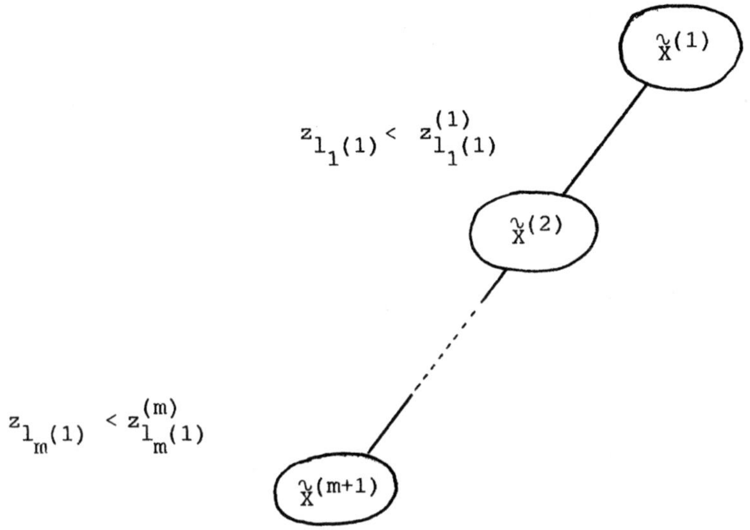

Fig. 1 : The depth first progression.

Different tests allow to terminate this first step.
Fathoming tests : the node (m+1) is fathomed if :

(a) $D_{m+1} = \Phi$

(b) $m_{ks_k}^{(m+1)} - M_{ks_k}^{(m+1)} \leq \varepsilon_{ks_k} \qquad \forall\, (ks_k)$

(c) The vector $\hat{X}$ of the incumbent values (values of the criteria for the best compromise already determined) is preferred to the new ideal point (of components $M_{ks_k}^{(m+1)}$).

Stopping tests : the first step of the procedure is stopped if :

(d) More than q successive iterations do not bring an improvement of the incumbent point $\hat{X}$;

(e) More than Q iterations have been performed.

The parameters ε_{ks_k}, q and Q are fixed in the agreement with the decision maker.

3.2. It can be hoped that the appropriate choice of the criterion $z_{l_m(1)}$, at each level m, has been made so that at the end of the first step, a good compromise has been found.

Nevertheless, it is worth examining some other parts of the tree to confirm the decision maker's satisfaction. The complete tree is generated in the following manner : at each level, K+1 subnodes are introduced by successively adding the constraints :

1°) $z_{l_m(1)} < z^{(m)}_{l_m(1)}$

2°) $z_{l_m(2)} < z^{(m)}_{l_m(2)}$; $z_{l_m(1)} \geqq z^{(m)}_{l_m(1)}$

.
.
.

K+1) $z_{l_m(K+1)} < z^{(m)}_{l_m(K+1)}$; $z_{l_m(k)} \geqq z^{(m)}_{l_m(k)}$ k=1,...,K

where $l_m(k) \in \{ks_k \mid k=1,\ldots,K+1;\ s_k = 1,\ldots,S_k\}$ is a set of K+1 criteria - only one scenario being chosen for each objective. At each level m, they are ordered according to the priorities of the decision maker (see figure 2).

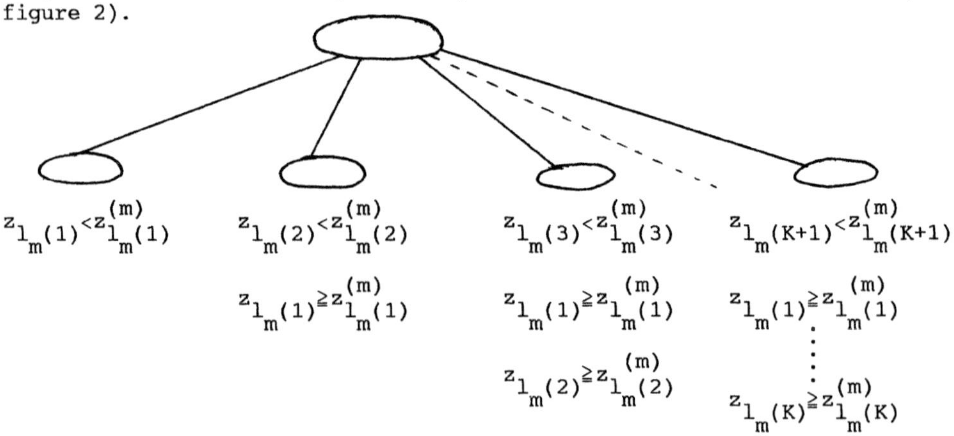

Fig. 2 : The branching procedure.

In this way, the feasible regions of the K+1 subnodes form a partition of the feasible region of the preceding node in the tree.

The usual backtracking procedure is applied; yet it seems unnecessary to explore the whole tree. Indeed the subnodes $k > \bar{K}$ of each branching correspond to a simultaneous relaxation of those criteria $l_m(k)$, $k \leq \bar{K}$, the decision maker wants to improve in priority ! Therefore, these subnodes do almost certainly not bring any improved solutions. It appears in practice that $\bar{K}$ could be fixed to 2 or 3 : at most, the tree branchs of figure 2 will be considered.

The fathoming tests a), b) and c) are again applied in this second step; each time a node is not fathomed, problem (7) is actualized and solved to determine a new compromise. The stopping tests d) and e) - with possibly a choice of new values for the parameters q and Q - limit the number of iterations in this backtracking procedure.

4. SOME CONCLUSIONS.

STRANGE can be easily extended to the case of non linear objective functions : the Miller's method (see [19]) is applied using piece-wise linear functions in an extended version of the simplex method called "Restricted Basis" (see [2] for details on RB-STRANGE).

STRANGE, RB-STRANGE, STRANGE-MOMIX are integrated in a DSS which address a large scope of multiobjective investment planning problems, including non linearity of functions, integer variables and uncertainties. The measurement of risk resulting in not satisfying several uncertain constraints is made possible by using an additional criterion. The value of this criterion within its variation range gives a useful measure of risk. This technique is not only mathematically meaningful but it largely contributes to a better understanding of uncertainties as practical applications have shown (see chapter II-3 of this book).

The scenario approach is also favourably perceived by the users, as mean values and dispersions of results are easily apprehended. The methods exposed here have the considerable advantage of requiring easy-to-answer questions and comparisons to the users.

This interactive DSS has been first implemented on a mainframe VAX computer to be used for real problems. For smaller problems and with the aim of a larger diffusion, a micro-computer version of STRANGE has been developed (see [13]) : it is user friendly, menu-driven decision tool making extensive use of graphical techniques in the interactive phases.

Let us add that recently STRANGE has been proved to be useful to easily solve fractional linear programming problem (FLP). Effectively, a optimal solution of this type of problem is nothing else that an efficient extreme solution of the bi-criterion problem naturally associated to the FLP problem. With its parametric analyses, STRANGE generates all the set of efficient solutions in at most two phases, in particular it generates the efficient extreme solutions, including the optimal solution of FLP problem.

REFERENCES.

[1] BENAYOUN R., de MONTGOLFIER J., TERGNY J. and LARICHEV O., "Linear programming with multiple objective functions : Step method (STEM)", Mathematical Programming 1(2) (1971) 366-375.

[2] DELHAYE C., "Système interactif d'aide à la décision multicritère", Travail de fin d'Etudes, Faculté Polytechnique de Mons (1987).

[3] DESPONTIN M., "Interactive economic policy formulation with multiregional econometric models", in : M. Despontin, P. Nijkamp and J. Spronk (eds.), Macro-Economic Planning with Conflicting Goals, Springer, Berlin, 1984.

[4] GOICOECHA A., HANSEN D. and DUCKSTEIN L., Introduction to Multi-Objective Analysis with Engineering and Business Application, John Wiley, New York, 1982.

[5] KUNSCH P., TEGHEM J., "Nuclear fuel cycle optimization using multi-objective integer linear programming", European Journal of Operations Research 31 (1987), 240-249.

[6] MARCOTTE O., SOLAND R., "An interactive branch and bound algorithm for multiple criteria optimization", Management Science, 32, 1 (1986), 61-75.

[7] SLOWINSKI R., TEGHEM J., "Fuzzy versus Stochastic Approaches to multicriteria linear programming under uncertainty", Naval Research Logistics, 35 (1988), 673-695.

[8] STANCU-MINASIAN I.M., "Recent results in stochastic programming with multiple objective functions", in M. Grauer, A. Lewandowski and A.P. Wierzbicki (eds.), Multi-Objective and Stochastic Optimization , IIASA Collaborative Proceedings Series CP-S12, 1982.

[9] TEGHEM J., "Multiobjective and stochastic linear programming", Foundations of Control Engineering, 8, 3-4 (1983), 225-232.

[10] TEGHEM J. Jr and KUNSCH P., "Multi-Objective decision making under uncertainty : An example for power systems", in Y.Y. Haimes and V. Chankong (eds.), Decision Making with Multiple Objectives, Springer, Berlin, 1985.

[11] TEGHEM J., DUFRANE D., THAUVOYE M., KUNSCH P., "STRANGE : an interactive method for multi-objective linear programming under uncertainty", European Journal of Operational Research, 26 (1986), 65-82.

[12] TEGHEM J., KUNSCH P., "An interactive DSS for multiobjective investment planning" in D. Mitra et al. (eds.), "Mathematical models for Decision Support", Springer-Verlag (1988).

[13] TEGHEM J., DELHAYE C., BOURGEOIS F., KUNSCH P., "STRANGE on micro" in C. Bana e Costa (ed.), "Multicriteria decision making : software and applications", Springer Verlag (to appear in 1990).

[14] TEGHEM J., KUNSCH P., "Survey of techniques to determine the efficient solutions for multi-objective integer linear programming", Asia-Pacific Journal of Operations Research, vol. 3 (1986), 95-108.

[15] TEGHEM J., KUNSCH P., "Interactive method for multiobjective integer linear programming" in Fandel et al. (eds.), "Large scale modelling and interactive decision analysis", Springer Verlag (1986), 75-87.

[16] TEGHEM J., KUNSCH P., "MOMIX : a method for multi-objective mixed linear programming", submitted for publication.
[17] TEGHEM J., L'HOIR C., "Portfolio selection by multi-objective linear programming with the MOMIX method", submitted for publication.
[18] TEGHEM J., "The use of the method STRANGE to fractional linear programming", submitted for publication.
[19] ZANGWILL W.I., "Non linear programming", Prenctice Hall (1967).

Part II
Chapter 3

APPLICATION OF "STRANGE" TO ENERGY STUDIES

P.L. Kunsch
BELGONUCLEAIRE
25 rue du champ de Mars
B-1050 BRUSSELS (Belgium)

ABSTRACT. STRANGE is an interactive multiple criteria decision aid system taking into account uncertainties on objective functions and on constraints by using the techniques of stochastic linear programming. It has been mainly used for applications in the field of energy planning. The paper presents some of the published case studies applicable to nuclear and renewable energies. The paper also shows how extensions of the method towards non-linear functions and mixed integer variables increase the usefulness of the approach in real cases.

1. Background

In the years of economic growth in the sixties, decision making within private or public corporations or economic units was based primarely on the maximization of profit as being the main objective of these groups.

With the increase of the economic uncertainties in the seventies, the economic decision process started to take into account multiple objectives such as balance of payments, safe supply of raw material, environment, employment etc.

This is particularly true in planning energy production and supply. The first oil crisis has indeed shown that cheap energy in unlimited supply is no longer possible in a finite world.

The high risk of interrupted energy supply might no longer be felt to be so acute as in those days, due to the diversification and more rational use of resources. However the long term energy issue is not solved. National economies have to live with the constraint of limited resources and of dependence with regard to outside suppliers. In more recent days they become more and more conscious of their endangered natural environment.

Although they are many energy models in use, mainly in the hands of large national bodies or of supranational institutions, like the European Economic Community (EEC) or the Organization for Economic Cooperation and Development (OECD), it is surprising to see how little account is taken of the trade-offs between conflicting objectives relevant to planning problems.

Most of these models are demand driven and, like in the days of

R. Slowinski and J. Teghem (eds.),
Stochastic versus Fuzzy Approaches to Multiobjective Mathematical Programming under Uncertainty, 117–130.

unlimited economic growth, they exclusively optimize the total production costs (e.g. see **Ref. [1]**)

2. STRANGE and Energy Planning

The use of the multiple criteria approach **STRANGE** (see Chapter II-2) as a decision aid is a modest attempt to add this multiple objective dimension as well as uncertainties to the important field of energy planning.

The method has been used in the context of an engineering and manufacturing company developing and building energy systems, based mainly on nuclear fuels but also on renewable fuels, like solar energy.

This original approach which has been discussed in chapter II-2 from the mathematical aspect has been coded into a computer code for several energy study models.

STRANGE addresses time-dependent medium to long-range planning problems with a continuous range of possible actions. Its originality lies in the use of multi-objective stochastic linear programming and in its interactive search for acceptable compromise solutions.

The approach is particularly suitable for evaluating the time-dependent deployment of new techniques and the penetration of energy systems : advanced nuclear reactors, renewable energies, co-generation of heat and electricity etc.

The name is derived for letters describing the first code which had been developed for planning " **STRA**tegy for **N**uclear **G**eneration of **E**lectricity" although its applicability is of course not limited to nuclear energy.

3. Optimal use of nuclear fuel

This first application of the **STRANGE** method has been developed within the activities of the company mentioned. This company is manufacturing fuel to be used in nuclear reactors. It has, for the sake of developing its activities in advanced fuel technologies to convince its customers- i.e. the electricity producers - of opportunities derived from the optimal use of the fuel.

To start with, one has to understand that nuclear fuel is behaving quite differently from classical fossile fuel. The latter is burnt once and for all in the thermal plant during a classical combustion process generating heat energy.

In contrast, the nuclear material contained in so called fuel assemblies is only partially burnt in the reactor during the process of fission. Fissile atoms, usually uranium atoms, are split to produce heat energy. After several years in the reactor, the fuel assemblies are no longer usable and they have to be removed and replaced by fresh ones. The spent fuel assembly contains "ashes" - nuclear wastes - but also still a lot of fissile atoms, which can be possibly reused- about 95% of the initial quantity. **Reprocessing** is the chemical process in which the fissile material still present in the spent assemblies can be

recuperated by separating it from the wastes. Reprocessed material can then be **recycled**, it means reused for the manufacturing of fresh fuel assemblies.
The decision problem which has to be solved is to know how to proceed with the spent fuel assemblies after they are removed from the reactor. Several decision branchings are possible leaving a lot of open strategic questions like the following ones:

Decision 1: Reprocessing or final storage of the fuel assemblies ? In the latter case, the still available fissile atoms cannot be reused.
Decision 2: Use of reprocessed material: storage strategy, recycling, use in advanced reactors?
Decision 3: Disposal of the wastes?

In our societies the answer to these question has often been the matter of controversial political and public debates.

The corresponding decision model can be handled completely with continuous variables in a linear programming system:

The equations of the constraints represent the material inventories and the transformations during the so-called nuclear fuel cycle starting with the mining of natural uranium and ending with the final storage of spent fuel assemblies or of waste material.
Other constraints represent technological limits on the different fuel types, or restrictions on the lead time related to some operations on the fuel, like the cooling time of burnt fuel before further handling, or on the manufacturing time of fuel assemblies. There are constraints on the capacities of storage facilities and of the production plants in which the fuel undergoes some transformation, like assembly manufacturing, or spent fuel reprocessing.
Political and economic constraints are related to commitments with regard to suppliers of raw material or of fuel cycle services.

In the case study thoroughly described in Ref. [2,3] four criteria have been optimized :

1. The **Fuel Cycle costs**, i.e. that variable part of the electricity production costs related to the burning of fuel. As they are many uncertainties in the per-unit costs of the different steps making up the nuclear fuel cycle - one being for instance the market price of the raw material, the coefficients of the cost objective function are affected by important uncertainties;

2. The **assured raw material supply** . Uranium is a commodity like fossile fuel. Uranium supply can be guaranteed by long-term contracts, or it can be purchased on a short time basis on the spot market. As in the case of oil, diversication is important for measuring the independence from the suppliers. A hoarding strategy to constitute strategic stocks might also be considered in certain cases. On the

other hand Uranium has limited resources recoverable at a given mining cost;

3. The **commmercial balance** is related to the fuel cycle services which are needed and which cannot always be provided by national companies because of the large investment and technology requirements neeeded. An example is given by the reprocessing of spent fuel which is in Europe almost completely taken over by French companies;

4. The **employment** has been considered as an important criterion, as the nuclear industry has been increasingly unpopular in the last decade, and there is a risk for this industry with high technological level to become unactractive for young professionals. This would have negative consequences however, including for the safe operating of nuclear plants, due to the lack of qualified people;

These criteria, which all except for the first one, go beyond a purely cost-benefit indicator, need to be represented by adequate ratings, not only currency units like francs or dollars.
It is clearly one of the difficult tasks of the analyst when preparing the model to define representative functions, which in the further steps of comparing criteria - i.e. really pears and apples- make sense to the decision-makers.

This comparaison of heterogeneous concepts: cost, employment etc. is provided by the scalarisation of the vector problem. In **STRANGE** the weights are given a limited importance. Instead a parametric analysis is proposed, which is a projection on the efficient frontier in the space of the objective functions.

One of the objectives is selected as the parametric variable. By solving the dual problem, the efficient solutions in the direction represented by this objective can be determined and plotted. Such plots display at one glance multiple efficient decisions, from which the decision maker extracts acceptable compromises.

Fig. 1 from **Ref. [2]** shows the result of a parametric analysis of the objective representing the assurance of raw material (Uranium) supply. This 3-D surface gives the amount of recycled material which will be reused as fuel in the reactors in the dependence of time and as a function of the level of safe uranium supply which is accepted. A modified compromise will be accepted as the decision-makers sees the evolution of all objective functions plotted against this objective as a relaxation variable.

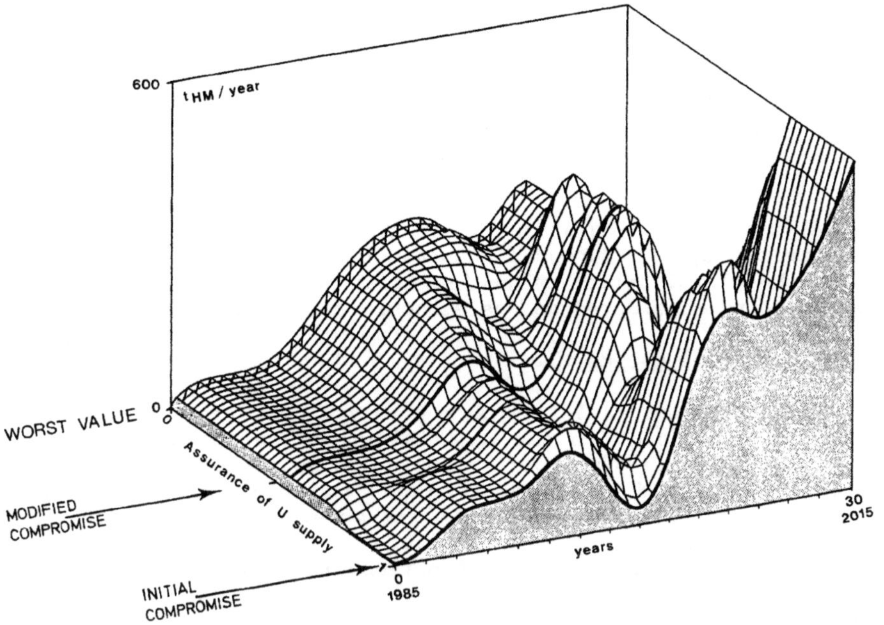

Fig. 1 Parametric analysis on the assurance of Uranium supply. The quantities of recycled material - in tons of heavy metal per year are plotted against time, starting in the year 1985.

The uncertainties in this model are considered to affect mainly the coefficients of the objective functions: the time evolution of costs, or different evolution in the safe supply of raw material. **STRANGE** handles this by considering discrete scenarios in the objective functions. All those scenarios are taken into account in the final compromise solution and in the parametric analyses, performed one at the time on one objective function represented in one scenario.

In the fuel cycle exercises, no uncertainty in the Right-Hand-Side of constraints has been considered, although they can be handled as well by introducing an additional objectives, representing the penalties for violated constraints (see below in the power system case study). For example, uncertainties in the capacities of fuel assembly manufacturing plants is represented by the possible loss of electricity production caused by the non-availability of the fuel to be loaded into the reactors.

This analysis of the logistics of the nuclear fuel cycle is part of a study sponsored by the European Economic Community with the purpose of analysing the global impact of nuclear fuel strategies on the countries of the community considered as one economic entity (**Ref. [2]**)

4. Generalized Fuel Cycles

The optimization of fuel use is a purely linear problem, as long as only terms proportional to the material quantities in their different forms are present in the coefficient of constraints and objectives. This remains true if for example only variable costs make up the production costs. In more realistic cases however, e.g. the presence of fixed costs or drops in the variable costs due to large serie manufacturing introduce non linear functions into the problem.

STRANGE, although at start a technique based on mathematical linear programming, could however be extended to the optimization of this kind of problems. The Miller's method (see **Ref. [4]**) is applied using piece-wise linear functions in an extended version of the simplex method called "Restricted Basis" (RB) (see Chapter II-2).
This technique has shown its influence on the fuel cycle optimization in the following way:

A recycling of fuel is first non-economic as large capital costs are needed to start operating a reprocessing pilot plant with a limited throughput. Per-unit costs drop however as a result of increasing load of the plant . Moreover investment costs of future plants are getting smaller thanks to the mass production and to the gained experience.
The resulting optimization therefore brings results which can be very different from the purely linear case in which the coefficient of the objectives are always constant whatever the level of the production variable. Therefore in the linear case, the take-off of some new technique or development can be made difficult (although the multiple criteria aspects can mitigate the purely economic impediments). In contrast the generalization to piece-wise linear functions makes the results much more realistic and permits smoother developments of techniques.

5. Power System Planning under uncertainties

5.1 Power-mix calculation

Power-mix calculations describe dynamically the shifting shares of the different energy production means in function of time, and assuming several evolution scenarios in the future for the basic data, like the costs.
An former version of STRANGE had been used for power-mix calculations (**Ref [5]**), and it was based on an aggregation techniques with weights derived from the pay-off matrix. This approach was abandoned as being non-interactive and it gave solutions quite sensitive to the definition and number of the scenarios describing the uncertainties.

This difficulty was avoided with the approach applied in the MOSLP method - Multiobjective Stochastic LP - as it is applied in the present **STRANGE** method presented in Chapter II-2 (see also **Ref. [6]**). It transforms the problem into an equivalent deterministic Multi-objective LP by :

- defining the situations (k,s_k) as new objectives, where s_k=1, S_k are the scenarios associated to objective k=1,K.

- using the 'stochastic programming with recourse' described in Chapter II-2 (**Ref. [6]**) to solve the uncertainties in the constraints. This comes out to the definition of an additional objective K+1, measuring the violation of the constraints across various scenarios for the realization of the Right-Hand-Sides.

5.2 Solar power system versus Diesel engines

An application of this technique to a power-mix problem has been extensively described in the literature (see **Ref. [6,7]**). It is based originally on a real case in which the contenders are classical Diesel engines and solar power systems.

The aim is to assure the electricity supply of small village communities in a Third World country, having no access to the electricity grid and purely domestic needs. An important aspect is the night production which represents the peak demand.
The merits of both systems are compared in the analysis. The main advantages of a solar generator is its long life-time, its easy maintenance - while the Diesel engine needs frequent maintenance and refitting, and of course its complete independence of fuel, as long as the sun is shining (This can be normally assumed during day-time); the main drawback with regard to a Diesel engine is the cash drain caused by the high investment costs, and also the difficulties related to the storage of energy in the form of heat for generating electricity after dark.
In a first stage two objectives are used for the analysis:

1. The **total production costs** is calculated from the investment, the variable operating costs, which include the variable fuel costs when applicable, the maintenance costs and the storage costs of heat for the production after dark.

2. The **dependence on suppliers** from outside the Third World country is measured as the sum of all expenditures, occurring during the life-time of the engines, which have to be paid to foreign manufacturers or suppliers. This includes for instance all of the Diesel fuel, imported to the country. On the other hand, it is assumed that an increasing part of the components used in the solar power plant can be manufactured locally, according to a learning curve of the indigeneous craftsmen.

The uncertainties attached to this problem are of two different kinds:

- First, some elements of the costs are uncertain, because of external economic conditions. This affects the cost coefficients in the two first objective functions. In order to represent these uncertainties, three future situations of the world are imagined in which the cost coefficients correspond to low, average and high side assumptions. This applies to both the Diesel fuel price and to the solar investment costs. In all there are therefore nine future scenarios affecting some coefficients of objectives 1 and 2;

- A second kind of uncertainty is related to the perturbations that can appear in the energy production for the following two reasons:
(a) Diesel fuel shortages are stochastic events. Several situations of fuel scarcity are possible. Each of them has a subjective probability of occurence, assessed from historical data, or from projections. Each situation corresponds to the non-satisfaction of a fraction of the normal demand going from slight fuel shortages to severe embargos.
(b) Solar energy production also fails if insolation is too poor. Statistics on the past years are available to define the probability, duration and importance of some typical situations of insufficient insolation.

Such uncertainties introduce discrete random elements in the vector of the Right-Hand-Side of the constraints. As described before in section 5. and in Chapter II-2, the technique of stochastic programming is applied:
A third objective is constructed representing the violation for non-satifaction of the demand. The original stochastic optimization problem with two objectives is then transformed into an equivalent deterministic optimization problem with three objectives.

3. The third objective is called the **failure for non-satisfying the electricity demand**. It is defined as the total of the failed energy units (kWh) not available because of the possible diesel fuel shortage situations, or because of the situations of unsufficient insolation . It is calculated as the expected value of the stochastic variable

representing this energy deficiency in the scarcity situations just described.

In mathematical programming the approach for calculating these stochastic variables is to introduce slack variables by default and by excess, which are measuring the violation of the constraints in each deficiency situation and are used in the definition of the third objective.

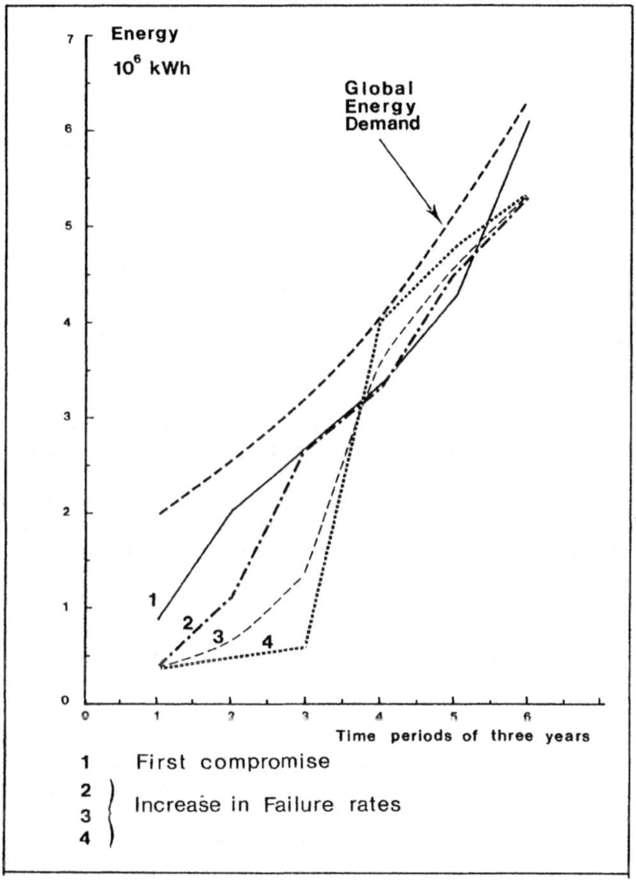

Fig. 2 Electrosolar energy production after dark in function of time for severable acceptable failure rates in the electricity production.

The aim of the multiobjective decision problem is to derive a possible long-term planning for the partial replacement of the traditional Diesel generators by the new solar power systems, to meet best all requirements represented by these objectives.
Fig. 2 shows various production schedules during the night - it means using the storage of heat coming from solar production. These production schedules for the new solar technology will depend on the importance of the acceptable failure rate, i.e. the fraction of non-satisfied electricity demand as described in the third objective. The calculation shows that the larger this rate will be, the later in time will the solar system be breaking through. In addition for the peak demand production and for back-up purposes, some additional Diesel capacity might still be required .

6. An extended Power System Planning with Mixed Integer Variables

Increments of production in an power-mix calculation can be handled continuously as long as the total size of the energy system is large enough.
For small-size systems, the incremental steps are too large with regard to the total size and they can therefore no longer be considered as being continuous variables.
Investment planning of power system is therefore by essence a mixed integer problem (MIP).

In the case study just considered in section 5, these discrete aspects have been ignored, because only long-term trends are needed, but in practical medium-range planning, MIP cannot be avoided.
The presence of integer variables in a Multiobjective Linear Programming system introduces supplementary difficulties in both the characterization of the efficient solutions and in the interactive steps of the method (see **Ref. [8]** for the details).

The first consequence is that the parametric analyses used in the interactive steps of **STRANGE** cannot be extended. Instead a completely new interactive approach, called **STRANGE-MOMIX**, has been developed in **Ref. [9]** (see also Chapter II-2).
It is based on a Branch and Bound tree which is explored "in-depth" first, and later on if necessary using a backtracking procedure.
The in-depth nodes of the tree are generated as being first compromise solutions found in the MIP and with the additional constraint imposed to a selected criterion to be better than an imposed limit. The tree can be explored as in the traditional BB-method in integer programming including fathoming and stopping tests.
The backtracking procedure is applied in a second stage to explore more parts of the tree. Subnodes are generated as a partition of the feasible region of the parent node by adding constraints on the criteria ranked according to their degree of preference.
STRANGE-MOMIX has been applied successfully to port-folio analyses. Its use in power mix calculations is required to have realistic

implementation schedules of producing units.
No results have been published so far although they were attempts to generalize the case study presented in the previous section and reactor mix calculation as in **Ref [3]**.
The reason for a limited use of these methodologies is not only the CPU effort requested for quite large problems- with hundreds of decision and dependent variables of with classically 20 to 50 are integer variables. It is also explained by the abscence of decision-makers ready to indulge in the quite time-consuming interactive steps.

7. The codes STRANGE for Energy System Decision Aid

A flow sheet of the computer code is given in **Fig. 3**. It shows the general organization of the Multiobjective Decision Aid (MCDA) code and the interface with the user, analyst or decision maker.
The modular structure makes the coding as little problem-dependent as possible. Some parts of the system however still have to be modified according to the specific problem which has to be solved.
The central components specific to the problem are (1) the database gathering all the input data, their uncertainties and their time evolution ;(2) the matrix generator calculating from the database the coefficients of the technical matrix and of the objective functions for various scenarios.

Other parts of the system are problem-independent, like the multiple criteria technique, the processors for preparing the data and the report generator of the calculation results.
All the variable and objective function values derived from the successive parametric analyses are stored on disk and can be displayed on the screen using the graphical tools. In that way, the interactive steps can be pursued as long as the decision-maker wishes, with possible backtracking to a previous solution.
Several versions of **STRANGE** presently exist, actually one for each of the different problems. They have been implemented on a minicomputer VAX 8200, and they are using LP packages, like XMP **(Ref. [10])**, adapted for using the restricted basis extension of the method.

Typical problems had 15 to 30 time steps representing time horizons of about the same number of years; there were typically two to three economic scenarios, about ten decision variables, 20 dependent variables and about 20 constraints per time step.

In the future developments, it can be thought of having a commercial package with matrix and report generator and LP (including MIP) organized in a completely problem-independent way to use the MCDA approach. The set-up time for preparing a new operational version of the code would be only represented by the time of the model analysis. This would improve the present conditions in which much time is spent with programming tasks, and even minor changes to the analysis translate into an significant coding effort.

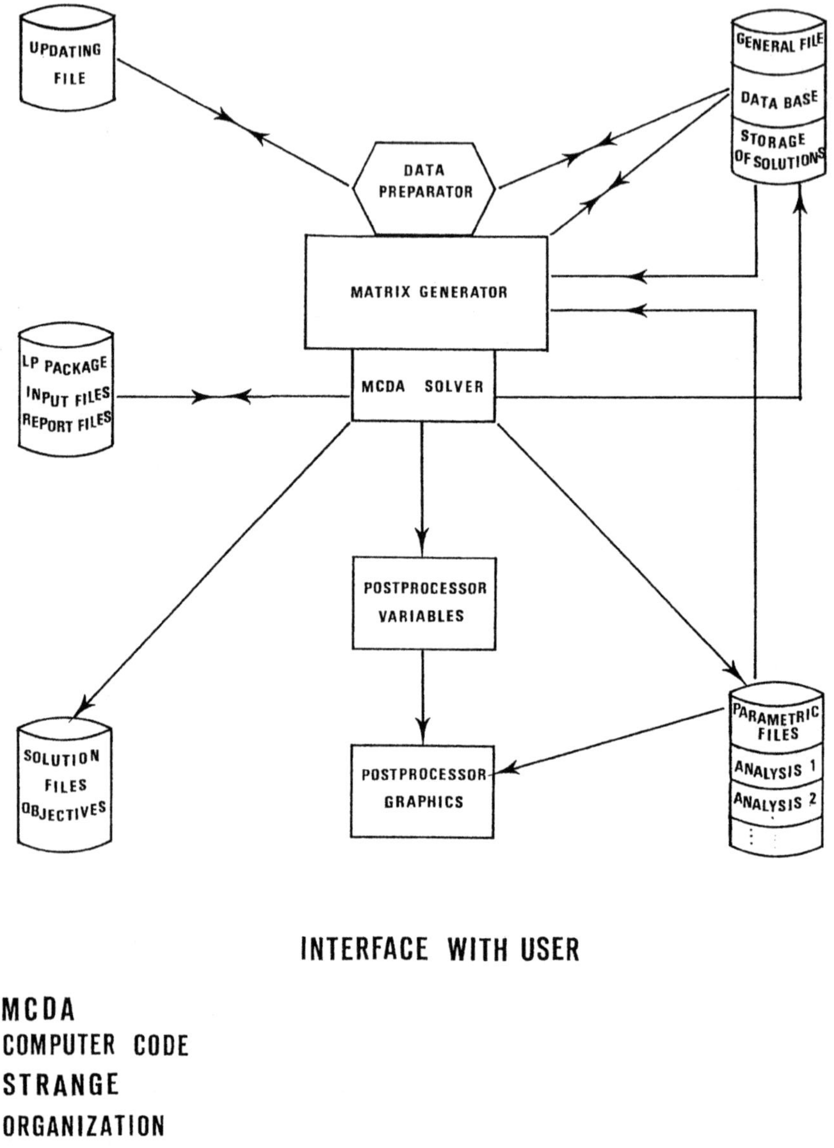

Fig. 3 The organization of the MCDA computer code **STRANGE**

8.Conclusions

STRANGE has been successfully applied up to the stage of parametric analyses in energy planning problems.

Its basic approach has been:

- purely continuous constraints;
- uncertainties in the linear objectives and in the Right-Hand-Side of constraints, the latter being in general capacity constraints;

The extension to piece-wise linear functions has been requested whenever the large variation range of the variables makes it impossible to assume constant coefficients in the objective functions.
Uncertainties in the left-hand-side of the constraints can be handled, although it is less frequent .
The burden of integer variable can be a considerable one, but the problem can be solved if enough time and computing efforts are affordable. The **STRANGE-MOMIX** approach addressing this type of problem has not yet been applied in full-scale MIP problems.
Exploratory steps in energy calculations are often satisfactorily described accurately enough with continuous models.
When it comes to details however, the MIP results specially on a short time scale can considerably depart from the continuous case. Considering the vast amounts of money spent by government and energy producing companies, and looking at the important and often very long-term impact of a power programm on the life of a nation and of its people, the cost of such calculations are still absolutely negligible. It is therefore worth making the effort, even then if the only result would be not to miss any of the important aspects of the decision problem.

This will help better comparing the pears and the apples, that is to say the antinomies of economy and society.

REFERENCES:

[1] EFOM-12c Energy Supply Modelling Package (1985)
Report EUR 8896 EN/II
EEC Luxembourg

[2] Kunsch, P.L., van Steenberghe, T. and Decressin, A. (1987)
A Multicriteria Study On LWR Fuel Management in Countries of The European Economic Community
IAEA Vienna Proceedings of an International Symposium on The Back-End Of The Nuclear Fuel Cycle IAEA-SM-294/11, 143-152

[3] Kunsch, P.L. and Teghem, J. Jr.
Nuclear Fuel Cycle Optimization using Multiobjective Linear Programming
EJOR, 31 ,240-249

[4] Zangwill, W.I. (1967)
Non Linear Programming
Prentice Hall

[5] van Dievoet J., Verraver J. and Kunsch P.L. (1981)
Multivariant Strategy Study For The Use Of Plutonium In A Nuclear Programme: The Case Of The large And Of The Small country Fast Reactor Fuel Cycle BNES Conference London, 55-62

[6] Teghem, J. Jr. Dufrane, D., Thauvoye , M. and Kunsch, P.L. (1986)
STRANGE: An Interactive Method For Multiobjective Linear Programming Under Uncertainty
EJOR 26,1, 65-82

[7] Teghem, J. Jr. and Kunsch P.L. (1985)
Multiobjective Decision Making Under Uncertainty, An Exemple For Power Systems.
Chankung, Haimes Eds, Decision Making With Multiple Objectives
Springer Verlag ,443-456

[8] Teghem, J. Jr. and Kunsch, P.L. (1986)
Complete Characterization of Efficient Solutions for Multiobjective Integer Linear Programming
Asia-Pacific Journal of Operations Research 3,2 95-108

[9] Teghem, J. Jr. and Kunsch P.L. (1988)
Interactive DSS For Multiobjective Linear Planning in,
Nato Asi Series Volume F48 Mathematical For Decision Support,
Editor G. Mitra, Springer Verlag, 123-133

[10] Marsten R.E. (1981)
The Design of the XMP Linear Programming Library
University of Arizona

Part II
Chapter 4

MULTIOBJECTIVE STOCHASTIC LINEAR PROGRAMMING WITH INCOMPLETE INFORMATION: A GENERAL METHODOLOGY

Bruno URLI
UQAT, Rouyn, Québec
Canada, J9X

Raymond NADEAU
Sciences de l'administration
Université Laval, Québec
Canada, G1K 7P4

ABSTRACT.

Numerous multiobjective linear programming methods have been proposed in the last two decades for contexts where the parameters are deterministic. In many real situations, however, parameters of a stochastic nature arise and the analyst is, as a result, confronted with a stochastic multiobjective linear programming problem. Recently, some methods have been developed to deal with this kind of problems; in most of them, the decision maker is supposed to be placed in a risky situation, i.e. a situation where he can associate probability distributions to the stochastic parameters. In many cases, we believe that it would be more realistic to suppose that the decision maker is in a situation of partial uncertainty, i.e. a situation where he possesses only an incomplete information about the stochastic parameters; for example, he could be able to precise only the bounds of variation of the parameters and eventually, their central values. For such situations, we propose a general multiobjective stochastic linear programming methodology which includes many modes of transformation of the stochastic objectives and constraints in order to obtain an equivalent multiobjective deterministic linear programming problem. Finally, this deterministic equivalent program is solved by an interactive method derived from STEM. Our methodology will be illustrated through a didactical example.

1. Introduction

For decades, specialists in Operational Research (O.R.) have been developing new methods aimed at modelling problems to be solved with more and more realism and precision. This research towards a greater adequacy of models with the reality has become particularly obvious in the basic O.R. model, that of linear programming. With this in mind, it

R. Slowinski and J. Teghem (eds.),
Stochastic versus Fuzzy Approaches to Multiobjective Mathematical Programming under Uncertainty, 131–161.

should be noted that the development of stochastic linear programming (SLP) has enabled a better focusing of one aspect of the reality (until then neglected) by taking into account the uncertainty. More recently, there has been a development of multiobjective linear programming (MOLP) methods in which the actions are no longer evaluated by are single objective function, but rather by several objective functions which the decision maker (D.M.) wishes to optimize simultaneously. Finally, in the last few years, a particular interest has arisen to simultaneously take into account these two types of generalizations which has led to multiobjective stochastic linear programming (MOSLP).

Up to now the MOSLP methodologies has been few in number, undoubtedly due to the addition of uncertainty to the difficulty of multicriteria; but there are nevertheless some which are referred to by Stancu-Minasian [1984]. Generally, from appropriate transformations of the objective functions and the stochastic constraints, the initial MOSLP problem is changed into an equivalent multicriteria deterministic problem. In order to obtain a best compromise solution, it is generally necessary to obtain supplementary information about the structure of preferences from the D.M.. With the classification of Hwang and Masud [1979], it is possible to distinguish two approaches within the MOSLP methods: that using an a priori articulation of preferences and that aiming at a progressive articulation of preferences also called interactive methods. Methods with a priori articulation of preferences are characterized by a certain mathematical refinement, by the very precise specification of hypotheses and the use of standard SLP approaches which enable the transformation of the initial stochastic problem to an equivalent deterministic quadratic or fractional program. In these methods, it should be noted that the uncertainty is in general located only in the objective functions.

To our knowledge, there exist today only two interactive MOSLP methods: the PROTRADE method (Goicoechea and al. [1982]) and the STRANGE method (Teghem and al. [1986]). These methods were developed to help the D.M. in the framework of concrete applications and deal with problems where the uncertainty lies in the constraints as well as in the objective functions.

Generally, the preceding MOSLP methods are aimed at problems where the set of actions is infinite and where the modelling of the undetermined data (as much for objective functions as for constraints) is probabilistic. In most of these methods, the modelling of the stochastic parameters is done by assuming that the intervals of variation of those parameters are known. The arbitrariness of each of these methods consists mainly of the measure of occurrence of some or all of the values of these intervals of uncertainty. If documentation about this judgement quantification relative to the possible values of these stochastic parameters exists (Huber [1974]), numerous studies have also shown the biases that make up these different techniques of elicitation (Kahneman, Slovic and Tversky, [1982]). Consequently, in a situation where there is little information about some parameters, if it seems reasonable to consider the uncertain nature of those parameters, wanting to associate a probability distribution to them is a hypothesis which we prefer to avoid. In many real problems, the D.M. is confronted with such situations, and in particular that is often the case in the framework of medium or long term planning problem.

In short, we are interested in a situation with uncertainty in which the D.M.'s information is sufficient to associate to each stochastic parameter a variation interval (i.e. a pessimistic value and an optimistic value) and eventually a most probable value. Furthermore, this information is incomplete in the sense that it does not permit the assignment of probability distributions to the stochastic parameters. Such a situation which we call uncertainty with incomplete information has never been tackled in the field of MOLP. In order answer to this problematic, we propose a general MOSLP methodology with incomplete information.

2. The problem and a look at the proposed methodology

The problem to be solved is as follows:

$$\left\{\begin{array}{ll} \text{Max} & Z_k(\underline{x}) = \sum_{j=1}^{m} c_{kj}\, x_j . \\ \text{Subject to} & \sum_{j=1}^{n} a_{ij}\, x_j \le b_i \ , \ x_j \ge 0 \\ i = 1,\dots, m;\ j = 1,\dots, n;\ k = 1,\dots, K & \end{array}\right. \tag{1}$$

where Z_k, $k = 1,\dots, K$, are real value functions defined on R^n, $\underline{x} \in R^n$, and the parameters c_{kj}, a_{ij} and b_i, $i = 1,\dots, m$, $j = 1,\dots, n$ and $k = 1,\dots, K$, are random variables (defined on some probability space) for which the probability distributions are unknown but for which the following hypotheses should be verified:

- for each coefficient c_{kj}, the D.M. knows the limits of the variation c^-_{kj} and c^+_{kj} (and eventually a central value c^o_{kj});
- for each coefficient a_{ij} and b_i, the D.M. knows the limits of the variation a^-_{ij}, a^+_{ij}, b^-_i and b^+_i.

To deal with problem (1), the proposed multicriteria methodology offers many options based on different modelling of uncertainty relative to the objective functions and the constraints. Two approaches are proposed with regard to modelling the uncertainty of the objective functions. The first one is based on the use of the D.M.'s. degree of satisfaction relative to some stochastic inequalities. In the most general case, that is where only the extreme values of the stochastic parameters in the objective functions are known, the deterministic equivalent program is non-linear. However, if the most probable values of these parameters are also known, this approach then generates equivalent linear objective functions. The other approach (called "mean-dispersion") has recourse to an equivalent deterministic formulation by corresponding to each criterion two measures, a measure of central tendency and a measure of dispersion. It is necessary in this approach to know the stochastic parameter's central values of the objective functions. However, if the D.M. has difficulty coming up with these values, it is always possible to define them automatically as middle points of the variation intervals. This automatic procedure is justified by the fact that (not possessing information about the probability distributions of the stochastic parameters) we determine their central values as if, by virtue of the maximal entropy principle, uniform probability distributions were defined on the variation intervals of these parameters. The equivalent objective functions generated by this second approach are also linear.

With regard to the modelling of uncertainty in the constraints of the problem (1), two approaches are also proposed: a "constraints with recourse" approach and a "satisfaction thresholds on the constraints" approach. In both cases, the equivalent deterministic constraints thusly obtained are linear. Subsequently, according to the type of situation to be dealt with, by combining one of the two proposed approaches to deal with the uncertainty in the objective functions with one of the two proposed approaches to deal with the uncertainty in the constraints, we obtain a multiobjective deterministic program (most of cases, linear) equivalent to (1). Finally, this deterministic equivalent program is solved by an interactive method derived from STEM. The first phase of this algorithm consists of obtaining a first compromise. Afterwards, through several interactive phases, new compromises are presented to the D.M. until a compromise considered to be satisfactory is obtained.

The main steps in our methodology are resumed in Figure 1. In this figure, each step is numbered: these numbers correspond to different sections of this work in which these steps are explained in more detail.

Figure 1

Schema of our general MOLSP methodology

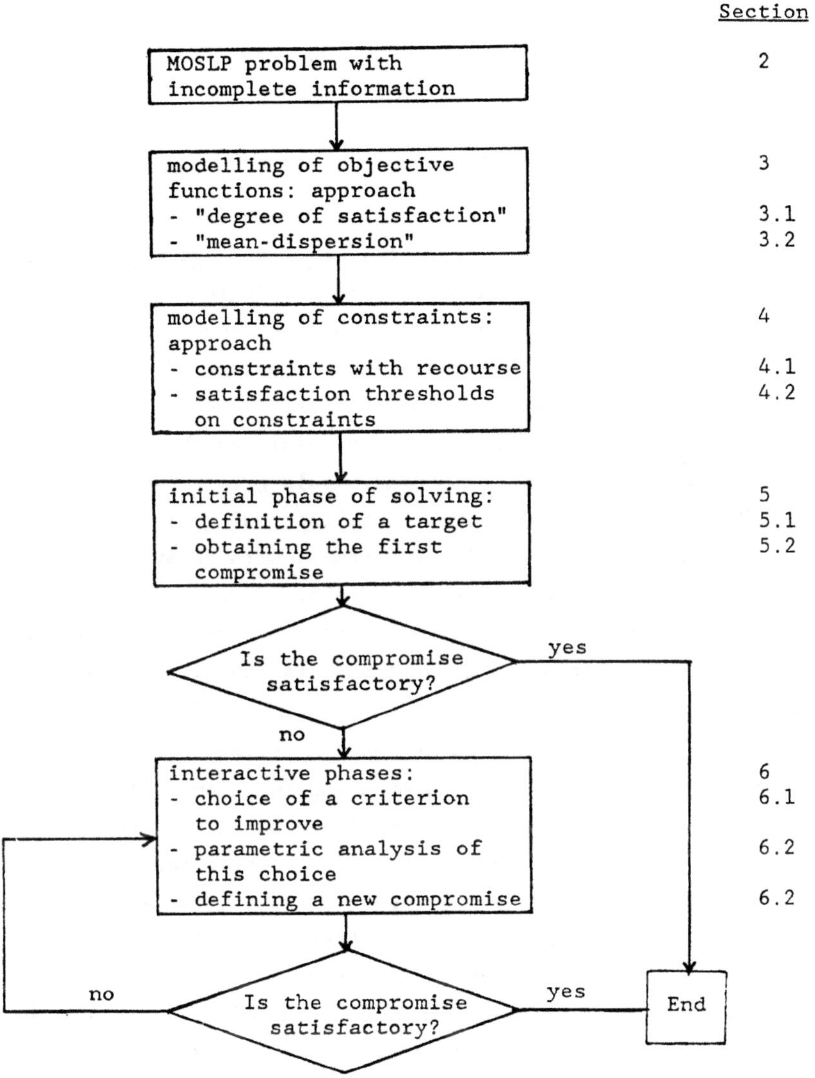

3. Modelling stochastic objective functions

3.1. THE "DEGREE OF SATISFACTION" APPROACH

To deal with the stochastic objective functions $Z_k(x)$, $k = 1, \ldots, K$, we will turn to a "Goal Programming" approach by transforming the initial problem (1) into a new problem modelled by means of stochastic inequalities in the form

$$Z_k(x) \geq g_k, \tag{2}$$

where g_k represents the goal to be attained relatively to the objective Z_k. It will be supposed that $c_{kj} \in [c_{kj}^-, c_{kj}^+]$ and $g_k \in [g_k^-, g_k^+]$; g_k^+ represents the goal that the D.M. wishes to attain for Z_k, whereas g_k^- represents the inferior bound which the D.M. is ready to accept for this goal g_k. In other words, the problem then consists of looking for an admissible solution x^* such that the attained values on the different objective functions Z_k are as close as possible to the goals g_k^+ while at the same time satisfying the minimal goals g_k^-. For each objective function Z_k, the solution is compared to the goal g_k in such a way that the D.M. is asked to express his preferences with regard to the deviations relative to these goals, that is relative to $(g_k - Z_k)$. In this context, the greater this deviation is, the less satisfying it is to the D.M. and the weaker his degree of satisfaction is for such a solution.

This degree of satisfaction of the D.M. relative to the attainment of the goal g_k is expressed by a decreasing functions $P_k(g_k - Z_k.)$ Taking into account that the objective functions $Z_k(x)$ and goals g_k are not defined with certainty, this degree of satisfaction translates the following assertion: "the D.M. is even less satisfied that $Z_k^-(x) = c_k^-.x$ is close to g_k^- and even more satisfied that $Z_k^+(x) = c_k^+.x$ is close to g_k^+ ". To do this, two slack variables $d_k^+ = g_k^+ - Z_k^+$ and $d_k^- = g_k^- - Z_k^-$ are used, so the degree of satisfaction $P_k(d_k^+, d_k^-)$ is expressed by the linear function (3) which is represented graphically by the Figure 2.

$$P_k(d_k^+, d_k^-) = \begin{cases} 0 \quad , & \text{if } d_k^- \geq 0 \\ 1 - d_k^+ / (d_k^+ - d_k^-) \; , & \text{if } d_k^+ \geq 0 \geq d_k^- \\ 1 \quad , & \text{if } d_k^+ \leq 0 \end{cases} \tag{3}$$

Figure 2

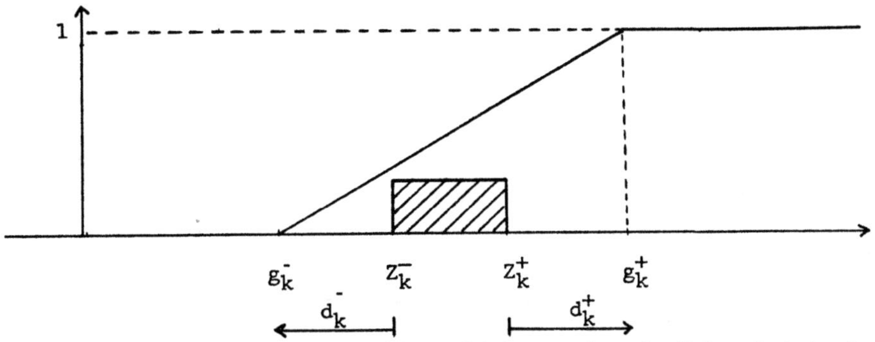

Consequently, the initial problem which consisted of "maximizing" the objective $Z_k(x)$ is replaced by the problem of maximizing the D.M.'s degree of satisfaction P_k relative to the attainment of goals g_k. Thus, the MOSLP problem (1) is transformed into the following equivalent multiobjective program which, except for the constraints $\sum_{j=1}^{n} a_{ij}\, x_j \le b_i$, is deterministic:

$$\begin{cases} \text{Max} \quad P_k(d_k^+, d_k^-) = 1 - d_k^+ / (d_k^+ - d_k^-) \\ \text{Subject to} \\ \qquad d_k^+ = g_k^+ - Z_k^+ \;, \quad d_k^- = g_k^- - Z_k^- \;, \\ \qquad d_k^+ \ge 0 \;, \quad d_k^- \ge 0 \;, \\ \qquad \sum_{j=1}^{n} a_{ij}\, x_j \le b_i,\; x_j \ge 0,\; i = 1, \ldots, m;\; j = 1, \ldots, n \end{cases}$$

However, it is necessary in this approach to determine the goals g_k. The bounds g_k^- and g_k^+ for g_k can either be obtained from the D.M. or by means of an automatic procedure. Through this last procedure, these goals are determined from what we identify as a goal matrix and which, in a way, consists of a generalization of the classic payoff matrix in order to take into account the uncertainty in the constraints of problem (1). For each objective function Z_k, we resolve the single-criterion deterministic programs obtained by replacing the stochastic parameters of problem (1) by their best values, i.e.

$$\max Z_k^+(x)$$
$$\text{s.t. } a_i^-.x \leq b_i^+$$

and by their worst values, i.e.

$$\max Z_k^-(x)$$
$$\text{s.t. } a_i^+.\bar{x} \leq b_i$$

The least constraining situation will be identified by $\alpha = 0$ and the most constraining by $\alpha = 1$ and the optimal solution of the corresponding program will be designed by x_k^α, $\alpha = 0,1$, $k = 1,\ldots, K$.

Then the goals will be chosen as follows

$$g_k^+ = Z_k^+(x_k^o) \quad \text{and}$$
$$g_k^- = \min_\alpha Z_k^-(x_k^\alpha) \quad , \alpha = 0,1;\ k = 1,\ldots, K$$

The goal g_k^+ is the ideal point because it is, in the space of objective functions, the point corresponding, for each objective function, to the best possible values of $Z_k(x)$ in the best of cases. The value g_k^- corresponds, for each objective, to the worst value of the goal matrix.

From this general model, a particular model corresponding with the case where, in addition to the extreme values, the D.M. also knows the most probable values c_k^o of the stochastic parameters c_{kj}, can be developed. Then the stochastic objective function Z_k is transformed by using the inequality (2) where Z_k^+ and Z_k^- are replaced by $Z_k^o = c_k^o.x$ and where $g_k \in [g_k^-, g_k^+]$. This inequality translates the following assertion: "the D.M. is less satisfied that $Z_k^o(x)$ is close to g_k^- and more satisfied that $Z_k^o(x)$ is close to g_k^+. So, as it has been done previously, taking into account the fact that here $d_k^+ - d_k^- = g_k^+ - g_k^-$, the D.M.'s degree of satisfaction $P_k(d_k^+)$ defined by (3) becomes:

$$P_k(d_k^+) = \begin{cases} 0, \text{ if } d_k^+ \geq g_k^+ - g_k^- \\ 1 - d_k^+ / (g_k^+ - g_k^-), \text{ if } 0 \leq d_k^+ \leq g_k^+ - g_k^- \\ 1, \text{ if } d_k^+ \leq 0 \end{cases}$$

This modelling of uncertainty in the objective functions consequently permits transforming of the initial stochastic program (1) into a multiobjective equivalent program where the objective functions are deterministic and linear; this equivalent program takes the form:

$$\begin{cases} \text{Max} & P_k(d_k^+) = 1 - d_k^+ / (g_k^+ - g_k^-) \\ \text{Subject to} & Z_k^o(x) + d_k^+ \geq g_k^+ \\ & d_k^+ \leq (g_k^+ - g_k^-) \\ & \sum_{j=1}^{n} a_{ij} \, x_j \leq b_i \, , \; x_j \geq 0, \\ & i = 1,\dots, m; \; j = 1,\dots, n; \; k = 1,\dots, K. \end{cases}$$

3.2. THE "MEAN-DISPERSION" APPROACH

As the parameters of the objective functions are specified by the means of intervals, the problem (1) can be viewed as a multicriteria multiparametric one. This approach (Steuer [1981]) brings one to transform the problem into a deterministic equivalent one where each stochastic function $Z_k(x)$ is replaced by the 2^n objective functions obtained by giving successively to c_{kj} the extreme values c_{kj}^- and c_{kj}^+. This modeling quickly becomes very bogged down as it leads to a multiplying of criteria which are at times very close to one another. Furthermore, in this modelling of uncertainty in the objective functions, all available information is not explicitly used, namely the central values c_k^o (if they are known).

Another approach, that of scenarios, also implies a multiplying of basic criteria by corresponding a criterion to each scenario. This is the approach used in the STRANGE method. This modeling implies the generation of a smaller number of criteria than the preceding one. Moreover, as the scenarios are chosen among all the possible combinations of c_{kj}^-, c_{kj}^o, c_{kj}^+, the central values are explicitely considered within the scenarios and are only kept the scenarios corresponding to the combinations sufficiently different from each other. Nevertheless, if the number of initial objective functions and/or the size of the decision vector increases, the multiparametric approach, as well the scenario approach, becomes cumbersome to handle.

In order to transform the stochastic objective functions of the problem (1), we propose to have recourse to two measures aimed at resuming the information relative to uncertainty in the parameters c_{kj}:

a measure of "central tendency" and one of "dispersion". Using a measure of dispersion jointly with a measure of central tendency permit a modeling which considers all the available information and avoids undesirable consequences (Nadeau [1983]) which result in a modeling of stochastic coefficients only by their central values as in PROTRADE.

Consequently, for each criterion Z_k two criteria are defined: a "central tendency" criterion Z_k^o and a "dispersion" criterion Z_k^Δ. The central tendency criterion can be defined as $Z_k^o = c_k^o.x$. In order to define Z_k^Δ, we turn to the general family of distance functions D_p which express the weighted distance between the values of $Z_k(x)$ for the different possible combinaisons of c_{kj}^- and c_{kj}^+ and the value Z_k^o of Z_k corresponding to the central value c_{kj}^o; more precisely, Z_k^Δ is defined by

$$Z_k^\Delta = D_p\left(Z_k^s, Z_k^o\right) = \left(\sum_s \lambda_s^p . |Z_k^s - Z_k^o|^p\right)^{1/p},$$

where s (s = 1, S (S ≤ 2^n)) is the index of the combinations of the extreme values of c_{kj}, λ_s is the weight of each of these combinations and p is a positive integer which expresses the importance that the D.M. places on the maximal deviation between Z_k^o and Z_k.

Even if in theory all the positive integer values for p are possible, only three values will be kept for this parameter: p = 1, p = 2 and p = ∞, which corresponds respectively to the geometric, euclidean and the Tchebycheff norm. These different measures are characterized by a growing weighting of the largest deviations between Z_k^s and Z_k^o. Thus, for the different values of p, we have different measures of dispersion, namely:

$$p = 1 \Rightarrow Z_k^\Delta = \sum_s \lambda_s |c_k^s.x - c_k^o.x|$$

$$p = 2 \Rightarrow Z_k^\Delta = \left(\sum_s \lambda_s^2 |c_k^s.x - c_k^o.x|^2\right)^{1/2}$$

$$p = \infty \Rightarrow Z_k^\Delta = \max_s \lambda_s |c_k^s.x - c_k^o.x|$$

Note that the measures D_1 and D_∞ are advantageous in that they permit to conserve the linearity of the initial problem (1), whereas measure D_2, which corresponds to the familiar concept of variance, leads to a quadratic program.

Adopting the preceding modelling of the uncertainty relative to the objective functions, the initial stochastic program (1) is replaced by an equivalent program in which the objective functions are deterministic and which takes the following form:

$$\begin{cases} \min\ (-\overset{o}{Z}_1,\ \overset{\Delta}{Z}_1, \ldots,\ -\overset{o}{Z}_K,\ \overset{\Delta}{z}_K) \\ \text{Subject to} \\ \qquad \Sigma\, a_{ij}\, x_j \le b_i\ ,\ x_j \ge 0 \\ \qquad i = 1, \ldots, m;\ j = 1, \ldots, n. \end{cases}$$

4. Modelling stochastic constraints

Now that the objective functions of the initial problem (1) have been transformed in order to possess a real mathematic signification, the second step consists of modifying the mathematically ill-defined stochastic constraints to well-defined mathematical entities. To do this, two types of constraints have been proposed within the single-criterion stochastic programming framework: constraints with recourse and constraints with probability thresholds. As the choice of one or the other of these approaches depends upon the concrete situation which is dealt with, it seems to us to be pertinent, in the development of a multiple-choice methodology, to take these two types of reasonings into account.

4.1. THE "CONSTRAINTS WITH RECOURSE" APPROACH.

In practice, faced with a program with stochastic constraints, if the D.M. is able to satisfactorily specify the penalties associated to the violations of the constraints and if the size of the associated deterministic problem remains reasonable, it is suitable to use a recourse type approach. But what does such an approach mean when the probability distributions of the stochastic coefficients of the constraints are not known?

In the same way as for the stochastic parameters of the objective functions, the different combinations (indexed by r, $r = 1, \ldots, R$, $R \le 2^{n+1}$) of the extreme values of a_{ij} and b_j, $i = 1, \ldots, m$, $j = 1, \ldots, n$, can be considered. Consequently, for each stochastic constraint $a_i.x \le b_i$, where $a^-_{ij} \epsilon [a^+_{ij}, a_{ij}]$ and $b^-_i \epsilon [b^+_i, b_i]$, R deterministic constraints of the form $a^r_i.x + v^r_i - w^r_i = b^r_i$, are associated at the most where w^r_i represents the violation of this i-th constraint for the r-th combination of the extreme values of a_{ij} and b_i and v^r_i is a classic slack variable.

If β_i^r expresses the penalty associated to the violation w_i^r of the i-th constraint, in order to minimize the cost of violations of the stochastic constraints, a recourse can be carried out by means of a supplementary objective function as in STRANGE. This supplementary objective functions (noted by Z_{K+1}) which considers the economic consequences of the violation of the constraints, could be defined as the sum of violations of the different combinations $(\Sigma_i \Sigma_r \beta_i^r w_i^r)$ or even the maximal violation of all these combinations$(\max_{ir} \beta_i^r w_i^r)$. In that choice, it must be remembered that the sum is compensatory whereas the maximum is not. Consequently, following the nature of the consequences resulting from the violations of the constraints, the recourse to any one of these possibilities could be justified.

Therefore, with this approach, the initial stochastic program (1) becomes

$$"\max" \; (Z_1(x), \ldots, Z_K(x), \; - Z_{K+1})$$
$$\text{subject to}$$
$$a_i^r.x + v_i^r - w_i^r = b_i^r, \; x_j \geq 0, \; v_i^r \geq 0, \; w_i^r \geq 0,$$
$$i = 1, \ldots, m; \; j = 1, \ldots, n; \; r = 1, \ldots, R,$$

in which the constraints are deterministic. This approach calls for the calculating of the violations for all the possible combinations of the extreme values of the stochastic parameters and then the obtained deterministic program is larger than the initial stochastic program (1).

Another possibility of a recourse consists of coming back exactly to the idea of recourse used in single-criterion stochastic programming by adding the cost of the violations of the constraints to each objective funtion Z_k. In this case, the initial stochastic program (1) would be reduced to a new program where each objective function Z_k of (1) would have been replaced by $Z_k + \sum_i \sum_r \beta_i^r \, w_i^r$ and where the constraints would be deterministic and would take the form $a_i^r.x + v_i^r - w_i^r = b_i^r$. When only the parameters b_i of the constraints of the program (1) are stochastic, the equivalent program associated to (1) by using the idea of recourse simply leads to doubling of the initial stochastic constraints according to whether b_i is equal to b_i^- or to b_i^+.

4.2. THE "SATISFACTION THRESHOLDS ON THE CONSTRAINTS" APPROACH.

It has been pointed out that the "constraints with recourse" approach can be used in so far as the D.M. is in a position to satisfactorily specify the penalties associated to the violations of the constraints and in so far as the size of the deterministic equivalent program remains reasonable. Furthermore, when both sides of the constraints are stochastic, the decomposition of the stochastic constraints in a system of deterministic constraints built from the possible combinations of the extreme values of the stochastic parameters of those constraints, becomes quite cumbersome. In that case, we propose to transform the stochastic constraints of problem (1) by using the idea of "satisfaction thresholds on the constraints", this approach having the noted advantage of not undulay increasing the size of the initial problem.

In order to transform the set of stochastic constraints:

$$a_i.x \le b_i \tag{4}$$

where $a_i \epsilon [a_i^-, a_i^+]$ and $b_i \epsilon [b_i^-, b_i^+]$, i = 1,..., m, j = 1,..., n, each one of these constraints is interpreted as it follows:

> "the D.M. hopes that $a_i^-.x$ will not be larger than b_i^+ and his satisfaction level will be even bigger than $a_i^+.x$ will be closer to b_i^-".

Consequently, we introduce the D.M.'s degree of satisfaction relative to the stochastic constraint (4); this degree of satisfaction is designed by μ and is defined by the following linear function:

$$\mu(a_i.x-b_i) = \begin{cases} 0, & \text{if } a_i^-.x \ge b_i^+ \\ 1, & \text{if } a_i^+.x \le b_i^- \\ (b_i^+ - a_i^-.x) \,/\, (b_i^+ - b_i^-) + (a_i^+ - a_i^-).x, & \text{elsewhere} \end{cases} \tag{5}$$

Some values of this function are represented graphically in Figure 3. The fact of expressing the D.M.'s degree of satisfaction by a linear function indirectly translates the choice of uniform probability distributions on the intervals of variation of the stochastic parameters.

Figure 3

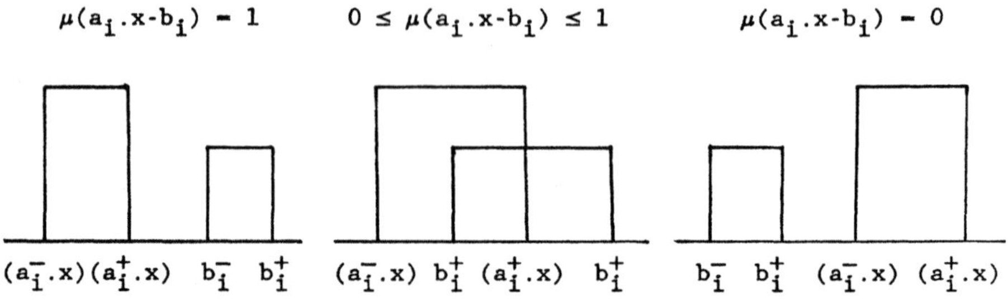

Therefore, in order to transform the stochastic constraints (4) into deterministic constraints, we turn to the idea of satisfaction threshold on the constraints. This threshold may take two forms. In its first form, the solution is required to jointly satisfy the set of the constraints with some satisfaction threshold fixed in advance. This threshold called "a joint satisfaction threshold on constraints" is noted by α; it corresponds to the idea of a joint probability threshold in chance-constrained programming. Using this type of threshold, the stochastic constraints (4) are transformed into the following deterministic constraint

$$\prod_i \mu(a_i.x \leq b_i) \geq \alpha,$$

which means that the set of the initial constraints is required to be jointly satisfied with a minimal satisfaction level α. If $\mu(a_i.x \leq b_i)$ is defined by (5), this equivalent deterministic constraint takes the form

$$\prod_i \{(b_i^+ - a_i^-.x) / (b_i^+ - b_i^-) + (a_i^+ - a_i^-).x\} \geq \alpha$$

which is a non-linear constraint.

The satisfaction threshold on constraints may take a second form: then the solution is required to individually satisfy each of the stochastic constraints (4) with some satisfaction threshold, a particular satisfaction threshold being defined on each constraints. This threshold called "individual satisfaction threshold on constraints" is noted by α_i; it corresponds to the idea of an individual probability threshold in chance-constrained programming. Using this type of threshold, each stochastic constraint (4) is replaced by a deterministic constraint of the form

$$\mu(a_i.x \le b_i) \ge \alpha_i$$

where $0 \le \alpha_i \le 1$. From the relation (5), this deterministic constraint takes the form

$$(a_i^- + \alpha_i(a_i^+ - a_i^-)) .x \le b_i^+ - \alpha_i (b_i^+ - b_i^-)$$

Figure 4

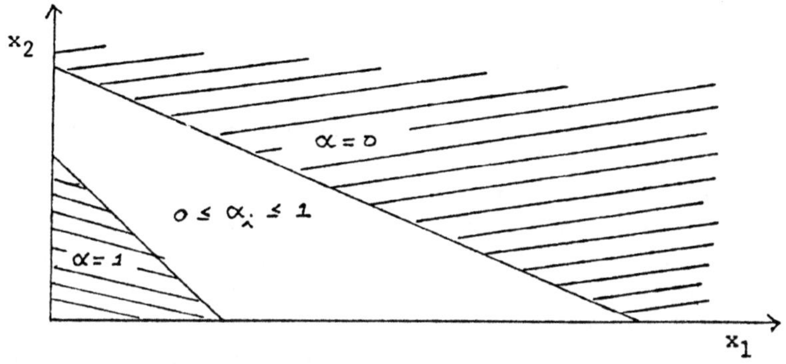

An illustration of the values taken by this inequality in the case where $x = (x_1, x_2)'$ is given in Figure 4. Furthermore, if only the parameters b_i of the constraints are uncertain, the equivalent deterministic constraint takes the form of

$$(b_i^+ - a_i.x) \ge \alpha_i (b_i^+ - b_i^-)$$

This last form is advantageous, in that it easily allows for a parametric analysis on the threshold α_i and thus allows for the analysis of the repercussions of changing the threshold α_i on the compromise solutions. Indeed, in the context of a methodology using this approach, these α_i thresholds would be determined in an interactive manner with the D.M.. Through these α_i, the D.M. can modify his requirements relative to his satisfaction level on the constraints. Thus, in the framework of an interactive approach, these α_i thresholds will be set to high values at the beginning of the procedure in order to be slackened in the subsequent steps and, at the same time, to allow for improvement of the obtained values for Z_k, $k = 1, \ldots, K$. Consequently, during these iterations, the D.M. may come to a satisfactory compromise between the quality and the reliability of the obtained compromise solution.

In this "satisfaction thresholds on the constraints" approach, the used information about the uncertain parameters of the constraints only consist of the intervals of variation of those parameters. If the D.M. could also determine central values of those parameters, how this supplementary information be used? The idea of partitioning the polyhedral space (Nadeau [1980]) defined by the constraints could be taken advantage of along with the different approaches previously presented, however but this idea is not developed in this paper.

5. The interactive solution algorithm

Briefly, facing the stochastic problem (1), we have proposed both approaches in order to transform the stochastic objective functions into deterministic ones and also both approaches in order to transform the stochastic constraints into deterministic ones. If one combines one of the two approaches for the objective functions with one of the two approaches for the constraints, one obtains a multiobjective deterministic program equivalent to (1) which, in most cases, is linear. Furthermore, if the equivalent program is not linear, it is always possible to turn to linearization methods to bring back a linear by parts program, in such a way that in every case, we arrive at a deterministic MOLP problem. In order to resolve this equivalent deterministic MOLP problem thusly obtained, the STEM method is proposed (Benayoun and al, [1971] to which the following modifications are added:

1° the defining of a target point 0 from the payoff table which enables the obtaining of the first compromise by means of information given by the D.M., whereas in the STEM method this first compromise is obtained automatically, without the intervention of the D.M.;

2° the addition of a parametric analysis, similar to the one used in the STRANGE method, which explores the consequences of an improvement asked for by the D.M. as well as the use of procedures which brings up the eventual indeterminations in the payoff table;

3° the addition (in methods using the "satisfaction thresholds on the constraints" approach) of a sensitivity analysis of the obtained compromises in relation to the satisfaction thresholds α_i.

The steps of this interactive algorithm are the defining of a target, the obtaining of a first compromise and a sequence of interactive phases, each of them generating a set of new compromises from a parametric analysis which explores the repercussions of the D.M.'s intervention following the proposed compromise in the previous step, and so on.

In what follows, that interactive algorithm will be briefly described supposing that, at the level of objective functions, the "mean-dispersion" approach has been chosen and that, at the level of constraints, the "constraints with recourse" approach has been chosen. In this case, the deterministic MOLP problem equivalent to the stochastic problem (1) takes the form of

$$\begin{cases} \min \{\overset{o}{Z}_1, \overset{\Delta}{Z}_1, \overset{o}{Z}_2, \overset{\Delta}{Z}_2, \ldots, \overset{o}{Z}_K, \overset{\Delta}{Z}_K, Z_{K+1}\} \\ \text{subject to} \\ a_i^r.x + v_i^r - w_i^r = b_i^r,\ x_j \geq 0,\ v_i^r \geq 0,\ w_i^r \geq 0 \\ i = 1, \ldots, m;\ j = 1, \ldots, n;\ r = 1, \ldots, R. \end{cases} \tag{6}$$

To simplify the writing of (6), this program will be written in the form of

$$\begin{cases} \min Z_k(y), \quad k = 1, 2, \ldots, 2K+1 \\ \text{subject to} \quad y \in D \end{cases} \tag{7}$$

where $Z_k(y)$ successively stands for $Z_1^o, Z_1^\Delta, Z_2^o, Z_2^\Delta, \ldots, Z_K^o, Z_K^\Delta, Z_{K+1}$ and where $D = \{y = (x, v, w) \simeq a_i^r.x + v_i^r - w_i^r = b_i^r,\ x_j \geq 0,\ v_i^r \geq 0,\ w_i^r \geq 0, i = 1, \ldots, m,\ j = 1, \ldots, n,\ r = 1, \ldots, R\}$

5.1. THE INITIAL PHASE

5.1.1. Defining a target

To solve (7), at first a payoff table must be built. To do this, for each criterion Z_k of (7) and for each combination r of the extreme values of constraints, the optimal value of Z_k^r is obtained by solving the program

$$\max Z_k^r(x), \text{ subject to } a_i^r.x \leq b_i^r.$$

This comes down to calculating the best solutions of $Z_k(y)$ for each of the polyhedrons formed by the R combinations of the extreme values of the coefficients of the constraints. Each of these optimizations gives an optimal solution y_k^r and the best one, y_k^*, is kept as the optimal solution for the criterion Z_k and noted by M_k, i.e.

$$Z_k(y_k^*) = \max_r Z_k(y_k^r) = M_k$$

Therefore, M_k represents the ideal point. When there is indetermination about y_k^*, we turn to the solution of the problem

$$\max_\ell Z_\ell(y),\ \ell \neq 1, \ldots, 2K+1,\ \ell \neq k$$
$$\text{subject to } y \in D,\ Z_k(y) = Z_k(y_k^*)$$

Similarly, the worst value of Z_k in the payoff table is noted by m_k and defined by

$$m_k = \min_{\ell} Z_k(y_{\ell}^*), \ \ell = 1,\ldots, \underset{*}{2K+1}, \ \ell \neq k$$

With this information concerning the best and the worst possible value of the objective functions $Z_k(y)$, we ask the D.M. to set an aspiration level 0_k on each criterion Z_k, $k = 1,\ldots, 2K + 1$. Afterwards, the distance, in the space of the criteria, with regard to this target $0 = \{0_1,\ldots, 0_{2K+1}\}$ will be minimized. As the target is set by the D.M., it is possible to come to a solution where all the deviations relative to this target are negative or zero. This situation would reveal the fact that this 0 objective is dominated, in which case it would be worthwhile to suggest to the D.M. to set a new target point 0' and begin the procedure again.

5.1.2. Obtaining a first compromise

As within the STEM method, the weights Π_k relative to each criterion Z_k have to be determined. These weights are defined by

$$\Pi_k = b_k \Big/ \sum_k b_k$$

where $b_k = [(M_k - m_k)/m_k] \cdot [1/\|c_k\|_2]$ and $\|c_k\|_2$ is the euclidian norm of the vector c_k.

A first compromise, denoted by y^1 is then obtained by solving the following program:

$$\left\{\begin{array}{l} \min M.\partial - \Sigma_k \epsilon_k \\ \text{subject to} \\ \quad \Pi_k\ (0_k - Z_k\ (y)) \leq \partial - \epsilon_k\ ,\ k = 1,\ldots, 2K + 1 \\ \quad y \in D \end{array}\right. \qquad (8)$$

In this program (8), the introduction of parameters M (a larger value than that of all the other coefficients of the program) and ϵ_k assure us of the efficiency of the first compromise y^1. This correction is due to Despontin [1984]. The subsequent compromises shall be written y^m where $m = 2, 3, 4,\ldots$

If, at the level of constraints, the "satisfaction thresholds on constraints" approach is adopted, then to obtain the first compromise, the values of α_i are at first chosen higher in order to be slackened later if necessary. Generally, the closer levels α_i are to 1.0, the lower the values of the objective functions are.

5.2. THE INTERACTIVE PHASES

With each new compromise, the D.M. receives the following information:

1° the obtained value for this compromise y^m (m being the number of the compromise) on each objective function Z_k of (7), i.e. $Z_k(y^m)$, $k = 1,\ldots,2K+1$. This information is given to the D.M. simultaneously with the range of variation $[m_k, M_k]$ of the considered criterion; in other words, the D.M. receives the information $(m_k, Z_k(y^m), O_k, M_k)$;

2° to facilitate the comparison of the obtained values for the different criteria, the relative values of $Z_k(y^m)$, noted by $Z_k'(y^m)$, are also calculated; this relative value, generally comprised between 0 and 1, translates the degree of attainment of the aspiration level O_k, i.e.

$Z_k'(y^m) = 1 - [(O_k - Z_k(y^m)) / (O_k - m_k)]$.

Consequently, the larger $Z_k'(y^m)$ is, the closer the D.M. is to his aspiration level O_k. If the value of $Z_k'(y^m)$ is more than 1.0, this means that for the considered objective, the D.M. has obtained a better value than the aspiration level O_k;

3° in order to permit an easier comparison of the variations relative to each objective function Z_k, the variation coefficient $C_k = Z_k^\Delta / Z_k^o$ corresponding to each Z_k^o are also presented to the D.M..

If, at the level of the constraints, the "satisfaction thresholds on the constraints" approach is used instead of the constraints with recourse" approach, a view of the evolution of the compromise solution for levels of satisfaction α_i situated in the proximity of those set a priori by the D.M., is also provided him with.

5.2.1. Choice of a criterion to improve

From the preceding information, the D.M. must state if he deems the compromise satisfactory (in which case the procedure ends) or if he wishes to try to obtain a better compromise. In the second situation, he must then specify an objective function Z_k^* that he wishes to improve, and, if possible, set the maximum level Δ_k^* of this improvement, which means that the maximum value desired by the D.M. for this objective function is $Z_k^*(y^m) + \Delta_k^*$. If the D.M. is incapable of evaluating Δ_k^*, this value can then be set at $\Delta_k^* = O_k - Z_k^*(y^m)$.

5.2.2. The parametric analysis

A parametric analysis allows to explore the consequences of the choices made by the D.M. in the preceding step. It consists of analyzing the repercussions on the different objective functions $Z_k(y)$ of the possible improvement level $\tau.\Delta_k^*$ of the objective function Z_k^*, $\tau\epsilon[0,1]$. Consequently, this parametric analysis is done by means of the following program:

$$\begin{cases} \min M.\partial - \Sigma_k \epsilon_k \\ \text{subject to} \\ \quad \Pi_k\ (O_k - Z_k\ (y)) \leq \partial - \epsilon_k \\ \quad \overset{*}{Z_k}(y) = \overset{*}{Z_k}(y^m) + \tau.\overset{*}{\Delta_k},\ 0 \leq \tau \leq 1 \\ \quad y \in D,\ k = 1,\ldots,\ 2K+1 \end{cases} \tag{9}$$

In resolving this program (9), a series of stability intervals of the form $[\tau_i, \tau_{i+1}]$ for the optimal basis of the problem are obtained. If the optimal solution of this parametric problem for a given τ is noted by y^τ, the value $Z_k(y^\tau)$ for each objective function Z_k, $k = 1,\ldots, 2K + 1$, can be calculated and presented to the D.M.. From the values $Z_k(y^\tau)$, the D.M. can opt an for improvement level τ^*.

To facilitate this choice, it may be worthwhile to graphically illustrate the incidence of the improvement level τ on the values $Z_k(y^\tau)$.To do this, the linear by parts curves $Z_k'(y^\tau) = f(\tau)$ can be traced.

Once τ^* is chosen, the D.M. is in front of a new compromise $y^{\tau*}$ which will simply noted by y^2, and the following equations are added to the problem (8):

$$\overset{*}{\Pi_k} = 0 \quad \text{and} \quad \overset{*}{Z_k}(y) \leq \overset{*}{Z_k}(y^{\tau^*})$$

This new compromise y^2 is treated like the preceding one and the procedure ends when the D.M. feels satisfied with the subsequent obtained compromise y^m. In any case, the procedure ends after 2K + 1 steps at the most because the objective function on which an improvement has been made is no longer considered (by adding the constraint $\Pi_k^* = 0$). The irrevocability of the improvements made to the different criteria is without a doubt a possible criticism of this procedure. However, a modification to the method allowing one to go back in the procedure is possible. It is a matter of being able to store the value Z_k^* and to re-establish D and Π_k^* upon demand.

6. Illustration of a particular method taken from the general methodology.

We shall illustrate one of the possible methods of our general MOSLP methodology from a didactic example taken from Slowinsky and Teghem [1986]. The presented method rests upon the modelling of objective functions by the "mean-dispersion" approach and the modelling of constraints by the "constraints with recourse" approach.

6.1. THE MOLSP PROBLEM WITH INCOMPLETE INFORMATION.

This example implies two decision variables (i.e. $\underline{x} = (x_1, x_2)$ and two criteria to minimize. These criteria are expressed as follows:

$$Z_1(x) = 15 + c_{11}\, x_1 + c_{12}\, x_2$$
$$Z_2(x) = 7 + c_{21}\, x_1 + c_{22}\, x_2$$

where the parameters c_{ij} are stochastic variables of which only the minimal, middle and maximal values are known: these values are respectively 2, 2.5 and 3 for c_{11}, - 4, - 3.5 and -3 for c_{12}, - 1, -0.875 and - 0.75 for c_{21} and 0.75, 0.875 and 1 for c_{22}. In this problem there are three constraints (other than those of non-negativity of the variables) and two of them are deterministic, namely:

(a) $-7x_1 + 12x_2 \leq 39$

(b) $x_1 - x_2 \leq 5$

The third constraint is under uncertainty and takes the following form:

(c) $a_{11}\, x_1 + a_{12}\, x_2 \leq b_1$

where the parameters a_{11}, a_{12} and b_1 are stochastic variables of which only the extreme values are known and are respectively [0,6] for a_{11}, [-2,4] for a_{12} and [13,19] for b_1.

6.2. THE EQUIVALENT DETERMINISTIC MOLP PROBLEM.

From each criterion Z_k of the initial problem (k = 1,2), we define two criteria Z_k^o and Z_k^Δ as follows:

$$\overset{o}{Z}_1 = 15 + 2.5x_1 - 3.5x_2$$

$$\overset{o}{Z}_2 = 7 - 0.875x_1 + 0.875x_2$$

and $\overset{\Delta}{Z}_k = (\overset{+}{c}_k - \overset{o}{c}_k).\underline{x}$, i.e.

$$\overset{\Delta}{Z}_1 = 0.5x_1 + 0.5x_2$$

$$\overset{\Delta}{Z}_2 = 0.125x_1 + 0.125x_2$$

In this example, for the dispersion criterion $\overset{\Delta}{Z}_k$, we have chosen the Tchebycheff norm and this dispersion then represents the most unfavorable deviation of Z_k in relation to $\overset{o}{Z}_k$.

Moreover, for this example, a "constraints with recourse" approach is adopted at the level of the constraints. This modelling requires a decomposition of the stochastic constraints according to the R combinations of the extreme values of the stochastic coefficients in the constraints. Consequently, the constraints {(a), (b), (c)} will be replaced by the set of the deterministic constraints $D_1 = \{(a), (b), (c_1) \text{ to } (c_8)\}$ where

$$\begin{array}{llll}
(c_1) & 6x_1 - 2x_2 \le 13 & (c_4) & 6x_1 - 2x_2 \le 19 \\
(c_2) & \quad\;\; + 4x_2 \le 13 & (c_5) & \quad\;\; + 4x_2 \le 19 \\
(c_3) & 6x_1 + 4x_2 \le 13 & (c_6) & 6x_1 + 4x_2 \le 19 \\
(c_7) & \quad\;\; - 2x_2 \le 13 & (c_8) & \quad\;\; - 2x_2 \le 19
\end{array}$$

Note that, here, the constraints (c_7) and (c_8) are always verified and that afterwards they can be eliminated. Finally a recourse is made by the intermediary of the supplementary criterion $Z_3 = \Sigma_r \beta^r w^r$. To simplify the example, we take $\beta^r = 1$, for all r. Consequently the equivalent deterministic MOLP problem takes the following form:

$$\left\{\begin{array}{l}
\text{"min" } \overset{o}{Z}_1 = 15 + 2.5x_1 - 3.5x_2 \\
\qquad \overset{o}{Z}_2 = 7 - .875x_1 + .875x_2 \\
\qquad \overset{\Delta}{Z}_1 = .5x_1 + .5x_2 \\
\qquad \overset{\Delta}{Z}_2 = .125x_1 + .125x_2 \\
\qquad Z_3 = \Sigma_r w^r \\
\text{subject to} \\
\qquad \begin{array}{lllllll}
(a) & -7x_1 & + 12x_2 & \le 39 & & & \\
(b) & x_1 & - x_2 & \le 5 & & & \\
(c_1) & 6x_1 & - 2x_2 & + v^1 & - w^1 & = & 13 \\
(c_4) & 6x_1 & - 2x_2 & + v^4 & - w^4 & = & 19 \\
(c_2) & & + 4x_2 & + v^2 & - w^2 & = & 13 \\
(c_5) & & + 4x_2 & + v^5 & - w^5 & = & 19 \\
(c_3) & 6x_1 & + 4x_2 & + v^3 & - w^3 & = & 13 \\
(c_6) & 6x_1 & + 4x_2 & + v^6 & - w^6 & = & 19
\end{array} \\
\qquad x_1, x_2, v^r, w^r \ge 0, \; r = 1, \ldots, 6
\end{array}\right. \tag{10}$$

6.3. SOLVING THE EQUIVALENT DETERMINISTIC MOLP PROBLEM.

6.3.1. The initial phase.

As it has been explained in section 5.1, the first step of the algorithm consists of building the payoff matrix and identifying a target point. This payoff matrix is given in the first columns of Table 1.

Table 1

	$\overset{o}{y}_1$	$\overset{\Delta}{y}_1$	$\overset{o}{y}_2$	$\overset{\Delta}{y}_2$	y_3	best M_k	goal O_k	worst m_k
$\overset{o}{z}_1$	3.62	15	27.5	15	3.62	3.62	8	27.5
$\overset{\Delta}{z}_1$	1.625	0	2.5	0	0	0	1	7.25
$\overset{o}{z}_2$	9.84	7	2.625	7	5.1	2.625	6	9.84
$\overset{\Delta}{z}_2$	0.41	0	0.625	0	0	0	0.4	1.81
z_3	0	0	56.0	0	0	0	10	56

Payoff matrix

Note that the diagonal of the payoff matrix gives the best possible values for each criterion; besides this value, in each line, the value of the other criteria for the optimal solution of the considered criterion is given. From the best value M_k and the worst value m_k given for each criterion, the D.M. can determine the goal O_k given in the last but one column of Table 1.

The subsequent step consists of calculating the weights Π_k to be associated to each criterion. Here, we obtain:

$$\Pi_1^o = 0.024,\ \Pi_1^\Delta = 0.072,\ \Pi_2^o = 0.17,\ \Pi_2^\Delta = 0.684,\ \Pi_3 = 0.05$$

Then the first compromise is obtained by solving deterministic program (11)

$$\begin{cases} \min 100.\delta - \Sigma_k \epsilon_k \\ \text{subject to} \\ \overset{o}{\pi}_1 (15 + 2.5x_1 - 3.5x_2 - 8) \le \delta - \epsilon_1 \\ \overset{o}{\pi}_2 (7 - .875x_1 + .875x_2 - 6) \le \delta - \epsilon_2 \\ \overset{\Delta}{\pi}_1 (.5x_1 + .5x_2 - 1) \le \delta - \epsilon_3 \\ \overset{\Delta}{\pi}_2 (.125x_1 + .125x_2 - .4) \le \delta - \epsilon_4 \\ \pi_3 (\Sigma_r w^r - 10) \le \delta - \epsilon_4 \\ y \in D_1 , x_1, x_2, w^r \ge 0, r = 1,\ldots, 6 \end{cases} \tag{11}$$

We obtain $y^1 = (x_1 = 1.3679,\ x_2 = 1.7976,\ w_3 = 2.3998)$ as the first compromise. From this first compromise, we enter into the interactive phases.

6.3.2. The interactive phases.

In order to evaluate the desirability of the first compromise y^1, the D.M. receives the information given in Table 2, i.e. the values of $Z_k(y^1)$, O_k, $Z_k'(y^1)$ and C_k.

Table 2

	$Z_k(y^1)$	O_k	$Z_k'(y^1)$	C_k
$\overset{o}{Z}_1$	12.13	8	0.789	0.13
$\overset{\Delta}{Z}_1$	1.58	1	0.907	
$\overset{o}{Z}_2$	7.37	6	0.642	0.054
$\overset{\Delta}{Z}_2$	0.396	0.4	1.003	
Z_3	2.4	10	1.165	

From the values of this table, the D.M. realizes that he has gone beyond his objectives for Z_2^{Δ} and Z_3 and that the worst obtained value is without a doubt that relative to Z_2^{o}. Based on this information, the D.M. judges the compromise as being unsatisfactory and wishes to obtain a better one. To do this, he decides to improve the criterion Z_2^o which is, in fact, the furthest away from the target 0.

A parametric analysis is made to examine the consequences on an improvement of Z_2^o. This analysis uses a single-criterion program (9) which here takes the following form:

$$\left\{\begin{array}{l}
\min 100.\delta - \Sigma_k \epsilon_k \\
\text{subject to} \\
\pi_1 \, (15 + 2.5x_1 - 3.5x_2 - 8) \le \delta - \epsilon_1 \\
\pi_2 \, (7 - .875x_1 + .875x_2 - 6) \le \delta - \epsilon_2 \\
\overset{\Delta}{\pi_1} \, (.5x_1 + .5x_2 - 1) \le \delta - \epsilon_3 \\
\overset{\Delta}{\pi_2} \, (.125x_1 + .125x_2 - .4) \le \delta - \epsilon_4 \\
\pi_3 \, (\Sigma_r \, w^r - 10) \le \delta - \epsilon_5 \\
-7x_1 + 12x_2 \le 39 \\
x_1 - x_2 \le 5 \\
6x_1 - 2x_2 + v^1 - w^1 = 13 \\
6x_1 + 4x_2 + v^4 - w^4 = 19 \\
\quad + 4x_2 + v^2 - w^2 = 13 \\
\quad + 4x_2 + v^5 - w^5 = 19 \\
6x_1 + 4x_2 + v^3 - w^3 = 13 \\
6x_1 + 4x_2 + v^6 - w^6 = 19 \\
-.875x_1 + .875x_2 - 1.37 \, \tau = -1 \\
\tau \le 1, \; x_1, \, x_2, \, v^r, \, w^r \ge 0, \; r = 1, \ldots, 6
\end{array}\right. \qquad (12)$$

By varying the value of the parameter τ $(0 \le \tau \le 1)$ included in the last constraint of (12), stability intervals for the optimal basis of the program are generated. Then, for example, if τ varies between 0.514 and 1, the variable v_6 enters in the base and the variable w_6 is taken out. The Table 3 sums up these stability intervals.

Table 3

Limits of intervals		Variables	
τ_i	τ_{i+1}	entering	taken out
1.0	0.514	v_6	w_6
0.514	0.027	v_6	w_6
0.027	0	v_1	w_1

The results of the parametric analysis are shown in Table 4. This table gives, for each criterion Z_k of (10) and for each improvement level τ, the value $Z_k(y)$ as well as its relative value $Z_k'(y)$ which is indicated in parentheses. Moreover, in order to help the D.M.

Table 4

τ	Z_1^o	Z_1^Δ	Z_2^o	Z_2^Δ	Z_3
1	12.13 (.788)	1.58 (.907)	7.37 (.642)	.39 (1.003)	2.39 (1.165)
.514	14.13 (.685)	1.86 (.861)	6.70 (.816)	.46 (.953)	5.99 (1.087)
.35	14.81 (.65)	1.96 (.846)	6.48 (.874)	.49 (.936)	8.42 (1.034)
.2	15.43 (.618)	2.05 (.832)	6.27 (.928)	.51 (.920)	10.65 (.986)
.027	16.15 (.582)	2.15 (.815)	6.03 (.99)	.53 (.902)	13.21 (.930)
0.0	16.26 (.576)	2.16 (.813)	6.00 (2.00)	.54 (.899)	13.85 (.916)

choose a satisfactory level τ^*, Figure 5 illustrates the evolution of the relative value Z_k' of the objective function Z_k of (10) relative to the value of τ.

Figure 5

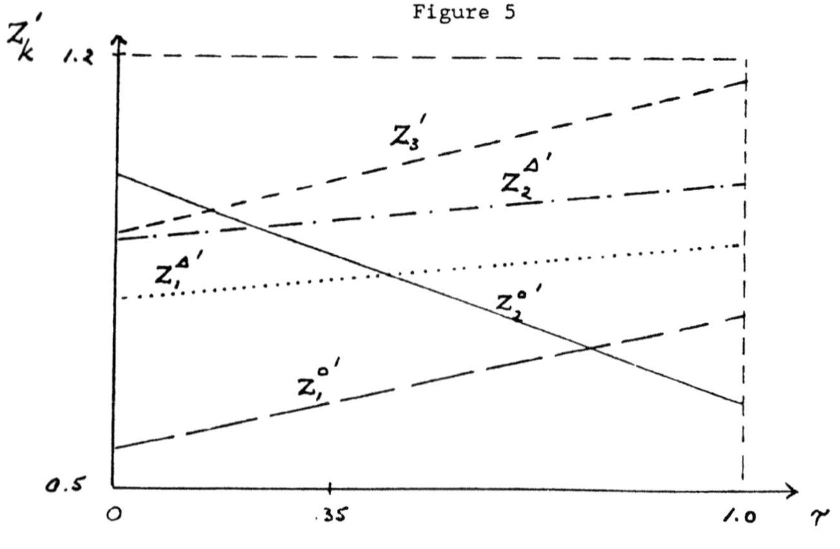

From data from Table 4 and Figure 5, suppose that the D.M. opts for the new compromise $y^2 = (x_1 = 2.258,\ x_2 = 1.665)$ corresponding to an improvement level $r^* = 0.35$. In fact, this decision implies that the value of the criterion Z_2^o will be set at $O_k + r^*.\Delta_k^* = 6 + (0.35)(1.37) = 6.48$ taking $\Delta_k^* = Z_k(y) - O_k$. The values of Z_k, Z_k', O_k and C_k corresponding to this new compromise are given in Table 5. In what follows,

Table 5

	$Z_k(y^2)$	O_k	$Z_k'(y^2)$	C_k
$\overset{o}{Z_1}$	14.815	8	0.65	0.13
$\overset{\Delta}{Z_1}$	1.962	1	0.846	
$\overset{o}{Z_2}$	6.48	6	0.874	0.075
$\overset{\Delta}{Z_2}$	0.49	0.4	0.936	
Z_3	8.428	10	1.034	

in programs (11) and (12), we now take $\Pi_2^o = 0$ and we add a constraint which fixes the worst value of Z_2^o to 6.48, namely the constraint

(13) $-0.875x_1 + 0.875x_2 \leq -0.52$.

From the information from Table 5, suppose that the D.M. now chooses to improve the value of the criterion Z_1^o. If the D.M. cannot set the maximal value for the improvement of this function, the value $\Delta_1 = Z_1(y^2) - O_1 = 6.815$ is taken. Then a parametric analysis is again made to examine the consequences of an improvement of Z_1^o by means of a program similar to (12). The results of the analysis are given in Table 6 and illustrated graphically in Figure 6. From this information, the D.M. then opts for an improvement level $r^* = 0.898$ for Z_1^o, which lead him to the new compromise $y^3 = (x_1, x_2) = (2.956, 3.363)$. For y^3, the values of Z_k', Z_k, O_k and C_k are given in Table 7.

Table 6

τ	Z_1^o	Z_1^Δ	Z_2^o	Z_2^Δ	Z_3
1.0	14.8 (.65)	1.96 (.84)	6.48 (.87)	.49 (.936)	8.40 (1.034)
0.957	14.5 (.65)	2.25 (.80)	6.48 (.87)	.56 (.883)	14.28 (.906)
0.898	14.12 (.69)	2.66 (.734)	6.48 (.87)	.66 (.812)	22.38 (.73)
0.768	13.24 (.73)	3.54 (.594)	6.48 (.87)	.88 (.656)	43.64 (.270)
0.678	12.61 (.76)	4.15 (.495)	6.48 (.87)	1.04 (.546)	60.7 (-.102)
0.548	11.72 (.81)	5.05 (.35)	6.48 (.87)	1.26 (.390)	89.16 (-.719)
0.30	9.96 (.89)	6.72 (.08)	6.48 (.87)	1.68 (.09)	149.28 (-2.04)
0.0	.800 (1.00)	8.77 (-.244)	6.48 (.87)	2.19 (-.272)	223.44 (-3.64)

Figure 6

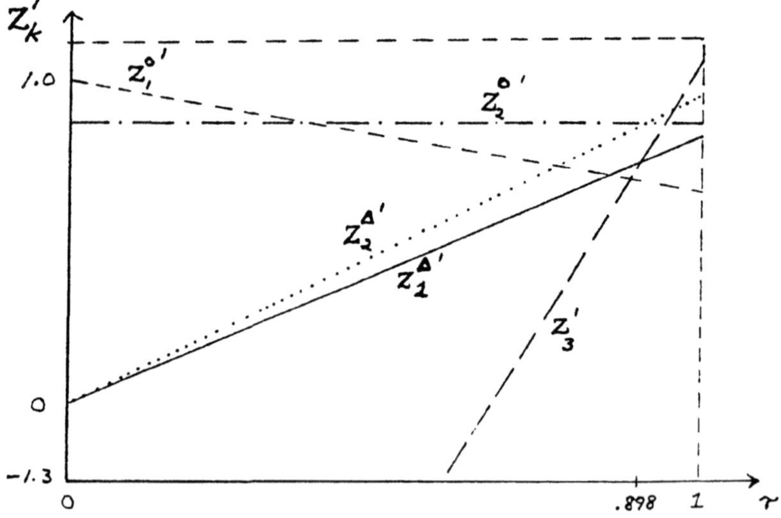

Table 7

	$Z_k(y^3)$	O_k	$Z'_k(y^3)$	C_k
$\overset{o}{Z}_1$	14.12	8	0.69	0.18
$\overset{\Delta}{Z}_1$	2.66	1	0.73	
$\overset{o}{Z}_2$	6.48	6	0.87	0.10
$\overset{\Delta}{Z}_2$	0.66	0.4	0.81	
Z_3	22.38	10	0.73	

In view of these results, the D.M. considers that this new solution y^3 is satisfactory and consequently he decides to keep it as the best compromise solution.

7. Conclusion

Given the fact that few MOSLP methodologies have been developed up to now and that this type of methodology is well adapted to many real decision problems, we think that the proposed MOSLP methodology can be an interesting decision support tool. It can be all the more useful as the context of uncertainty which it assumes seems very realistic. Moreover, this general methodology offers many options and it can be ajusted to different situations. Obviously, in this paper, we have only outlined those different possibilities which would have to be explored more thoroughly and to be tested in concrete problems, which we intend to do in future. Finally, we wish to develop micro-computer software based on this metholodogy, which would facilitate its application.

8. References

BENAYOUN R. et al., (1971), Linear programming with multiple objective functions: STEP Method (STEM), Mathematical Programming, vol. 1, no. 3.

CHARNES A.,W. COOPER, (1959),Chance-constrained programming, Management Sciences, 6.

DESPONTIN M., (1982), Regional multiple objective quantitative economic policy: a belgian model. EJOR, vol. 10, no. 1.

DESPONTIN M., (1984), Interactive economic policy formulation with multiregional econometric models, in: Despontin, Nijkam et Spronk, "Macro planning with conflicting goals", Springer Verlag, Berlin.

GOICOECHEA A., D.R. HANSEN and L. DUCKSTEIN, (1982), Multiobjective decision analysis with engineering and business applications, John Wiley, New York.

HUBER G., (1974), Methods for quantifying subjective probabilities and multiattribute utilities, Decision Sciences, vol. 5.

HWANG C.L., A.S. MASUD, (1979), Multiobjective decision making-methods and applications, Springer Verlag, Berlin.

KAHNEMAN, D., P. SLOVIC and A. TVERSKY, (1982), Judgment under uncertainty: heuristics and biases, Cambridge Univ. Press, Cambridge.

NADEAU, R. and R. THEODORESCU, (1980), Restricted Bayes strategies for programs with simple recourse, Operations Research, vol. 28, no. 3, part II.

NADEAU R., (1983), La programmation linéaire stochastique. Document du LAMSADE 26, Université de Paris-Dauphine.

SLOWINSKI R. and J. TEGHEM, (1988), Fuzzy vs stochastic approaches to multiciteria linear programming under uncertainty, Naval Research Logistics, vol. 35, p. 673-695.

SOMMER G. and M.A. POLLATSCHEK, (1978), A fuzzy programming approach to an air pollution regulation problem, in "Progress in cybernetics and systems research", vol. 3.

STANCU-MINASIAN I.M. and M.J. WETS, (1976) A research bibliography in stochastic programming, 1955-1975, Operations Research 24, p. 1078-1119.

STANCU-MINASIAN I.M., (1984), Stochastic programming with multiple objective functions, D. Reidel Publ. Company.

STEUER R., (1976), Multiple objective linear programming with interval criterion weights, Management Science, vol. 23, no. 3, p. 305-316.

STEUER R., Algorithms for linear programming problems with interval objective functions coefficients, Mathematics of O.R. 6, p. 333-348.

TEGHEM J. Jr., D. DUFRANE, M. THAUVOYE and P. KUNSCH, (1986), STRANGE: an interactive method for multi-objective linear programming under uncertainty, E.J.O.R., 26, 1, p. 65-82.

TEGHEM J. and P. KUNSCH, (1985), Multi-objective decision making under uncertainty: an example for power systems, in Haimes, Y., Chankong, V., "Decision making with multiple objective" Springer Verlag, Berlin, no. 242, p. 443-456.

Part II
Chapter 5

COMPUTATION OF EFFICIENT SOLUTIONS OF STOCHASTIC OPTIMIZATION PROBLEMS WITH APPLICATIONS TO REGRESSION AND SCENARIO ANALYSIS

K. MARTI
Universität der Bundeswehr München
Fakultät für Luft- und Raumfahrttechnik
Werner-Heisenberg-Weg 39
D-8014 Neubiberg/München
W-Germany

ABSTRACT

In engineering and economics often a certain vector x of inputs or decisions must be chosen, subject to some constraints, such that the expected costs (or loss) arising from the deviation between the output $A(\omega)x$ of a stochastic linear system $x \rightarrow A(\omega)x$ and a desired stochastic target vector $b(\omega)$ are minimal. Hence, one has the following stochastic linear optimization problem

$$\text{minimize } F(x) = Eu(A(\omega)x-b(\omega)) \quad \text{s.t. } x \in D, \qquad (1)$$

where u is a convex loss function on $\mathbb{R}^m$, $(A(\omega),b(\omega))$ is a random $(m,n+1)$-matrix, "E" denotes the expectation operator and D is a convex subset of $\mathbb{R}^n$. Concrete problems of this type are e.g. stochastic linear programs with recourse, error minimization and optimal design problems, acid rain abatement methods, problems in scenario analysis and non-least square regression analysis.

Solving (1), the loss function u should be exactly known. However, in practice mostly there is some uncertainty in assigning appropriate penalty costs to the deviation between the output $A(\omega)x$ and the target $b(\omega)$. For finding in this situation solutions "hedging against uncertainty" a set of so-called efficient points of (1) is defined and a numerical procedure for determining these compromise solutions is derived. Several applications are discussed.

R. Slowinski and J. Teghem (eds.),
Stochastic versus Fuzzy Approaches to Multiobjective Mathematical Programming under Uncertainty, 163–188.

1. Introduction

Many problems in stochastic optimization, as e.g. in stochastic linear programming with recourse, portfolio optimization, error minimization and optimal design, nominal value optimization, inductive inference, statistical prediction, statistical decision making, regression and scenario analysis, can be formulated, see e.g. [8], as a mean value minimization problem of the type

$$\text{minimize } F(x) \quad \text{s.t.} \quad x \in D, \tag{1}$$

where the objective function $F = F_u$ of (1) is defined by

$$F(x) = Eu(A(\omega)x - b(\omega)), \quad x \in \mathbb{R}^n. \tag{2}$$

Here, $(A(\omega), b(\omega))$ is a random $m\times(n+1)$ matrix, "E" denotes the expectation operator, D is a convex subset of $\mathbb{R}^n$ and $u: \mathbb{R}^m \longrightarrow \mathbb{R}$ designates a convex loss function measuring the loss arising from the deviation $z = A(\omega)x - b(\omega)$ between the output $A(\omega)x$ of the stochastic linear system $x \longrightarrow A(\omega)x$ and the random target $b(\omega)$.

Solving (1), the loss function u should be exactly known. However, in practice mostly there is some uncertainty about the appropriate selection of u, e.g. due to the difficulties in assigning appropriate penalty costs to the deviation $z = A(\omega)x - b(\omega)$ between the output $A(\omega)x$ and the target $b(\omega)$. We suppose that $u \in U$, where U is a **given** set of convex loss functions containing the true, but unknown loss function u_o.

A possible way out in this situation of uncertainty about u is to provide the decision maker with a certain set $E = E_{D,U} (\subset D)$ of compromise solutions or solutions "hedging against uncertainty" [18], called the **efficient** points or solutions of (1), being substitutes for an optimal solution $x^* \in \operatorname{argmin}_{x \in D} F(x)$. This set $E_{D,U}$, or at least its closed hull $\bar{E}_{D,U}$, should contain, of course, an optimal solution $x^* = x^*_u$ of (1) for every $u \in U$.

An important class $U = \mathbf{C}_m^J$ of loss functions u is the set of partially monotone increasing convex loss functions on $\mathbb{R}^m$ defined as follows:

Definition 1.1. Let J be a given subset of $\{1,2,...,m\}$. For $J = \phi$ we put $\mathbf{C}_n^{\phi} := \mathbf{C}_m$, where $\mathbf{C}_m$ is the set of all convex functions u on $\mathbb{R}^m$. If $J \neq \phi$, then $\mathbf{C}_m^J$ denotes the set of all convex functions $u: \mathbb{R}^m \longrightarrow \mathbb{R}$ having the following property:

$$z_I \le w_I, \; z_{II} = w_{II} \Longrightarrow u(z) \le u(w) \tag{3}$$

Here, $z_I \in \mathbb{R}^{|J|}$, $z_{II} \in \mathbb{R}^{m-|J|}$ is the partition of any $z \in \mathbb{R}^m$ into the two subvectors $z_I = (z_i)_{i \in J}$, $z_{II} = (z_i)_{i \notin J}$. Moreover $z_I \le w_I$ means that $z_j \le w_j$ for all $j \in J$.

Remark. In many cases one has loss functions $u \in \mathbf{C}_m^J$ with this additional strict partial monotonicity property:

$$z_I \le w_I, \; z_{II} = w_{II}, \; z_i < w_i \text{ for some } i \in J \Longrightarrow u(z) < u(w). \tag{3.1}$$

A related property of an element u of $\mathbf{C}_m^J$ is

$$z_I < w_I, \; z_{II} = w_{II} \Longrightarrow u(z) < u(w), \tag{3.2}$$

where $z_I < w_I$ means that $z_i < w_i$ for all $i \in J$.

We emphasize here that also in the (rare) most favorable case that loss function u in (1) is exactly known, in solving (1), a set $E_{D,U}$ of efficient solutions of (1) has the following important practical meaning:

* **Necessary optimality condition**: Each candidate $\tilde{x} \in D$ for an optimal solution of (1) with any $u \in U$ must fulfill the constraint

$$\tilde{x} \in E_{D,U} (\tilde{x} \in \bar{E}_{D,U}, \text{ resp.}), \tag{4}$$

hence, (4) yields informations about the location and the geometrical structure of an optimal solution of (1).

* **Sharper constraints in (1)**: Having a set $E_{D,U}$ of efficient solu-

tions of (1), the constraint "$x \in D$" in (1) may be replaced by the sharper restriction (4), i.e. (1) can be replaced by

$$\min Eu(A(\omega)x-b(\omega)) \quad \text{s.t.} \quad x \in E_{D,U} \; (x \in \bar{E}_{D,U}, \text{ resp.}) \tag{5}$$

1.1. A basic definition of efficient solutions. For a given set U of convex loss functions u containing the true, but unknown loss function u_o, a first definition of efficient solutions can be given as follows:

Definition 1.2. A point $x \in D$ is called a **nondominated, admissible** or **Pareto-optimal solution** of (1) if there is no vector $\tilde{x} \in D$, $\tilde{x} \neq x$, such that

$$F_u(\tilde{x}) \le F_u(x) \text{ for all } u \in U, \tag{6.1}$$

$$F_{\tilde{u}}(\tilde{x}) < F_{\tilde{u}}(x) \text{ for some } \tilde{u} \in U, \tag{6.2}$$

where $F_u(x) := Eu(A(\omega)x-b(\omega))$ is the objective function of (1).

Let $E_{D,U}^{(o)}$ denote the set of all nondominated solutions of (1) according to Definition 1.2.

1.2. Discretely distributed stochastic programs. In the following we consider the construction of a set $E_{D,U}$ of efficient solutions for (1) in the case that $(A(\omega),b(\omega))$ has a discrete distribution $P_{(A(\ ),b(\ ))}$, i.e. in the following we suppose that

$$P_{(A(\cdot),b(\cdot))} = \sum_{i=1}^{r} \alpha_i \, \epsilon_{(A^i,b^i)}, \tag{7}$$

where $r>1$ is an integer, $\alpha_i>0$, $i=1,\dots,r$, $\sum_{i=1}^{r} \alpha_i=1$ and $\epsilon_{(A^i,b^i)}$ denotes the one-point measure in the given $m\times(n+1)$ matrix (A^i,b^i), $i=1,2,\dots,r$.

Very often the random matrix $(A(\omega),b(\omega))$ can be represented by

$$(A(\omega),b(\omega)) = (\bar{A},\bar{b}) + \sum_{t=1}^{L} \xi_t(\omega)(A^{(t)},b^{(t)}), \tag{8}$$

where $L\geq 1$ is an integer such that

$$L < m(n+1) \text{ or even } L \ll m(n+1), \tag{8.1}$$

$(\bar{A},\bar{b}),(A^{(1)},b^{(1)}),\ldots,(A^{(L)},b^{(L)})$ are L+1 given fixed $m\times(n+1)$ matrices and $\xi_1(\omega),\ldots,\xi_L(\omega)$ are zero mean real valued random variables having the discrete distribution

$$P(\xi_1(\omega) = \xi_1^i, \xi_2(\omega) = \xi_2^i, \ldots, \xi_L(\omega) = \xi_L^i) = \alpha_i,\ i=1,\ldots,r. \tag{8.2}$$

By the above representation (8), (8.1), (8.2) there is expressed the fact that in practice the number L of generating random variables $\xi_1(\omega),\ldots,\xi_L(\omega)$ in $(A(\omega),b(\omega))$ is much smaller than the total number $m(n+1)$ of elements of the random matrix $(A(\omega),b(\omega))$.

Let Ξ_t denote the r-row vectors

$$\Xi_t = (\xi_t^1,\ldots,\xi_t^i,\ldots,\xi_t^r),\ t=1,\ldots,L, \tag{9}$$

of realizations of the random variables $\xi_t(\omega)$, $t=1,\ldots,L$. Moreover, let Ξ be the $L\times r$ matrix having the rows $\Xi_1,\ldots,\Xi_L$. By 1_r and α we designate the r-vectors

$$1_r = (1,1,\ldots,1)',\ \alpha = (\alpha_1,\alpha_2,\ldots,\alpha_r)'.$$

Since $\xi_1(\omega),\ldots,\xi_L(\omega)$ have zero means, we obviously have

$$\Xi\alpha = 0. \tag{10}$$

If $\alpha_1=\alpha_2= \ldots =\alpha_r = \frac{1}{r}$ and we still assume that $\xi_1(\omega),\ldots,\xi_L(\omega)$ are uncorrelated random variables, then 1_r, $\Xi_1',\ldots,\Xi_L'$ are mutually orthogonal r-vectors.

Before we start the construction of efficient solutions of (1), in the next section 2 some concrete applications are discussed.

2. Examples

2.1. Regression and curve-fitting problems. The functional dependence of a variable Y on certain other variables $Z_1, Z_2, \dots, Z_\ell$ is often modelled by the l.i.p.-(linear in parameter-) equation

$$Y = \sum_{k=o}^{\nu} \beta_k \varphi_k(Z_1, Z_2, \dots, Z_\ell) + R. \tag{11}$$

Here $\varphi_o(z), \varphi_1(z), \dots, \varphi_\nu(z)$, $z \in \mathbb{R}^\ell$, denote given functions on $\mathbb{R}^\ell$, β_o, $\beta_1, \dots, \beta_\nu$ are unknown coefficients to be estimated and R designates an error or noise term.

2.1.1. Regression of a sample. Let $Z = (Z_1, Z_2, \dots, Z_\ell)$ be the ℓ-vector of the so-called predictor variables Z_j, $j=1,\dots,\ell$. For given values z_i, $i=1,2,\dots,r$, of Z, let $y_1, y_2, \dots, y_r$, resp., be the observed values of the response variable Y related to the input values $z_1, z_2, \dots, z_r$, respectively.

Estimators $\hat{\beta}_k$, $k=0,1,\dots,\nu$, of the unknown coefficients $\beta_k, k=0,1,\dots,\nu$, are then determined usually by solving the minimization (or regression) problem

$$\min_{\beta \in D} \sum_{i=1}^{r} \delta(y_i - \sum_{k=o}^{\nu} \beta_k \varphi_k(z_i)), \tag{12}$$

where $\beta = (\beta_o, \beta_1, \dots, \beta_\nu)'$, $\delta = \delta(w)$ is a convex distance or error function and the condition "$\beta \in D$" represents possible constraints for $\beta_o, \beta_1, \dots, \beta_r$ in the model (11). A very popular error criterion is, of course, the Least-Square-Criterion given by

$$\delta(w) = w^2 \tag{LSQ}$$

However, several disadvantages of the LSQ-criterion are discussed - also recently - by many authors, as e.g. Davies (1967), Appa et al. (1973), Gentl et al. (1977), Watson (1980), Narula (1982), Zanakis et al. (1982), Dielman et al. (1982), Bloomfield and Steiger (1983), Everitt (1987), Dodge (ed.) (1987).

Consequently, there are considered therefore several alternatives for the LSQ-criterion, e.g. the L_p-estimators represented by the error function

$$\delta(w) = |w|^p, \ p\geq 1.$$

As mentioned above, a very important case is the Least-Absolute-Value-Criterion given by

$$\delta(w) = |w|. \qquad \text{(LAV)}$$

Hence, since in the regression problem (12) not a unique error function $\delta(w)$ can be assigned, the regression analysis is a problem as described in section 1. Indeed we have only to define the quantities arising in section 1 as follows:

$$m = 1, \ n = \nu+1 \qquad (13.1)$$

$$x = (\beta_o, \beta_1, \ldots, \beta_\nu)' \qquad (13.2)$$

$$u(w) = \delta(-w), \qquad (13.3)$$

moreover, it is $P((A(\omega), b(\omega)) = (A^i, b^i)) = \alpha_i$, $i=1,2,\ldots,r$, where

$$A^i = (\varphi_o(z_i), \varphi_1(z_i), \ldots, \varphi_\nu(z_i)), \ b^i = y_i \qquad (13.4)$$

$$\alpha_i = \frac{1}{r}. \qquad (13.5)$$

2.1.2. Curve-fitting. We obtain exactly the same minimization problem (12) if a (l.i.p.-)parametric model

$$\hat{h}(z) = \sum_{k=o}^{\nu} \beta_k \varphi_k(z), \ z\in \mathbb{R}^\ell,$$

with given basis functions $\varphi_k: \mathbb{R}^\ell \longrightarrow \mathbb{R}$, $k=0,1,\ldots,\nu$, is fitted to measurements $y_1, y_2, \ldots, y_r$ of an unknown function $y=h(x)$, $x\in \mathbb{R}^\ell$, taken at r given points $z_1, z_2, \ldots, z_r \in \mathbb{R}^\ell$. The LSQ-criterion is also in curve-fitting problems a very popular criterion though there is, of course, no unique error function $\delta=\delta(w)$ in general. Because of the disadvantages of the LSQ-criterion mentioned in section 2.1.1 which are true also here, there are considered several alternatives, see e.g. [9],[24]. Hence, also curve-fitting can be interpreted as a problem of type (1),(2), where the corresponding quantities m,n,x,u etc. are again defined by (13.1)-(13.5).

2.2. Air quality planning problems. Presently the determination of acid rain abatement strategies or similar methods for improving the air quality is of utmost importance, see e.g. Turner (1979), Fortin and Mc Bean (1983), Fronza and Melli (1984), Ellis et al. (1985), (1986), Guldman (1986), Fuessle et al. (1987). The problem can be described as follows:

A certain pollutant, e.g. SO_2, is emitted from n sources or emission regions, e.g. power plants, $S_1, S_2, \ldots, S_k, \ldots, S_n$. Let E_k, $k=1,\ldots,n$, denote the corresponding emission levels (e.g. in tons/year). The pollutant is transfered then - depending on the weather conditions - to possible receptors $R_1, R_2, \ldots, R_i, \ldots, R_m$, e.g. sensitive areas as lakes, forests etc.. Suppose that G_i is the maximum level of the pollutant which may be deposited in the considered planning period (e.g. a year) at receptor R_i, $i=1,\ldots,m$. Let t_{ik}, $0 \le t_{ik} \le 1$, $i=1,\ldots,m$, $k=1,\ldots,n$, denote the so-called transfer coefficients representing the rate of pollutant deposition at receptor R_i that results from 1 unit of pollutant emission at source S_k. Because of

- weather stochasticity
- variability between different meteorological models or scenarios (e.g. Gaussian diffusion model, 3-D-hydrodynamical model etc.)

describing the transfer of the pollutant between sources and receptors,

the transfer coefficients t_{jk} must be considered a random variables $t_{ik} = t_{ik}(\omega)$ on a certain probability space $(\Omega, \mathfrak{A}, P)$.

Furthermore, let $B_i=B_i(\omega)$ be the level of noncontrollable pollutant deposition at receptor R_i, i=1,...,m, e.g. due to natural sources as volcanoes etc..

Denoting by x_k, k=1,...,n, the percentage of emission level reduction at source S_k, $E_k - x_kE_k = E_k(1-x_k)$ is the remaining emission level of the pollutant after reconstruction at source S_k. Adding the different amounts of the pollutant still transfered from controllable and noncontrollable sources to a receptor R_i, we get the constraint

$$\sum_{k=1}^{n} E_k(1-x_k)t_{ik}(\omega) + B_i(\omega) \leq G_i, \tag{14.1}$$

for each i=1,...,m. The violation of constraint (14.1) causes certain serious damages and therefore certain costs at receptor R_i, i=1,...,m. We assume that these costs are given by

$$u_i(\sum_{k=1}^{n} E_k(1-x_k)t_{ik}(\omega) + B_i(\omega)-G_i),$$

where $u_i: \mathbb{R} \longrightarrow \mathbb{R}$ designates a nondecreasing, convex cost function. If c_kx_k with a given cost factor $c_k>0$ are the costs for the reconstruction at source S_k, k=1,...,n, then

$$C = \sum_{k=1}^{n} c_kx_k + \sum_{i=1}^{m} u_i(\sum_{k=1}^{n} E_k(1-x_k)t_{ik}(\omega) + B_i(\omega)-G_i) \tag{14.2}$$

are the total cost arising from the reconstruction at sources $S_1,S_2,\ldots,S_n$ and due to the damages at receptors $R_1,R_2,\ldots,R_m$. It is easy to see that C can be represented by

$$C = c'x + \tilde{u}(T(\omega)x-q(\omega)), \qquad (14.2)'$$

where $c=(c_1,\ldots,c_n)'$, $x=(x_1,\ldots,x_n)'$,

$$\tilde{u}(z) = \sum_{i=1}^{m} u_i(z_i), \quad z\in \mathbb{R}^m, \tag{14.3}$$

and the i-th row T_i of T and the i-th element h_i of h are defined by

$$T_i(\omega) = -(E_1t_{i1}(\omega), E_2t_{i2}(\omega),\ldots,E_nt_{in}(\omega)) \tag{14.4}$$

$$h_i(\omega) = G_i - B_i(\omega) - \sum_{k=1}^{n} E_kt_{ik}(\omega). \tag{14.5}$$

For the decision variables $x_1,\ldots,x_n$ we have the constraints

$$x_k^{min} \le x_k \le x_k^{max} \text{ with given bounds } 0 \le x_k^{min} \le x_k^{max} \le 1. \tag{14.6}$$

Since there is no unique way to select a nondecreasing, convex cost function u_i for measuring the violation of constraint (14.1), we have again a stochastic optimization problem as described in section 1, where $\tilde{u} \in \mathbf{C}_m^{\{1,\ldots,m\}}$, cf. (14.3).

2.3. __Scenario analysis__. Given a certain planning problem, in scenario analysis [17],[19],[22],[29] the future evolution or development of the system to be considered is anticipated or explored by means of a (usually small) number r (e.g. r=3,4,5,6) of so-called scenarios s^1, $s^2,\ldots,s^i,\ldots,s^r$. Scenarios s^i, $i=1,2,\ldots,r$, are plausible alternative models of the future development given by "extreme points" of a set of certain basic or key variables. An individual scenario or a certain mixture of the scenarios $s^1,\ldots,s^r$ is assumed then to be revealed in the considered future time period.

We assume now that the planning problem can be described mathematically by the optimization problem

$$\min c'x \quad \text{s.t.} \quad Tx = (\le)h, \; x \in D, \tag{15}$$

where D is a given convex subset of $\mathbb{R}^n$ and the data (c,T,q) are given by

$$(c,T,h) = (c^i,T^i,h^i) \text{ for scenario } s^i, \; i=1,2,\ldots,r,$$

where c^i is an n-vector, T^i an m×n matrix and h^i an m-vector. Having written here the scenarios $s^1,s^2,\ldots,s^r$ by means of (15) and the data (c^i,T^i,h^i), $i=1,2,\ldots,r$, and facing therefore the subproblems

$$\min c^{i\prime}x \quad \text{s.t.} \quad T^ix = (\le)h^i, \; x \in D, \tag*{$(15)_{(i)}$}$$

$i=1,2,\ldots,r$, the decision maker has then to select an appropriate decision $x \in D$.

Since one is unable in general to predict with certainty which scenario s^i will occur, scenario analysts are looking for decisions x^o which are

"robust" with respect to the different scenarios or "scenario-independent",

cf. [17],[19],[22]. Obviously, this robustness concept is closely related to the idea of detecting "similarities" within the family of optimal solutions $x^*(s^i)$, $i=1,2,\dots,r$, of the individual subproblems $(15)_{(i)}$, $i=1,\dots,r$, cf. [25],[26].

Let $\alpha_1,\alpha_2,\dots,\alpha_r$ with $\alpha_i>0$, $i=1,\dots r$, $\sum_{i=1}^{r}\alpha_i=1$, be (subjective) probabilities for the occurence of $s^1,\dots,s^r$, resp., or weights reflecting the relative importance of $s^1,\dots,s^r$, resp., in the uncertain environment [18]. Considering loss functions $u\in\mathbf{C}_m^J$ for evaluating the violations $z^i=T^ix-h^i$ of the constraint T^ix-h^i, $T^ix\le h^i$, resp., in $(15)_{(i)}$, a class of robust or scenario-independent decisions are obviously the efficient solutions of

$$\min_{x\in D}\sum_{i=1}^{r}\alpha_i(c^{i\prime}x+u(T^ix-h^i)) \tag{16}$$

which is a stochastic optimization problem of the type (1),(2) having a discrete distribution, see section 1.3. Further robustness properties can be obtained by varying the probability or weight distribution $\alpha=(\alpha_1,\alpha_2,\dots,\alpha_r)'$.

3. A system of linear relations for the construction of descent directions

Fundamental for the computation of the set $E_{D,U}$ of efficient solutions of (1) is the following construction method for descent directions of the objective function F of (1), cf. [15].

We suppose now that the true, but unknown loss function u of (1) is, see Definition 1.1, an element of $\mathbf{C}_m^J$ for some known index set J {1,2,...,m}. We remember that for any vector $z \in \mathbb{R}^m$ the subvectors z_I, z_{II} are defined by

$$z_I = (z_i)_{i \in J}, \ z_{II} = (z_i)_{i \notin J}, \tag{17}$$

see (3). Of course, if $J=\emptyset$, then $z=z_{II}$ and z_I does exist. For any $m \times (n+1)$ matrix (A,b) let (A_I,b_I), (A_{II},b_{II}), resp., denote the submatrices of (A,b) having the rows (A_i,b_i) with $i \in J$, $i \in \{1,...,m\} \backslash J$, respectively.

Given an n-vector x, e.g. the t-th iteration point of an algorithm for solving (1), we consider, in extension of system (3.1)-(3.4b) in [15], the following system of linear relations for the unknowns (y,Π), where $y \in \mathbb{R}^n$ and Π is an auxiliary $r \times r$ matrix:

(18.1) $\sum_{j=1}^{r} \pi_{ij} = 1, \ \pi_{ij} \geq 0, \quad i,j=1,2,...,r$

(18.2) $\alpha_j = \sum_{i=1}^{r} \alpha_i \pi_{ij}, \quad j=1,2,...,r$

(18.3.1) $A_I^j y - b_I^j \leq \sum_{i=1}^{r} \frac{\alpha_i \pi_{ij}}{\alpha_j} (A_I^i x - b_I^i), \quad j=1,2,...,r$

(18.3.2) $A_{II}^j y - b_{II}^j = \sum_{i=1}^{r} \frac{\alpha_i \pi_{ij}}{\alpha_j} (A_{II}^i x - b_{II}^i), \quad j=1,2,...,r$

(18.4a) The transition probability measure

$$K^j = \sum_{i=1}^{r} \beta_{ij} \epsilon_{z^i}, \ \beta_{ij} = \frac{\alpha_i \pi_{ij}}{\alpha_j}, \ z^i = A^i x - b^i,$$

is not a one-point measure for at least one j, $1 \leq j \leq r$

(18.4b) There exists at least one j, $1 \leq j \leq r$, such that K^j is not a one-point measure and $\pi_{ij} > 0$ for all j=1,2,...,r

(18.4c) At least one inequality in (18.3.1) holds with "<".

In order to take into consideration also the constraint "$x \in D$" in (1), we may add to (18.1)-(18.4c) the condition

(18.5) $x \in D$.

Note. a) By ϵ_z we denote the one-point measure in a point $z \in \mathbb{R}^m$. b) According to (18.1), Π is a stochastic matrix. System (18.1)-(18.3) has always the trivial solution $(y,\pi)=(x,I)$, where I is the r×r identity matrix. c) If (y,Π) solves (18.1)-(18.3), then

$$\bar{A}_I y \le \bar{A}_I x \tag{19.1}$$

$$\bar{A}_{II} y = \bar{A}_{II} x, \tag{19.2}$$

where $\bar{A}_I = EA_I(\omega)$, $\bar{A}_{II} = EA_{II}(\omega))$. d) If $J = \emptyset$, then (18.3.1) vanishs and (18.3.2) reads

$$A^j y - b^j = \sum_{i=1}^{r} \frac{\alpha_i \pi_{ij}}{\alpha_j} (A^i x - b^i), \quad j=1,2,\ldots,r. \tag{18.3.2)'$$

In the special case

$$(A_I^j, b_I^j) = (\bar{A}_I, \bar{b}_I) \text{ for all } j=1,2,\ldots,r, \tag{20}$$

i.e. if $(A_I(\omega),\ b_I(\omega))$ is constant with probability one, then (18.3.1) is obviously reduced to the condition

$$\bar{A}_I y \le \bar{A}_I x,$$

cf. (19.1).

The meaning of (18.3.1)-(18.3) and the additional conditions (18.4a)-(18.4c) for the basic mean value minimization problem (1) with objective function F is summarized in the next result.

Theorem 3.1. Let J be any fixed subset of $\{1,2,\ldots,m\}$. a) If (y,Π) is a solution of (18.1)-(18.3.2), then $F(y) \le F(x)$ for every $u \in \mathbf{C}_m^J$. For $J=\emptyset$ also the converse holds: If there is a vector y such that $F(y) \le F(x)$ for all $u \in \mathbf{C}_m (= \mathbf{C}_m^{\emptyset})$, then there exists an r×r matrix Π such that (y,Π) satisfies (18.1), (18.2) and (18.3.2)'. b) If (y,Π) is a solution of (18.1)-(18.3.2) and (18.4a), then $F(y) < F(x)$ for every $u \in \mathbf{C}_m^J$ which is strictly convex on $\mathrm{conv}\{z^i : 1 \le i \le r\}$. c) If (y,Π) is a solution of (18.1)-(18.3.2) and (18.4b), then $F(y) < F(x)$ for every $u \in \mathbf{C}_m^J$ which is not affine-linear on $\mathrm{conv}\{z^i : 1 \le i \le r\}$. d) If (y,Π) fulfills (18.1)-(18.3.2) and (18.4c), then $F(y) < F(x)$ for every $u \in \mathbf{C}_m^J$ satisfying (3.1).

A simple, but important consequence of the above theorem is stated next:

Corollary 3.1. For given $x \in \mathbb{R}^n$ or $x \in D$ let (y,Π) be any solution of (18.1)-(18.3.2) such that $y \neq x$, $y \in D \setminus \{x\}$, respectively. a) In this case $h=y-x$ is a descent direction, a feasible descent direction, resp., of F at x for every $u \in \mathbf{C}_m^J$ such that F is not constant on the line segment $\overline{xy}$ joining x and y. b) If (y,Π) fulfills also (18.4a),(18.4b),(18.4c), resp., then $h=y-x$ is a (feasible) descent direction of F at x for every $u \in \mathbf{C}_m^J$ which is strictly convex on $\mathrm{conv}\{z^i: 1 \le i \le r\}$, is not affine-linear on $\mathrm{conv}\{z^i: 1 \le i \le r\}$, fulfills (3.1), respectively.

4. Efficient solutions of (1)

In the following we suppose that the unknown loss function u is an element of $\mathbf{C}_m^J$, see Definition 1.1, where J is a given subset of $\{1,2,\ldots,m\}$.

For a given point $x \in D$, the descent direction-finding procedure described in section 3 only can fail completely if for each solution (y,Π) of (18.1)-(18.3.2) with $x \in D$ it holds

$$A^j y = A^j x \text{ for each } j=1,2,\ldots,r. \tag{21}$$

Indeed, in this case we either have $y=x$ or for arbitrary loss functions u the objective function F of (1) is constant on the whole line through the points x,y. This observation suggests the following basic efficiency concept.

Definition 4.1. A point $x \in D$ is called a ($\mathbf{C}_m^J$-) efficient point or a ($\mathbf{C}_m^J$-) efficient solution of (1) if and only if for each solution (y,Π) of (18.1)-(18.3.2) with $y \in D$ it is $A^j y = A^j x$ for each $j=1,2,\ldots,r$, i.e. $A(\omega)x = A(\omega)y$ w.p.1.

Let $E_{D,J}$ denote the set of all ($\mathbf{C}_m^J$-) efficient points of (1).

For deriving parametric representations of $E_{D,J}$, we need the following definitions and lemmata:

For a given n-vector x and $z^i = A^i x - b^i$, $i=1,\dots,r$, let $S=S_x \subset \{1,2,\dots,r\}$ with $|S| =: s$ be an index set such that

$$Z_1 := \{z^i : 1 \le i \le r\} = \{z^i : i \in S\}, \tag{22.1}$$

where $z^i \neq z^j$ for $i \neq j$. Defining $\tilde{\alpha}_i$, τ_{ij}, $\tilde{\beta}_{ij}$, $i \in S$, $j=1,2,\dots,r$, by

$$\tilde{\alpha}_i = \sum_{z^t = z^i} \alpha_t \tag{22.2}$$

$$\tau_{ij} = \frac{1}{\tilde{\alpha}_i} \sum_{z^t = z^i} \alpha_t \pi_{tj} \tag{22.3}$$

$$\tilde{\beta}_{ij} = \frac{\tilde{\alpha}_i \tau_{ij}}{\alpha_j}, \tag{22.4}$$

we find that relations (18.1)-(18.3.2) can be represented by

$$\sum_{j=1}^{r} \tau_{ij} = 1,\ \tau_{ij} \ge 0,\ i \in S,\ j=1,2,\dots,r \tag{23.1}$$

$$\alpha_j = \sum_{i \in S} \tilde{\alpha}_i \tau_{ij},\ j=1,2,\dots,r \tag{23.2}$$

$$A_I^j y - b_I^j \le \sum_{i \in S} \tilde{\beta}_{ij} z_I^i,\ j=1,2,\dots,r \tag{23.3}$$

$$A_{II}^j y - b_{II}^j = \sum_{i \in S} \tilde{\beta}_{ij} z_{II}^i,\ j=1,2,\dots,r. \tag{23.4}$$

Consider now the closed convex hull

$$C_1 = \overline{\text{conv}}\, Z_1 = \text{conv}\, Z_1$$

of Z_1, cf. [15], and let Z_1^e denote the set of extreme points of C_1. It is Z_1^e Z_1, see [15]. For $J \neq \emptyset$ we further consider the set $Z_1^{e,u}$ of upper extreme points of C_1 which are defined as follows.

<u>Definition 4.2.</u> An extreme point of C_1, i.e. an element $z^i \in Z_1^e$, is called an upper extreme point of C_1 if and only if there is no vector $\tilde{z} \in C_1$ such that

$$z^i \neq \tilde{z},\ z_I^i \le \tilde{z}_I,\ z_{II}^i = \tilde{z}_{II}. \tag{24}$$

The existence of upper extreme points is guaranteed by the next result.

Lemma 4.1. C_1 has at least one upper extreme point.

For the next lemma we still need the s×r matrix $\mathbf{T}^o = (\tau^o_{ij})$ defined by

$$\tau^o_{ij} = \begin{cases} 0, & \text{if } z^i \neq z^j \\ \frac{\alpha_j}{\alpha_i}, & \text{if } z^i = z^j \end{cases} \quad \text{for } i \in S,\ j=1,2,\dots,r. \tag{25}$$

Lemma 4.2. Let (y,Π) be a solution of (18.1)-(18.3.2). Let $\mathbf{T}=\mathbf{T}(\Pi)=(\tau_{ij})$ be the s×r matrix having the elements τ_{ij} given by (22.3). If (21) holds, i.e. $A^j y = A^j x$ for all $j=1,2,\dots,r$, then $\mathbf{T}(\Pi)=\mathbf{T}^o$ and (18.3.1) holds with "=".

Lemma 4.2 implies the following important property of efficient solutions of (1):

Corollary 4.1. Let $x \in D$ be an efficient solution of (1). If (y,Π) is any solution of (18.1)-(18.3.2) with $y \in D$, then $\mathbf{T}(\Pi)=\mathbf{T}^o$ and (18.3.1) holds with "=".

For $J=\emptyset$ we obtain the set $E_D := E_{D,\emptyset}$ of all $\mathbf{C}_m$- efficient solutions of (1). This set is studied intensively in [15]. An important relationship between E_D and $E_{D,J}$ for any $J \subset \{1,2,\dots,m\}$ is shown in the following.

Lemma 4.3. It holds $E_{D,J} \subset E_D$ for every $J \subset \{1,2,\dots,m\}$.

4.1. Comparison of Definitions 4.1 and 1.2. Comparing the efficient solutions according to Definition 4.1 and the nondominated solutions according to Definition 1.2, first for $J=\emptyset$, i.e. $U=\mathbf{C}_m$ we find the following correspondence.

Theorem 4.1. $E_{D,\emptyset} = E^{(0)}_{D,\mathbf{C}_m}$.

The next corollary follows immediately from the above theorem and Lemma 4.3.

Corollary 4.2. $E_{D,J} \subset E_{D,\emptyset} = E^{(o)}_{D,\mathbb{C}_m}$ for every $J \subset \{1,2,\ldots,m\}$.

Considering now $U = \mathbb{C}^J_m$, we find the next inclusion.

Theorem 4.2a. $E_{D,J} \supset E^{(o)}_{D,\mathbb{C}^J_m}$ for every $J \subset \{1,2,\ldots,m\}$.

The following inclusion follows from Corollary 4.2 and Theorem 4.2a, cf. also (9.2) and (11.3).

Corollary 4.3. $E^{(o)}_{D,\mathbb{C}^J_m} \subset E_{D,J} \subset E^{(o)}_{D,\mathbb{C}_m}$ for every $J \subset \{1,2,\ldots,m\}$.

A converse statement to Theorem 4.2a can be obtained for case (20), i.e. if $(A_I(\omega),b_I(\omega)) = (\bar{A}_I,\bar{b}_I)$ w.p.1. This holds in many applications, see section 2.

Theorem 4.2b. If $(A_I(\omega),b_I(\omega)) = (\bar{A}_I,\bar{b}_I)$ w.p.1, then $E_{D,J} = E^{(o)}_{D,\mathbb{C}^J_m}$ for each $J \subset \{1,2,\ldots,m\}$.

4.2. Further characterizations of $E_{D,J}$.

The ($\mathbb{C}^J_m$-) efficiency of a point $x \in D$ can also be described in the following way.

Theorem 4.3. A point $x \in D$ is ($\mathbb{C}^J_m$-) efficient if and only if for every solution (y,Π) of (18.1)-(18.3.2) it holds $A^j y = A^j x$ for all $j=1,2,\ldots,r$, or $h=y-x$ is not a feasible direction for D at x.

A special case. In the important special case $b^i = b_o$ for all $i=1,2,\ldots,r$, where b_o is a fixed m-vector, system (18.1)-(18.3.2) is reduced to the following equivalent system for (y,Π):

$$\sum_{j=1}^{r} \pi_{ij} = 1, \quad \pi_{ij} \geq 0, \quad i,j=1,2,\ldots,r$$

$$\alpha_j = \sum_{i=1}^{r} \alpha_i \pi_{ij}, \quad j=1,2,\ldots,r$$

$$A^j_I y \leq \sum_{i=1}^{r} \frac{\alpha_i \pi_{ij}}{\alpha_j} A^i_I x$$

$$A^j_{II} y = \sum_{i=1}^{r} \frac{\alpha_i \pi_{ij}}{\alpha_j} A^i_{II} x.$$

4.3. Necessary optimality conditions without (sub-)gradients

If $x \in D$ is efficient, then, according to Theorem 4.3, the descent direction-finding method described in section 3 fails at x. Since especially in any optimal solution x^* of (1) there does not exist a feasible descent direction, obviously, efficient points must be candidates for optimal solution x^* of (1):

Theorem 4.4. Suppose that for every $x \in D$ and every solution (y,Π) of (18.1)-(18.3.2) with $y \in D$ the objective function F of (1) with a loss function $u \in C_m^J$ is constant on the line segment xy if and only if $A^j y = A^j x$ for every $j=1,2,\dots,r$. If x^* is an optimal solution of (1), then $x^* \in E_{D,J}$.

Note. The assumption in Theorem 4.4 concerning F is fulfilled e.g. if $u \in C_m^J$ is strictly convex on the convex hull $\mathrm{conv}\{(A^j y - b^j)(A^j x - b^j):$ $x,y \in D,\ 1 \le j \le r\}$ generated by the line segments $(A^j y - b^j)(A^j x - b^j)$ joining $A^j y - b^j$ and $A^j x - b^j$.

If the assumption in Theorem 4.4 concerning F does not hold, then it may happen that the objective function F of (1) is constant on a certain line segment xy though $A^j y \neq A^j x$ for at least one index j, $1 \le j \le r$. Hence, Theorem 4.4 can not be applied then directly. However, in this case the following modification of Theorem 4.4 holds true.

Theorem 4.5. Let u be an arbitrary loss function from $\mathbf{C}_m^J$ for some $J \subset \{1,2,\dots,m\}$. If D is a compact convex subset of $\mathbb{R}^n$, then there exists at least one optimal solution x^* of (1) lying in the closure $\bar{E}_{D,J}$ of the set $E_{D,J}$ of efficient solutions of (1).

According to Theorem 4.4 and 4.5 the efficient solutions determine the locations of the optimal solutions x^* of (1) for arbitrary loss functions $u \in C_m^J$. Hence, if the loss function $u \in C_m^J$ can not be specified exactly in a practial situation, then the set $E_{D,J}$ of efficient solutions of (1) contains possible substitutes for an optimal solution x^* of (1).

5. Parametric representation of $E_{D,J}$

In the following the considerations in [15] are generalized to the case of partly monotone loss functions u, i.e. in the following we suppose that $u \in C_m^J$ for some index set $J \subset \{1,2,\ldots,m\}$, see Definition 1.1.

5.1. Characterization of efficient solutions by optimality conditions.

For solving the descent direction-generating relations (18.1)-(18.3.2) and (18.4a) or (18.4b) or (18.4c) we can use, see Theorem 3.1 and Corollary 3.1, the quadratic pogram

$$\text{minimize} \sum_{j=1}^{r} \theta_j \left(\eta'\left(A_I^j y - b_I^j - \sum_{i\in S} \tilde{\beta}_{ij} z_I^i\right) + \sum_{i\in S} \tilde{\beta}_{ij}^{\,2}\right) \qquad (\tilde{P}_{x,D}^Q)$$

s.t.

$$\sum_{j=1}^{r} \tau_{ij} = 1, \quad \tau_{ij} \ge 0, \quad i,j=1,2,\ldots,r$$

$$\alpha_j = \sum_{i\in S} \tilde{\alpha}_i \tau_{ij}, \quad j=1,2,\ldots,r$$

$$A_I^j y - b_I^j \le \sum_{i\in S} \tilde{\beta}_{ij} z_I^i, \quad j=1,2,\ldots,r$$

$$A_{II}^j y - b_{II}^j = \sum_{i\in S} \tilde{\beta}_{ij} z_{II}^i, \quad j=1,2,\ldots,r$$

$$y \in D$$

which generalizes a corresponding quadratic program in [15].

On the other hand, also by means of $(\tilde{P}_{x,D}^Q)$, efficient solutions of (1) can be characterized as follows:

Lemma 5.1. A vector $x \in D$ is an efficient solution of (1) if and only if $(\tilde{P}_{x,D}^Q)$ has an optimal solution $(y^*, \mathbf{T}^*)$ such that $A^j y^* = A^j x$ for all $j=1,2,\ldots,r$.

Note. According to Lemma 4.2 we then also have that $\mathbf{T}^* = \mathbf{T}^o$ and (18.3.1) holds with "=".

5.2. The Kuhn-Tucker conditions for $(\tilde{P}_{x,D}^Q)$.

Based on Lemma 5.1, we consider now the Lagrangian Ψ of $(\tilde{P}_{x,D}^Q)$, where we suppose that the feasible domain D of (1) is given by

$$D = \{x \in \mathbb{R}^n : g_k(x) \le 0, \ k=1,2,\ldots,\kappa\}. \qquad (26)$$

Here, $g_1, g_2, \ldots, g_\kappa$ are diffferentiable, convex functions. Moreover, we suppose that $(\tilde{P}^Q_{x,D})$ has a feasible solution $(y, \mathbf{T})$ such that

$$g_k(y) < 0 \text{ for each non-affine linear } g_k. \tag{26.1}$$

No constraint qualifications are needed in the important special case

$$D = \{x \in \mathbb{R}^n : Gx \le g\}, \tag{26.2}$$

where (G,g) is a given $\kappa \times (n+1)$ matrix.

Putting $\theta_j = 1$ for all $j=1,2,\ldots,r$, the Lagrangian Ψ of $(\tilde{P}^Q_{x,D})$ is then given, with $R := \{1,2,\ldots,r\}$, by

$$\begin{aligned}
\Psi &= \Psi(y, \mathbf{T}; (\mu_i)_{i\in S}, (\lambda_j)_{j\in R}, (\gamma_{jI})_{j\in R}, (\gamma_{jII})_{j\in R}, \rho) = \\
&= \sum_{j=1}^{r} \Big(\eta'(A^j_I y - b^j_I - \sum_{i\in S} \frac{\tilde{\alpha}_i \tau_{ij}}{\alpha_j} z^i_I) + \sum_{i\in S} \Big(\frac{\tilde{\alpha}_i \tau_{ij}}{\alpha_j}\Big)^2\Big) + \\
&+ \sum_{i\in S} \mu_i \Big(\sum_{j=1}^{r} \tau_{ij} - 1\Big) + \sum_{j=1}^{r} \lambda_j \Big(\sum_{i\in S} \tilde{\alpha}_i \tau_{ij} - \alpha_j\Big) + \\
&+ \sum_{j=1}^{r} \gamma_{jI}'\Big(A^j_I y - b^j_I - \sum_{i\in S} \frac{\tilde{\alpha}_i \tau_{ij}}{\alpha_j} z^i_I\Big) + \\
&+ \sum_{j=1}^{r} \gamma_{jII}'\Big(A^j_{II} y - b^j_{II} - \sum_{i\in S} \frac{\tilde{\alpha}_i \tau_{ij}}{\alpha_j} z^i_{II}\Big) + \sum_{k=1}^{\kappa} \rho_k g_k(y)
\end{aligned} \tag{27}$$

with Lagrangian multipliers $\mu_i, \lambda_j \in \mathbb{R}, \gamma_{jI} \in \mathbb{R}^{|J|}, \gamma_{jII} \in \mathbb{R}^{m-|J|}, i\in S, j\in R,$ $\rho \in \mathbb{R}^\kappa$. Defining

$$\hat{\gamma}_j = \begin{pmatrix} \eta + \gamma_{jI} \\ \gamma_{jII} \end{pmatrix}, \quad j=1,2,\ldots,r, \tag{28}$$

the partial derivatives of Ψ with respect to y and $\tau_{ij}, i\in S, j\in R$, are given by

$$\begin{aligned}
\frac{\partial\Psi}{\partial y} &= \sum_{j=1}^{r} A^{j\prime}_I \eta + \sum_{j=1}^{r} A^{j\prime}_I \gamma_{jI} + \sum_{j=1}^{r} A^{j\prime}_{II} \gamma_{jII} + \sum_{k=1}^{\kappa} \rho_k \nabla g_k(y) \\
&= \sum_{j=1}^{r} A^{j\prime} \hat{\gamma}_j + \sum_{k=1}^{\kappa} \rho_k \nabla g_k(y),
\end{aligned} \tag{29.1}$$

$$\begin{aligned}
\frac{\partial\Psi}{\partial \tau_{ij}} &= -\frac{\tilde{\alpha}_i}{\alpha_j} \eta' z^i_I + 2\frac{\tilde{\alpha}_i}{\alpha_j} \tilde{\beta}_{ij} + \mu_i + \lambda_j \tilde{\alpha}_i + \\
&\quad - \frac{\tilde{\alpha}_i}{\alpha_j} \gamma_{jI}' z^i_I - \frac{\tilde{\alpha}_i}{\alpha_j} \gamma_{jII}' z^i_{II} \\
&= 2\frac{\tilde{\alpha}_i}{\alpha_j} \tilde{\beta}_{ij} + \mu_i + \lambda_j \tilde{\alpha}_i - \frac{\tilde{\alpha}_i}{\alpha_j} \hat{\gamma}_j' z^i.
\end{aligned} \tag{29.2}$$

The necessary and sufficient conditions for an optimal solution (y^*,T^*) of $(\tilde{P}^Q_{x,D})$ read then as follows:

$$0 = \frac{\partial\Psi}{\partial y} = \sum_{j=1}^{r} A^{j\,\prime}\hat{\gamma}_j + \sum_{k=1}^{\kappa} \rho_k \nabla g_k(y) \tag{30.1}$$

$$0 \le \frac{\partial\Psi}{\partial\tau_{ij}} = 2\,\frac{\tilde{\alpha}_i}{\alpha_j}\,\tilde{\beta}_{ij} + \mu_i + \lambda_j\,\tilde{\alpha}_i - \frac{\tilde{\alpha}_i}{\alpha_j}\,\hat{\gamma}_j{}' z^i,\ i\in S,\ j\in R \tag{30.2}$$

$$\tau_{ij} \ge 0,\ \tau_{ij}\,\frac{\partial\Psi}{\partial\tau_{ij}} = 0,\ i\in S,\ j\in R \tag{30.3}$$

$$A^j_I y - b^j_I \le \sum_{i\in S} \frac{\tilde{\alpha}_i\,\tau_{ij}}{\alpha_j}\, z^i_I,\ j\in R \tag{30.4}$$

$$0 = \gamma_{jI}{}'\,\frac{\partial\Psi}{\partial\gamma_{jI}} = \gamma_{jI}{}'\,(A^j_I y - b^j_I - \sum_{i\in S}\tilde{\beta}_{ij}\, z^i_I),\ j\in R \tag{30.5}$$

$$\gamma_{jI} \ge 0,\ j\in R \tag{30.6}$$

$$A^j_{II} y - b^j_{II} = \sum_{i\in S}\tilde{\beta}_{ij}\, z^i_{II},\ j\in R \tag{30.7}$$

$$g_k(y) \le 0,\ k=1,2,\ldots,\kappa \tag{30.8}$$

$$\rho_k\, g_k(y) = 0,\ \rho_k\ge 0,\ k=1,2,\ldots,\kappa. \tag{30.9}$$

Using the characterization of efficient solution of (1) given by Lemma 5.1, and having the necessary and sufficient optimality conditions (30.1)-(30.9) for $(\tilde{P}^Q_{x,D})$, we can characterize now efficient solutions of (1) as follows, see also the note to Lemma 5.1.

Theorem 5.1. Let D be given by (26), where it is supposed that $(\tilde{P}^Q_{x,D})$ fulfills the constraint qualification (26.1) for each $x\in D$. A point $x\in D$ is efficient if and only if there exists $y\in \mathbb{R}^n$, an $s\times r$ matrix $\mathbf{T}$, $s=|S_x|$, numbers μ_i, λ_j, ρ_k and vectors $\gamma_{jI}\in \mathbb{R}^{|J|}$, $\gamma_{jII}\in \mathbb{R}^{m-|J|}$ for $i\in S_x$, $j=1,2,\ldots,r$, $k=1,2,\ldots,\kappa$, such that the tuple $((y,\mathbf{T}), (\mu_i)_{i\in S}, (\lambda_j)_{j\in R}, (\gamma_{jI})_{j\in R}, (\gamma_{jII})_{j\in R}, \rho)$ satisfies

i) the optimality conditions (30.1)-(30.9)

ii) $A^j y = A^j x$ for all $j=1,2,\ldots,r$

iii) $\mathbf{T} = \mathbf{T}^o$.

Let the rm×n matrix $\mathbb{A}$ be defined by

$$\mathbb{A} = \begin{pmatrix} A^1 \\ A^2 \\ \vdots \\ A^r \end{pmatrix}.$$

The relations in Theorem 5.1 read then - explicitly - as follows:

$$A^j y = A^j x \text{ for all } j=1,2,\dots,r, \tag{31.1}$$

which simply means that

$$y = x, \tag{31.1)'}$$

provided that rank $\mathbb{A}$=n,

$$0 = \sum_{j=1}^{r} A^{j'} \hat{\gamma}_j + \sum_{k=1}^{\kappa} \rho_k \nabla g_k(y) \tag{31.2}$$

$$\gamma_{jI} \geq 0, \ j=1,2,\dots,r \tag{31.3}$$

$$g_k(x) \leq 0, \ k=1,2,\dots,\kappa \tag{31.4}$$

$$g_k(y) \leq 0, \ \rho_k g_k(y) = 0, \ \rho_k \geq 0, \ k=1,2,\dots,\kappa \tag{31.5}$$

$$2\frac{\tilde{\alpha}_i}{\alpha_j} + \mu_i + \lambda_j \tilde{\alpha}_i - \frac{\tilde{\alpha}_i}{\alpha_j} \hat{\gamma}_j{}' z^i = 0, \text{ if } z^i = z^j, \ i \in S_x, \ j \in R \tag{31.6}$$

$$\mu_i + \lambda_j \tilde{\alpha}_i - \frac{\tilde{\alpha}_i}{\alpha_j} \hat{\gamma}_j{}' z^i \geq 0, \text{ if } z^i \neq z^j, \ i \in S_x, \ j \in R \tag{31.7}$$

5.3. <u>Parametric representation of efficient solutions.</u> By elimination of μ_i in (31.6),(31.7), we get the equivalent conditions

$$\lambda_j - \lambda_i - \left(\frac{\hat{\gamma}_j}{\alpha_j} - \frac{\hat{\gamma}_i}{\alpha_i}\right)' z^i = 2\left(\frac{1}{\alpha_i} - \frac{1}{\alpha_j}\right), \text{ if } 1 \leq i,j \leq r, z^i = z^j \tag{31.6a}$$

$$\lambda_j - \lambda_i - \left(\frac{\hat{\gamma}_j}{\alpha_j} - \frac{\hat{\gamma}_i}{\alpha_i}\right)' z^i \geq \frac{2}{\alpha_i}, \text{ if } 1 \leq i,j \leq r, z^i \neq z^j. \tag{31.7a}$$

Thus, Theorem 5.1 yields the following **parametric representation** of $E_{D,J}$:

<u>Corollary 5.1.</u> Let D be given by (26), and assume that the constraint qualification (26.1) holds for every x∈D. An n-vector x is an efficient solution of (1) if and only if x satisfies the linear relations

$$\lambda_j - \lambda_i - \left(\frac{\hat{\gamma}_j}{\alpha_j} - \frac{\hat{\gamma}_i}{\alpha_i}\right)' z^i = 2\left(\frac{1}{\alpha_i} - \frac{1}{\alpha_j}\right), \text{ if } z^i = z^j \tag{32.1}$$

$$\lambda_j - \lambda_i - \left(\frac{\hat{\gamma}_j}{\alpha_j} - \frac{\hat{\gamma}_i}{\alpha_i}\right)' z^i \geq \frac{2}{\alpha_i}, \text{ if } z^i \neq z^j, \tag{32.2}$$

where $\lambda_1, \lambda_2, \ldots, \lambda_r$ are arbitrary real parameters, and the parameter m-vectors $\gamma_1, \gamma_2, \ldots, \gamma_r$ and further parameter vectors $\rho \in \mathbb{R}^\kappa$, $y \in \mathbb{R}^n$ must be selected such that

$$\sum_{j=1}^{r} A^{j'} \hat{\gamma}_j + \sum_{k=1}^{\kappa} \rho_k \nabla g_k(y) = 0 \tag{32.3}$$

$$\gamma_{jI} \geq 0, \ j=1,2,\ldots,r \tag{32.4}$$

$$g_k(x) \leq 0, \ k=1,2,\ldots,\kappa \tag{32.5}$$

$$g_k(y) \leq 0, \ \rho_k g_k(y) = 0, \ \rho_k \geq 0, \ k=1,2,\ldots,\kappa \tag{32.6}$$

$$A^j y = A^j x, \ j=1,2,\ldots,r. \tag{32.7}$$

References

[1] Appa, G., Smith, C.: On L_1- and L_∞- estimation. Math. Programming 5, 73-87 (1973)

[2] Bloomfield. P., Steiger. W.L.: Least Absolute Deviations. Boston-Basel-Stuttgart, Birkhäuser 1983

[3] Davies, M.: Linear approximation using the criterion of least total deviation. J. Roy. Statist. Soc. Ser. B, 29, 101-109 (1967)

[4] Dielman, T., Pfaffenberger, R.: LAV Estimation in Linear Regression: A Review. In [27], pp. 31-52

[5] Dodge, Y. (ed.): Statistical Data Analysis Based on the L_1-norm and Related Methods. Amsterdam-New York-Oxford-Tokyo, North Holland 1987

[6] Ellis, J.H., McBean, E.A., Farquhar, G.J.: Chance-constrained/ stochastic linear programming model for acid rain abatement-I. Complete colinearity and noncolinearity. Atmospheric Environment 19, 925-937 (1985)

[7] Ellis, J.H., McBean, E.A., Farquhar, G.J.: Chance-Constrained/ Stochastic Linear Programming Model for Acid Rain Abatement-II. Limited Colinearity. Atmospheric Environment 20, 501-511 (1986)

[8] Ermoliev, Yu., Wets, R.J-B (Eds.): Numerical Techniques for Stochastic Optimization. Springer Series in Computational Mathematics, Vol. 10. Berlin-Heidelberg-New York, Springer 1988

[9] Everitt, B.S.: Introduction to Optimization Methods and their Application in Statistics. London-New York, Chapman and Hall 1987

[10] Fortin, M., McBean, E.A.: A Management Model for Acid Rain Abatement. Atmospheric Environment 17(11), 2331-2336 (1983)

[11] Fronza, G., Melli, P.: Assignment of emission abatement levels by stochastic programming. Atmospheric Environment 18, 531-535 (1984)

[12] Fuessle, R.W., Brill, E.D., Liebman, J.C.: Air Quality Planning: A General Chance-Constrained Model. J. Envir. Engrg., ASCE, 113(1), 106-123 (1987)

[13] Gentl, J.E., Kennedy, W.J., Sposito, V.A.: On least absolute value estimation. Communications in Statistics, Vol.6, 1977

[14] Guldman, J.-M.: Interactions between Weather Stochasticity and the Locations of Pollution Sources and Receptors in Air Quality Planning: A Chance-Constrained Approach. Geographical Analysis 18, 198-214 (1986)

[15] Marti, K.: Descent Directions and Efficient Solutions in Discretely Distributed Stochastic Programs. LN in Economics and Mathematical Systems, Vol. 299. Berlin-Heidelberg-New York, Springer 1988

[16] Narula, S.C.: Optimization Techniques in Linear Regression. In [27], pp. 11-29

[17] Reibnitz, U. von: Scenario Techniques. Hamburg, McGraw-Hill Company, GmbH, 1988

[18] Rockafellar, R.T., Wets, R.J-B: Scenarios And Policy Aggregation In Optimization Under Uncertainty. IIASA Working Paper WP-87-119, IIASA Laxenburg, Austria, December 1987

[19] Rotach, M., et.al.: Szenarien künftiger Entwicklungen. Institut für Verkehrsplanung und Transporttechnik der ETH Zürich, November 1982

[20] Sherman, S.: On a theorem of Hardy, Littlwood, Polya and Blackwell. Proc. Nat. Acad. Sci. USA 37, 826-831 (1951)

[21] Sinitsyn, I.N.: Statistical Linearization. Automation and Remote Control 35, 765-776 (1974)

[22] Steinsiek, E., Knauer, P.: Szenarien als Instrument der Umweltplanung. Angewandte Systemanalyse 2, 10-19 (1981)

[23] Turner, D.B.: Atmospheric Dispersion Modeling: A Critical Review. J. of the Air Pollution Control Association 29, 502-519 (1979)

[24] Watson, G.A.: Approximation Theory and Numerical Methods. New York, J. Wiley 1980

[25] Wets, R.: The Aggregation Principle in Scenario Analysis and Stochastic Optimization. Research supported in part by grants from the National Science Foundation and the Air Force Office of Scientific Research 1988

[26] Wets, R.: Stochastic Programming. In: Nemhauser et.al. (Eds.): Optimization Vol.1, pp. 573-623. Amsterdam, North Holland 1989

[27] Zanakis, S.H., Rustagi, J.S. (Eds.): Optimization in Statistics. TIMS Studies in the Management Sciences, Vol. 19, Amsterdam, North Holland 1982

[28] Zanakis, S.H., Rustagi, J.S.: Introduction to contributions in Regression and Correlation. In [27], pp. 7-9

[29] Zentner, R.D.: Scenarios: A New Tool for Corporate Planners. Chemical and Engineering News, Int. Edition, vol. 53, October 1975, pp. 22-34

Part III.
THE FUZZY APPROACH

Part III
Chapter 1

INTERACTIVE DECISION MAKING FOR MULTIOBJECTIVE PROGRAMMING PROBLEMS WITH FUZZY PARAMETERS

MASATOSHI SAKAWA
Department of Computer Science, Faculty of Engineering, Iwate University, Ueda, Morioka 020 Japan

HITOSHI YANO
Nagoya Municipal Women's College, Nagoya 464 Japan

ABSTRACT. This paper presents interactive decision making methods for multiobjective linear, linear fractional and nonlinear programming problems with fuzzy parameters. On the basis of the α-level sets of the fuzzy numbers, the concept of α-multiobjective programming and (local) M-α-Pareto optimality is introduced. Through the interaction with the decision maker (DM), the fuzzy goals of the DM for each of the objective functions in α-multiobjective programming are quantified by eliciting the corresponding membership functions. After determining the membership functions, in order to generate a candidate for the (local) satisficing solution which is also (local) M-α-Pareto optimal, if the DM specifies the degree α of the α-level sets and the reference membership values, the (augmented) minimax problem is solved and the DM is supplied with the corresponding (local) M-α-Pareto optimal solution together with the trade-off rates among the values of the membership functions and the degree α. Then by considering the current values of the membership functions and α as well as the trade-off rates, the DM responds by updating his/her reference membership values and/or the degree α. In this way the (local) satisficing solution for the DM can be derived efficiently from among an M-α-Pareto optimal solution set. Based on the proposed methods for multiobjective linear, linear fractional and nonlinear programming problems with fuzzy parameters, interactive computer programs are developed and an illustrative numerical example for nonlinear case is demonstrated.

1. INTRODUCTION

In multiobjective programming problems, multiple objectives are usually noncom-

R. Slowinski and J. Teghem (eds.),
***Stochastic versus Fuzzy Approaches to Multiobjective Mathematical Programming under Uncertainty*, 191–228.**

mensurable and cannot be combined into a single objective. Moreover, the objectives often conflict with each other in that any improvement of one objective function can be achieved only at the expense of another. Accordingly, the aim is to find the compromise or satisficing solution of the decision maker (DM) which is also Pareto optimal based on his/her subjective value-judgement (e.g.,[3,29,37] etc.). However, when formulating the multiobjective programming problem which closely describes and represents the real decision situation, various factors of the real system should be reflected in the description of the objective functions and the constraints. Naturally these objective functions and constraints involve many parameters whose possible values may be assigned by the experts. In the conventional approach, such parameters are fixed at some values in an experimental and/or subjective manner through the experts' understanding of the nature of the parameters.

In most real-world situations, however, it may be reasonable to assume that the possible values of these parameters are often only imprecisely or ambiguously known to the experts. In this case, it would certainly be more appropriate to interpret the experts' understanding of the parameters as fuzzy numerical data which can be represented by means of fuzzy subsets of the real line known as fuzzy numbers [5]. The resulting multiobjective programming problem involving fuzzy parameters would be viewed as the more realistic version of the conventional one.

Recently, Tanaka and Asai [31,32] formulated the multiobjective linear programming problems with fuzzy parameters. Following the fuzzy decision or the minimum operator proposed by Bellman and Zadeh [1] together with triangular membership functions for fuzzy parameters, they considered two types of fuzzy multiobjective linear programming problems; one is to decide the nonfuzzy solution and the other is to decide the fuzzy solution.

More recently, Orlovski [14] formulated general multiobjective nonlinear programming problems with fuzzy parameters. He presented two approaches to the formulated problems by making systematic use of the extension principle of Zadeh [36] and demonstrated that there exist in some sense equivalent nonfuzzy formulations.

In this paper, for the multiobjective linear, linear fractional and nonlinear programming problems with fuzzy parameters characterized by fuzzy numbers, the concept of α-multiobjective programming and (local) M-α-Pareto optimality are introduced on the basis of the α-level sets of the fuzzy numbers. Then by assuming that the fuzzy goals of the DM for each of the objective functions in α-multiobjective programming can be quantified by eliciting the corresponding membership functions, interactive decision making methods for multiobjective linear, linear fractional and nonlinear programming problems with fuzzy parameters to derive the (local) satisficing solution of the DM efficiently from among an (local) M-α-Pareto optimal solution set are presented following our recent research results [15-28,34,35].

2. INTERACTIVE DECISION MAKING FOR MULTIOBJECTIVE LINEAR PROGRAMMING PROBLEMS WITH FUZZY PARAMETERS

2.1. Problem statement and solution concept

Consider multiobjective linear programming (MOLP) problems of the following form:

$$\begin{array}{ll} \min & (c_1 x, c_2 x, \cdots, c_k x) \\ \text{subject to} & x \in X = \{x \in E^n \mid a_j x \leq b_j, j = 1, \ldots, m, x \geq 0\}, \end{array} \tag{2.1}$$

where x is an n-dimensional column vector of decision variables, $c_1, \ldots, c_k$ are n-dimensional cost factor row vectors, $a_1, \ldots, a_m$ are n-dimensional constraint row vectors and $b_1, \ldots, b_m$ are constants.

Fundamental to the MOLP is the Pareto optimal concept, also known as a noninferior solution. Qualitatively, a Pareto optimal solution of the MOLP is one where any improvement of one objective function can be achieved only at the expense of another.

Definition 2.1 (Pareto optimal solution)
$x^* \in X$ is said to be a Pareto optimal solution to the MOLP, if and only if there does not exist another $x \in X$ such that $c_i x \leq c_i x^*, i = 1, \ldots, k$ with strict inequality holding for at least one i.

In practice, however, it would certainly be more appropriate to consider that the possible values of the parameters in the description of the objective functions and the constraints usually involve the ambiguity of the experts' understanding of the real system. For this reason, in this section, we consider the following multiobjective linear programming problem involving fuzzy parameters (MOLP-FP) [20,25] :

$$\begin{array}{ll} \min & (\tilde{c}_1 x, \tilde{c}_2 x, \ldots, \tilde{c}_k x) \\ \text{subject to} & x \in X(\tilde{a}, \tilde{b}) = \{x \in E^n \mid \tilde{a}_j x \leq \tilde{b}_j, j = 1, \ldots, m; x \geq 0\}. \end{array} \tag{2.2}$$

Here $\tilde{c}_i = (\tilde{c}_{i1}, \ldots, \tilde{c}_{in})$, and $\tilde{a}_j = (\tilde{a}_{j1}, \ldots, \tilde{a}_{jn}), \tilde{b}_j$ represent respectively fuzzy parameters involved in the objective function $\tilde{c}_i x$ and the constraint $\tilde{a}_j x \leq \tilde{b}_j$.

We now assume that $\tilde{c}_{i1}, \ldots, \tilde{c}_{in}, \tilde{a}_{j1}, \ldots, \tilde{a}_{jn}$ and $\tilde{b}_j$ in the MOLP-FP are fuzzy numbers whose membership functions are denoted by $\mu_{\tilde{c}_{i1}}(c_{i1}), \ldots, \mu_{\tilde{c}_{in}}(c_{in})$, $\mu_{\tilde{a}_{j1}}(a_{j1}), \ldots, \mu_{\tilde{a}_{jn}}(a_{jn})$, and $\mu_{\tilde{b}_j}(b_j)$ respectively.

For simplicity in the notation, define the following vectors :

$$\begin{array}{llll} c = (c_1, \ldots, c_k), & \tilde{c} = (\tilde{c}_1, \ldots, \tilde{c}_k), & a = (a_1, \ldots, a_m), & \tilde{a} = (\tilde{a}_1, \ldots, \tilde{a}_m), \\ b = (b_1, \ldots, b_m), & \tilde{b} = (\tilde{b}_1, \ldots, \tilde{b}_m). & & \end{array}$$

Then we can introduce the following α-level set or α-cut [5] of the fuzzy numbers $\tilde{a}_{jr}, \tilde{b}_j$ and $\tilde{c}_{ir}$.

Definition 2.2 (α-level set)
The α-level set of the fuzzy numbers $\tilde{a}_{jr}, \tilde{b}_j$ and $\tilde{c}_{ir}$ is defined as the ordinary set $L_\alpha(\tilde{a}, \tilde{b}, \tilde{c})$ for which the degree of their membership functions exceeds the level $\alpha \in [0,1]$:

$$L_\alpha(\tilde{a}, \tilde{b}, \tilde{c}) = \{(a,b,c) \mid \mu_{\tilde{a}_{jr}}(a_{jr}) \geq \alpha, \mu_{\tilde{b}_j}(b_j) \geq \alpha, \mu_{\tilde{c}_{ir}}(c_{ir}) \geq \alpha, \\ i = 1, \ldots, k, j = 1, \ldots, m, r = 1, \ldots, n\}. \tag{2.3}$$

For a certain degree α, the MOLP-FP (2.2) can be understood as the following nonfuzzy α-multiobjective linear programming (α-MOLP) problem [20,25].

$$\begin{array}{ll} \min & (c_1 x, c_2 x, \ldots, c_k x) \\ \text{subject to} & x \in X(a,b) = \{x \in E^n | a_j x \leq b_j, j = 1, \ldots, m; x \geq 0\}, \\ & (a,b,c) \in L_\alpha(\tilde{a}, \tilde{b}, \tilde{c}). \end{array} \tag{2.4}$$

It should be emphasized here that in the α-MOLP the parameters (a,b,c) are treated as decision variables rather than constants.

On the basis of the α-level sets of the fuzzy numbers, we introduce the concept of α-Pareto optimal solutions to the α-MOLP [20,25].

Definition 2.3 (α-Pareto optimal solution)
$x^* \in X(a^*, b^*)$ is said to be an α-Pareto optimal solution to the α-MOLP, if and only if there does not exist another $x \in X(a,b), (a,b,c) \in L_\alpha(\tilde{a}, \tilde{b}, \tilde{c})$ such that $c_i x \leq c_i^* x^*, i = 1, \ldots, k$, with strict inequality holding for at least one i, where the corresponding values of parameters (a^*, b^*, c^*) are called α-level optimal parameters.

As can be immediately understood from Definition 2.3, in general, α-Pareto optimal solutions to the α-MOLP consist of an infinite number of points and some kinds of subjective judgement should be added to the quantitative analyses by the decision maker (DM). Namely, the DM must select his/her compromise or satisficing solution from among α-Pareto optimal solutions based on his/her subjective judgement.

However, considering the imprecise nature of the DM's judgement, it is natural to assume that the DM may have imprecise or fuzzy goals for each of the objective functions in the α-MOLP. In a minimization problem, a goal stated by the DM may be to achieve "substantially less than some value A". This type of statement can be quantified by eliciting a corresponding membership function.

In order to elicit a membership function $\mu_i(c_i x)$ from the DM for each of the objective functions $c_i x, i = 1, \ldots, k$, in the α-MOLP, we first calculate the individual minimum and maximum of each objective function under the given constraints for $\alpha = 0$ and $\alpha = 1$. By taking account of the calculated individual minimum and maximum of each objective function for $\alpha = 0$ and $\alpha = 1$ together with the rate of increase of membership of satisfaction, the DM may be able to determine his/her membership function $\mu_i(c_i x)$ in a subjective manner which is a strictly monotone decreasing function with respect to $c_i x$.

So far we have restricted ourselves to a minimization problem and consequently assumed that the DM has a fuzzy goal such as "$c_i x$ should be substantially less than A_i". In the fuzzy approaches, we can further treat a more general case where the DM has two types of fuzzy goals, namely fuzzy goals expressed in words such as "$c_i x$ should be in the vicinity of C_i " (called fuzzy equal) as well as "$c_i x$ should be substantially less than A_i or greater than B_i " (called fuzzy min or fuzzy max). Such a generalized α-MOLP (Gα-MOLP) problem may now be expressed as :

$$\begin{array}{lll} \text{fuzzy min} & c_i x & (i \in I_1), \\ \text{fuzzy max} & c_i x & (i \in I_2), \\ \text{fuzzy equal} & c_i x & (i \in I_3), \\ \text{subject to} & & x \in X(a,b), \ (a,b,c), \in L_\alpha(\tilde{a},\tilde{b},\tilde{c}), \end{array} \tag{2.5}$$

where $I_1 \cup I_2 \cup I_3 = \{1, 2, \ldots, k\}, I_i \cap I_j = \phi, \ i, j = 1, 2, 3, i \neq j$.

In order to elicit a membership function $\mu_i(c_i x)$ from the DM for a fuzzy goal like "$c_i x$ should be in the vicinity of C_i ", it is obvious that we can use different functions to the left and right sides of C_i.

Concerning the membership functions of the Gα-MOLP, it is reasonable to assume that $\mu_i(c_i x), i \in I_1$ and the right side functions of $\mu_i(c_i x), i \in I_3$ are strictly monotone decreasing and continuous functions with respect to $c_i x$, and $\mu_i(c_i x), i \in I_2$ and the left side functions of $\mu_i(c_i x), i \in I_3$ are strictly monotone increasing and continuous functions with respect to $c_i x$. To be more explicit, each membership function $\mu_i(c_i x)$ of the Gα-MOLP for $i \in I_1, i \in I_2$ or $i \in I_3$ is defined as follows :

(1) $i \in I_1$:

$$\mu_i(c_i x) = \begin{cases} 1 \quad \text{or} \to 1 & \text{if } (c_i x)_R^1 \geq c_i x, \\ D_{iR}(c_i x) & \text{if } (c_i x)_R^1 < c_i x < (c_i x)_R^0, \\ 0 \quad \text{or} \to 0 & \text{if } c_i x \geq (c_i x)_R^0, \end{cases} \tag{2.6}$$

(2) $i \in I_2$:

$$\mu_i(c_i x) = \begin{cases} 0 \quad \text{or} \to 0 & \text{if } (c_i x)_L^0 \geq c_i x, \\ D_{iL}(c_i x) & \text{if } (c_i x)_L^0 < c_i x < (c_i x)_L^1, \\ 1 \quad \text{or} \to 1 & \text{if } c_i x \geq (c_i x)_L^1, \end{cases} \tag{2.7}$$

(3) $i \in I_3$:

$$\mu_i(c_i x) = \begin{cases} 0 \quad \text{or} \to 0 & \text{if} \quad c_i x \leq (c_i x)_L^0, \\ D_{iL}(c_i x) & \text{if} \quad (c_i x)_L^0 \leq c_i x \leq (c_i x)_L^1, \\ 1 & \text{if} \quad (c_i x)_L^1 \leq c_i x \leq (c_i x)_R^1, \\ D_{iR}(c_i x) & \text{if} \quad (c_i x)_R^1 \leq c_i x \leq (c_i x)_R^0, \\ 0 \quad \text{or} \to 0 & \text{if} \quad (c_i x)_R^0 \leq c_i x. \end{cases} \tag{2.8}$$

Here it is assumed that $D_{iR}(c_i x)$ or $D_{iL}(c_i x)$ is respectively a strictly monotone decreasing or increasing and continuous function with respect to $c_i x$ and may be linear or nonlinear, and $(c_i x)_L^0$ and $(c_i x)_R^0$ are unacceptable levels for $c_i x$ and $(c_i x)_L^1$ and $(c_i x)_R^1$ are totally desirable levels for $c_i x$. When fuzzy equal is included in the fuzzy goals of the DM, it is desirable that $c_i x$ should be as close to C_i as possible. Consequently, the notion of α-Pareto optimal solutions defined in terms of objective functions cannot be applied. For this reason, we introduce the concept of M-α-Pareto optimal solutions [18] which is defined in terms of membership functions instead of objective functions, where M refers to membership.

Definition 2.4 (M-α-Pareto optimal solution)
$x^* \in X(a^*, b^*)$ is said to be an M-α-Pareto optimal solution to the Gα-MOLP, if and only if there does not exist another $x \in X(a, b), (a, b, c) \in L_\alpha(\tilde{a}, \tilde{b}, \tilde{c})$ such that $\mu_i(c_i x) \geq \mu(c_i^* x^*), i = 1, \ldots, k$, with strict inequality holding for at least one i, where the corresponding values of parameters (a^*, b^*, c^*) are called α-level optimal parameters.

Observe that the concept of M-α-Pareto optimal solutions defined in terms of membership functions is a natural extension of that of α-Pareto optimal solutions defined in terms of objective functions, when fuzzy equal is included in the fuzzy goals of the DM.

Having elicited the membership functions $\mu_i(c_i x), i = 1, ..., k$ from the DM for each of the objective functions $c_i x, i = 1, \ldots, k$, if we introduce a general aggregation function

$$\mu_D(\mu_1(c_1 x), \mu_2(c_2 x), \ldots, \mu_k(c_k x), \alpha), \tag{2.9}$$

a general fuzzy α-multiobjective decision problem (Fα-MODP) can be defined by:

$$\begin{array}{ll} \max & \mu_D(\mu_1(c_1 x), \mu_2(c_2 x), \ldots, \mu_k(c_k x), \alpha), \\ \text{subject to} & (x, a, b, c) \in P(\alpha), \quad \alpha \in [0, 1], \end{array} \tag{2.10}$$

where $P(\alpha)$ is the set of M-α-Pareto optimal solutions and the corresponding α-level optimal parameters to the Gα-MOLP.

Probably the most crucial problem in the Fα-MODP is the identification of an appropriate aggregation function which well represents the human decision makers' fuzzy preferences. If $\mu_D(\cdot)$ can be explicitly identified, then the Fα-MODP reduces to a standard mathematical programming problem. However, this rarely happens and as an alternative, it becomes evident that an interaction with the DM is necessary.

Throughout this section, we make the following assumptions.

Assumption 2.1
The fuzzy goals of the DM can be quantified by eliciting the corresponding membership functions through the interaction with the DM.
Assumption 2.2
$\mu_D(\cdot)$ exists and is known only implicitly to the DM, which means the DM cannot specify the entire form of $\mu_D(\cdot)$, but he/she can provide local information concerning his/her preference. Moreover, it is strictly increasing and continuous with respect to $\mu_i(\cdot)$ and α.

2.2. Minimax Problems for MOLP-FP

For generating a candidate for the satisficing solution which is also M-α-Pareto optimal, in our decision making method, the DM is asked to specify the degree α of the α-level set and the reference levels of achievement of the membership functions, called reference membership values. Observe that the idea of the reference membership values first appeared in Sakawa and Yano [15] can be viewed as an obvious extension of the idea of the reference point of Wierzbicki [33]. Once the DM's degree α and reference membership values $\bar{\mu}_i, i = 1, \ldots, k$, are specified, the corresponding M-α-Pareto optimal solution, which is in a sense close to his/her requirement or better than that if the reference levels are attainable, is obtained by solving the following minimax problem.

$$\min_{\substack{x \in X(a,b) \\ (a,b,c) \in L_\alpha(\tilde{a},\tilde{b},\tilde{c})}} \max_{1 \le i \le k} (\bar{\mu}_i - \mu_i(c_i x)), \tag{2.11}$$

or equivalently:

$$\begin{aligned} \min \quad & v \\ \text{subject to} \quad & \bar{\mu}_i - \mu_i(c_i x) \le v, i = 1, \ldots, k, \\ & a_j x \le b_j, j = 1, \ldots, m, \quad x \ge 0, \quad (a,b,c) \in L_\alpha(\tilde{a},\tilde{b},\tilde{c}). \end{aligned} \tag{2.12}$$

However, with the strictly monotone decreasing or increasing membership function given by (2.6)-(2.8), which may be nonlinear, the resulting problem becomes a nonlinear programming problem. In order to solve the formulated problem on the basis of the linear programming method, we first convert each constraint $\bar{\mu}_i - \mu_i(c_i x) \le v, i = 1, \ldots, k$, of the minimax problem (2.12) into the following form using the strictly monotone property of $D_{iL}(\cdot)$ and $D_{iR}(\cdot)$.

$$c_i x \le D_{iR}^{-1}(\bar{\mu}_i - v) \qquad i \in I_1 \cup I_3. \tag{2.13}$$

$$c_i x \ge D_{iL}^{-1}(\bar{\mu}_i - v) \qquad i \in I_2 \cup I_3. \tag{2.14}$$

Now we introduce the following set-valued functions $S_{iR}(\cdot)$, $S_{iL}(\cdot)$ and $T_j(\cdot,\cdot)$.

$$S_{iR}(c_i) = \{(x, v) \mid c_i x \le D_{iR}^{-1}(\bar{\mu}_i - v)\} \qquad i \in I_1 \cup I_3. \tag{2.15}$$
$$S_{iL}(c_i) = \{(x, v) \mid c_i x \ge D_{iL}^{-1}(\bar{\mu}_i - v)\} \qquad i \in I_2 \cup I_3. \tag{2.16}$$
$$T_j(a_j, b_j) = \{x \mid a_j x \le b_j\} \qquad j = 1, \ldots, m. \tag{2.17}$$

Then, it can be verified that the following relations hold for $S_{iR}(\cdot)$, $S_{iL}(\cdot)$ and $T_j(\cdot,\cdot)$, when $x \ge 0$.

Proposition 2.1
(1) If $c_i^1 \le c_i^2$, then $S_{iR}(c_i^1) \supset S_{iR}(c_i^2)$ and $S_{iL}(c_i^1) \subset S_{iL}(c_i^2)$.
(2) If $a_j^1 \le a_j^2$, then $T_j(a_j^1, b_j) \supset T_j(a_j^2, b_j)$.
(3) If $b_j^1 \le b_j^2$, then $T_j(a_j, b_j^1) \subset T_j(a_j, b_j^2)$.

Now from the properties of the α-level sets for the vectors of the fuzzy numbers $\tilde{c}_i$, $\tilde{a}_j$ and the fuzzy numbers $\tilde{b}_j$, it should be noted here that the feasible regions for c_i, a_j and b_j can be denoted respectively by the closed intervals $[c_{i\alpha}^L, c_{i\alpha}^R]$, $[a_{j\alpha}^L, a_{j\alpha}^R]$ and $[b_{j\alpha}^L, b_{j\alpha}^R]$.

Consequently, by making use of the results in Proposition 2.1, we can obtain an optimal solution to (2.12) by solving the following problem.

$$\begin{array}{ll} \min & v \\ \text{subject to} & c_{i\alpha}^L x \le D_{iR}^{-1}(\bar{\mu}_i - v) \qquad i \in I_1 \cup I_3, \\ & c_{i\alpha}^R x \ge D_{iL}^{-1}(\bar{\mu}_i - v) \qquad i \in I_2 \cup I_3, \\ & a_{j\alpha}^L x \le b_{j\alpha}^R, j = 1, \ldots, m, \qquad x \ge 0. \end{array} \tag{2.18}$$

It is important to note here that in this formulation, if the value of v is fixed, it can be reduced to a set of linear inequalities. Obtaining the optimal solution v^* to the above problem is equivalent to determining the minimum value of v so that there exists an admissible set satisfying the constraints of (2.18). Since v satisfies $\bar{\mu}_{max} - 1 \le v \le \bar{\mu}_{max}$, where $\bar{\mu}_{max}$ denotes the maximum value of $\bar{\mu}_i, i = 1, \ldots, k$, we have the following method for solving this problem by combined use of the bisection method and the simplex method of linear programming.

Step 1: Set $v = \bar{\mu}_{max}$ and test whether an admissible set satisfying the constraints of (2.18) exists or not by making use of the phase one of the simplex method. If an admissible set exists, proceed. Otherwise, the DM must reassess his/her membership functions.

Step 2: Set $v = \bar{\mu}_{max} - 1$ and test whether an admissible set satisfying the constraints of (2.18) exists or not using the phase one of the simplex method. If an admissible set exists, set $v^* = \bar{\mu}_{max} - 1$. Otherwise go to the next step, since the minimum v which satisfies the constraints of (2.18) exists between $\bar{\mu}_{max} - 1$ and $\bar{\mu}_{max}$.

Step 3: For the initial value of $v = \bar{\mu}_{max} - 0.5$, update the value of v using the bisection method as follows :

(a) $v_{n+1} = v_n - 1/2^{n+1}$ if admissible set exists for v_n,

(b) $v_{n+1} = v_n + 1/2^{n+1}$ if no admissible set exists for v_n.

Namely, for each $v_n (n = 1, 2, \ldots)$, test whether an admissible set of (2.18) exists or not using the sensitivity analysis technique for the changes in the right hand side of the simplex method and determine the minimum value of v satisfying the constraints of (2.18).

In this way, we can determine the optimal solution v^*. Then the DM selects an appropriate standing objective from among the objectives $c_i x, i = 1, \ldots, k$. For notational convenience, in the following, without loss of generality, let it be $c_1 x$ and $1 \in I_1$. Then the following linear programming problem is solved for $v = v^*$.

$$\begin{array}{lll} \min & c_{1\alpha}^L x & \\ \text{subject to} & c_{i\alpha}^L x \leq D_{iR}^{-1}(\bar{\mu}_i - v^*) & i \in I_1 \cup I_3, \\ & c_{i\alpha}^R x \geq D_{iL}^{-1}(\bar{\mu}_i - v^*) & i \in I_2 \cup I_3, \\ & a_{j\alpha}^L x \leq b_{j\alpha}^R, j = 1, \ldots, m, & x \geq 0. \end{array} \tag{2.19}$$

For convenience in our subsequent discussion, we assume that the optimal solution x^* to (2.19) satisfies the following conditions:

$$\begin{array}{ll} c_{i\alpha}^L x^* = D_{iR}^{-1}(\bar{\mu}_i - v^*) & i \in I_1 \cup I_{3R}, \\ c_{i\alpha}^R x^* = D_{iL}^{-1}(\bar{\mu}_i - v^*) & i \in I_2 \cup I_{3L}, \end{array}$$

where $I_3 = I_{3L} \cup I_{3R}$ and $I_{3L} \cap I_{3R} = \phi$.

Then it is interesting to note that $c_{i\alpha}^L (i \in I_1 \cup I_{3R}), c_{i\alpha}^R (i \in I_2 \cup I_{3L}), a_{j\alpha}^L$ and $b_{j\alpha}^R, j = 1, \ldots, m$ are α-level optimal parameters for any M-α-Pareto optimal solution.

The relationships between the optimal solutions to (2.18) and the M-α-Pareto optimal concept of the Gα-MOLP can be characterized by the following theorems.

Theorem 2.1

(1) If x^* is a unique optimal solution to (2.18), then x^* is an M-α-Pareto optimal solution to the Gα-MOLP.

(2) If x^* is an M-α-Pareto optimal solution to the Gα-MOLP, then x^* is an optimal solution to (2.18) for some $\bar{\mu} = (\bar{\mu}_1 \ldots, \bar{\mu}_k)$.

It must be observed here that for generating M-α-Pareto optimal solutions using Theorem 2.1, uniqueness of solution must be verified. In the ad hoc numerical approach, however, in order to test the M-α-Pareto optimality of a current optimal solution x^*, we formulate and solve the following linear programming problem :

$$
\begin{array}{llllll}
\max & \sum_{i=1}^{k} \varepsilon_i & & & & \\
\text{subject to} & c_{i\alpha}^L x + \varepsilon_i = c_{i\alpha}^L x^*, & \varepsilon_i \geq 0, & i \in I_1 \cup I_{3R}, & & (2.20) \\
 & c_{i\alpha}^R x - \varepsilon_i = c_{i\alpha}^R x^*, & \varepsilon_i \geq 0, & i \in I_2 \cup I_{3L}, & & \\
 & a_{j\alpha}^L x \leq b_{j\alpha}^R, j = 1, \ldots, m, & x \geq 0. & & &
\end{array}
$$

Let $\bar{x}$ and $\bar{\varepsilon}$ be an optimal solution to (2.20). If all $\bar{\varepsilon}_i = 0$, then x^* is an M-α-Pareto optimal solution. If at least one $\bar{\varepsilon}_i > 0$, it can be easily shown that $\bar{x}$ is an M-α-Pareto optimal solution.

2.3. Interactive algorithm for MOLP-FP

Now given the M-α-Pareto optimal solution for the degree α and the reference membership values specified by the DM by solving the corresponding minimax problem, the DM must either be satisfied with the current M-α-Pareto optimal solution and α or update his/her reference membership values and/or the degree α. In order to help the DM express his/her degree of preference, trade-off information between a standing membership function and each of the other membership functions as well as between the degree α and the membership functions is very useful. Such a trade-off information is easily obtainable since it is closely related to the simplex multipliers of the problem (2.19).

To derive the trade-off information, define the following Lagrangian function L corresponding to the problem (2.19).

$$
\begin{aligned}
L = c_{1\alpha}^L x + & \sum_{i \in I_1 \cup I_{3R}} \Pi_{iR} \{ c_{i\alpha}^L x - D_{iR}^{-1}(\bar{\mu}_i - v^*) \} \\
+ & \sum_{i \in I_2 \cup I_{3L}} \Pi_{iL} \{ D_{iL}^{-1}(\bar{\mu}_i - v^*) - c_{i\alpha}^R x \} + \sum_{j=1}^{m} \lambda_j (a_{j\alpha}^L x - b_{j\alpha}^R) \qquad (2.21)
\end{aligned}
$$

where Π_{iL}, Π_{iR} and λ_j are simplex multipliers corresponding to the constraints of (2.19).

Here, we assume that the problem (2.19) has a unique and nondegenerate optimal solution satisfying the following conditions :

(1) $\Pi_{iR} > 0 \quad i \in I_1 \cup I_{3R}, i \neq 1,$
(2) $\Pi_{iL} > 0 \quad i \in I_2 \cup I_{3L}.$

Then, by using the results in Haimes and Chankong [7], the following expression holds.

$$-\frac{\partial(c_{1\alpha}^L x)}{\partial(c_{i\alpha}^L x)} = \Pi_{iR} \qquad i \in I_1 \cup I_{3R}, i \neq 1, \tag{2.22}$$

$$-\frac{\partial(c_{1\alpha}^L x)}{\partial(c_{i\alpha}^R x)} = -\Pi_{iL} \qquad i \in I_2 \cup I_{3L}. \tag{2.23}$$

Furthermore, using the strictly monotone decreasing or increasing property of $D_{iR}(\cdot)$ or $D_{iL}(\cdot)$ together with the chain rule, if $D_{iR}(\cdot)$ and $D_{iL}(\cdot)$ are differentiable at the optimal solution to (2.19), it holds that

$$-\frac{\partial\mu_1(c_{1\alpha}^L x)}{\partial\mu_i(c_{i\alpha}^L x)} = \frac{D_{1R}'(c_{1\alpha}^L x)}{D_{iR}'(c_{i\alpha}^L x)}\Pi_{iR} \qquad i \in I_1 \cup I_{3R}, i \neq 1, \tag{2.24}$$

$$-\frac{\partial\mu_1(c_{1\alpha}^L x)}{\partial\mu_i(c_{i\alpha}^R x)} = \frac{D_{1R}'(c_{1\alpha}^L x)}{D_{iL}'(c_{i\alpha}^R x)}\Pi_{iL} \qquad i \in I_2 \cup I_{3L}, \tag{2.25}$$

where $D_{iR}'(\cdot)$ and $D_{iL}'(\cdot)$ denote the differential coefficients of $D_{iR}(\cdot)$ and $D_{iL}(\cdot)$ respectively.

Regarding a trade-off rate between $\mu_1(c_{1\alpha}^L x)$ and α, the following relation holds based on the sensitivity theorem [6].

$$\frac{\partial\mu_1(c_{1\alpha}^L x)}{\partial\alpha} = D_{1R}'(c_{1\alpha}^L x)\left\{\frac{\partial(c_{1\alpha}^L)}{\partial\alpha}x + \sum_{i\in I_1\cup I_{3R}} \Pi_{iR}\frac{\partial(c_{i\alpha}^L)}{\partial\alpha}x - \sum_{i\in I_2\cup I_{3L}} \Pi_{iL}\frac{\partial(c_{i\alpha}^R)}{\partial\alpha}x \right.$$
$$\left. + \sum_{j=1}^{m} \lambda_j\left\{\frac{\partial(a_{j\alpha}^L)}{\partial\alpha}x - \frac{\partial(b_{j\alpha}^R)}{\partial\alpha}\right\}\right\} \tag{2.26}$$

It should be noted here that in order to obtain the trade-off rate information from (2.24)~(2.26), all the constraints of the problem (2.19) must be active for the current optimal solution. Therefore, if there are inactive constraints, it is necessary to replace $\bar{\mu}_i$ for inactive constraints by $D_{iR}(c_{i\alpha}^L x^*) + v^*$ or $D_{iL}(c_{i\alpha}^R x^*) + v^*$ and solve the corresponding problem (2.19) for obtaining the simplex multipliers.

Now, following the above discussions, we can present an interactive algorithm in order to derive the satisficing solution for the DM from among the M-α-Pareto optimal solution set. The steps marked with an asterisk involve interaction with the DM.

Step 0 : Calculate the individual minimum and maximum of each objective function under the given constraints for $\alpha = 0$ and $\alpha = 1$.

Step 1* : Elicit a membership function $\mu_i(c_i x)$ from the DM for each of the objective functions.

Step 2* : Ask the DM to select the initial value of α $(0 \leq \alpha \leq 1)$ and set the initial reference membership values $\bar{\mu}_i = 1, i = 1, \ldots, k$.

Step 3 : For the degree α and the reference membership values specified by the DM, solve the minimax problem and perform the M-α-Pareto optimality test.

Step 4* : The DM is supplied with the corresponding M-α-Pareto optimal solution and the trade-off rates between the membership functions and the degree α. If the DM is satisfied with the current membership function values of the M-α-Pareto optimal solution and α, stop. Otherwise, the DM must update his/her reference membership values and/or the degree α by considering the current values of the membership functions and α together with the trade-off rates between the membership functions and the degree α, and return to Step 3.

Here it should be stressed for the DM that

(1) any improvement of one membership function can be achieved only at the expense of at least one of the other membership functions for some fixed degree α, and

(2) the greater value of the degree a gives worse values of the membership functions for some fixed reference membership values.

3. INTERACTIVE DECISION MAKING FOR MULTIOBJECTIVE LINEAR FRACTIONAL PROGRAMMING PROBLEMS WITH FUZZY PARAMETERS

3.1. Problem statement and solution concept

Consider the multiobjective linear fractional programming (MOLFP) problems [9,10] of the following form:

$$\begin{array}{lrcl} \min & z_1(x) & = & p_1(x)/q_1(x), \\ \min & z_2(x) & = & p_2(x)/q_2(x), \\ \vdots & & & \vdots \\ \min & z_k(x) & = & p_k(x)/q_k(x), \\ \text{subject to} & \multicolumn{3}{l}{x \in X = \{x \in E^n \mid a_j x \leq b_j, j = 1, \cdots, m; x \geq 0\},} \end{array} \tag{3.1}$$

where x is an n-dimensional column vector of decision variables, a_j is an n-dimensional constraint row vector, b_j is a constant, $z_1(x), \cdots, z_k(x)$ are k distinct linear fractional objective functions and

$$\begin{aligned} p_i(x) &= c_{i1}x + c_{i2}x + \cdots + c_{in}x + c_{i,n+1}, \\ q_i(x) &= d_{i1}x + d_{i2}x + \cdots + d_{in}x + d_{i,n+1}. \end{aligned} \tag{3.2}$$

These objective functions are independent each other and cannot be combined into a single objective function. Here it is customary to assume that the $q_i(x) > 0$ for all $x \in X$.

Fundamental to the MOLFP is the Pareto optimal concept, also known as a noninferior solution. Qualitatively, a Pareto optimal solution of the MOLFP is one where any improvement of one objective function can be achieved only at the expense of another.

Definition 3.1 (Pareto optimal solution)

$x^* \in X$ is said to be a Pareto optimal solution to the MOLFP, if and only if there does not exist another $x \in X$ such that $z_i(x) \leq z_i(x^*), i = 1, \cdots, k$ with strict inequality holding for at least one i.

In practice, however, it would certainly be appropriate to consider that the possible values of the parameters in the description of the objective functions and the constraints usually involve the ambiguity of the experts' understanding of the real system . For this reason, in this paper, we consider the following multiobjective linear fractional programming problem involving fuzzy parameters (MOLFP-FP) [16,22] :

$$\begin{array}{ll} \min & z(x, \tilde{c}, \tilde{d}) = (z_1(x, \tilde{c}_1, \tilde{d}_1), z_2(x, \tilde{c}_2, \tilde{d}_2), \cdots, z_k(x, \tilde{c}_k, \tilde{d}_k)) \\ \text{subject to} & x \in X(\tilde{a}, \tilde{b}) = \{x \in E^n \mid \tilde{a}_j x \leq \tilde{b}_j, j = 1, \cdots, m; x \geq 0\}, \end{array} \tag{3.3}$$

where

$$z_i(x, \tilde{c}_i, \tilde{d}_i) = p_i(x, \tilde{c}_i)/q_i(x, \tilde{d}_i), \tag{3.4}$$

and

$$\begin{aligned} p_i(x, \tilde{c}_i) &= \tilde{c}_{i1}x_1 + \tilde{c}_{i2}x_2 + \cdots + \tilde{c}_{in}x_n + \tilde{c}_{i,n+1}, \\ q_i(x, \tilde{d}_i) &= \tilde{d}_{i1}x_1 + \tilde{d}_{i2}x_2 + \cdots + \tilde{d}_{in}x_n + \tilde{d}_{i,n+1}. \end{aligned} \tag{3.5}$$

Here $\tilde{c}_i = (\tilde{c}_{i1}, \ldots, \tilde{c}_{in}, \tilde{c}_{i,n+1})$, $\tilde{d}_i = (\tilde{d}_{i1}, \ldots, \tilde{d}_{in}, \tilde{d}_{i,n+1})$ and $\tilde{a}_j = (\tilde{a}_{j1}, \ldots, \tilde{a}_{jn})$, $\tilde{b}_j$ represent respectively fuzzy parameters involved in the objective function $z_i(x, \tilde{c}_i, \tilde{d}_i)$ and the constraint $\tilde{a}_j x \leq \tilde{b}_j$.

We now assume that these fuzzy parameters $\tilde{c}_{ir}, \tilde{d}_{ir}$ and $\tilde{a}_{js}, \tilde{b}_j$, $i = 1, \cdots, k$, $j = 1, \cdots, m$, $r = 1, \cdots, n+1$, $s = 1, \cdots n$, in the MOLP-FP are fuzzy numbers [5] whose membership functions are $\mu_{\tilde{c}_{ir}}(c_{ir})$, $\mu_{\tilde{d}_{ir}}(d_{ir})$ and $\mu_{\tilde{a}_{js}}(a_{js})$, $\mu_{\tilde{b}_j}(b_j)$ respectively.

For simplicity of notation, define the following vectors :

$$c_i = (c_{i1}, \ldots, c_{i,n+1}), \quad d_i = (d_{i1}, \ldots, d_{i,n+1}), \quad a_j = (a_{j1}, \ldots, a_{jn}).$$

We also introduce the following α-level set of the fuzzy numbers $\tilde{a}_{js}, \tilde{b}_j$, $\tilde{c}_{ir}$ and $\tilde{d}_{ir}$.

Definition 3.2 (α-level set)
The α-level set of the fuzzy numbers $\tilde{a}_{js}, \tilde{b}_j$ $\tilde{c}_{ir}$ and $\tilde{d}_{ir}$ is defined as the ordinary set $L_\alpha(\tilde{a}, \tilde{b}, \tilde{c}, \tilde{d})$ for which the degree of their membership functions is greater than or equal to the level $\alpha \in [0,1]$:

$$L_\alpha(\tilde{a}, \tilde{b}, \tilde{c}, \tilde{d}) = \{(a,b,c,d) \mid \mu_{\tilde{a}_{js}}(a_{js}) \geq \alpha, \mu_{\tilde{b}_j}(b_j) \geq \alpha, \mu_{\tilde{c}_{ir}}(c_{ir}) \geq \alpha, \mu_{\tilde{d}_{ir}}(d_{ir}) \geq \alpha,$$
$$i = 1, \ldots, k, j = 1, \ldots, m, r = 1, \ldots, n+1, s = 1, \ldots, n\}. \tag{3.6}$$

As can be seen from the definition, the α-level set $L_\alpha(\tilde{a}, \tilde{b}, \tilde{c}, \tilde{d})$ is the set of the closed intervals depending on the level α.

For a certain degree α, the MOLFP-FP can be understood as the following nonfuzzy α-multiobjective linear fractional programming (α-MOLFP) problem [16,22].

$$\begin{array}{ll} \min & z(x,c,d) = (z_1(x,c_1,d_1), z_2(x,c_2,d_2), \cdots, z_k(x,c_k,d_k)) \\ \text{subject to} & x \in X(a,b) = \{x \in E^n \mid a_j x \leq b_j, j = 1, \cdots, m; x \geq 0\}, \\ & (a,b,c,d) \in L_\alpha(\tilde{a}, \tilde{b}, \tilde{c}, \tilde{d}). \end{array} \tag{3.7}$$

where

$$z_i(x, c_i, d_i) = p_i(x, c_i)/q_i(x, d_i), \tag{3.8}$$

and

$$\begin{aligned} p_i(x, c_i) &= c_{i1}x_1 + c_{i2}x_2 + \cdots + c_{in}x_n + c_{i,n+1}, \\ q_i(x, d_i) &= d_{i1}x_1 + d_{i2}x_2 + \cdots + d_{in}x_n + d_{i,n+1}. \end{aligned} \tag{3.9}$$

For linear fractional programming, it is customary to assume that $q_i(x, d_i) > 0$, $i = 1, \cdots, k$, for all $x \in X(a,b)$. In this section, for simplicity, we further assume that $p_i(x, c_i) > 0$, $i = 1, \cdots, k$, for all $x \in X(a,b)$.

On the basis of the α-level sets of the fuzzy numbers, we introduce the concept of α-Pareto optimal solutions to the α-MOLFP [16,22].

Definition 3.3 (α-Pareto optimal solution)
$x^* \in X(a^*, b^*)$ is said to be an α-Pareto optimal solution to the α-MOLFP, if and only if there does not exist another $x \in X(a,b)$, $(a,b,c,d) \in L_\alpha(\tilde{a}, \tilde{b}, \tilde{c}, \tilde{d})$ such that $z_i(x, c_i, d_i) \leq z_i(x^*, c_i^*, d_i^*)$, $i = 1, \cdots, k$, with strict inequality holding for at least one i, where the corresponding values of parameters (a^*, b^*, c^*, d^*) are called α-level optimal parameters.

As can be seen from Definition 3.3, usually, α-Pareto optimal solutions consist of an infinite number of points, and the DM must select his/her compromise or satisficing solution from among α-Pareto optimal solutions based on his/her subjective value-judgement.

However, considering the imprecise nature of the DM's judgement, it is reasonable to assume that the DM may have fuzzy goals for each of the objective functions in the α-MOLFP. For example, a goal stated by the DM may be to achieve "substantially less than A". This type of statement can be quantified by eliciting a corresponding membership function.

In order to elicit a membership function $\mu_i(z_i(x, c_i, d_i))$ from the DM for each of the objective functions $z_i(x, c_i, d_i)$, $i = 1, \cdots, k$, in the α-MOLFP, we first calculate the individual minimum and maximum of each objective function under the given constraints for $\alpha = 0$ and $\alpha = 1$. By taking account of the calculated individual minimum and maximum of each objective function for $\alpha = 0$ and $\alpha = 1$ together with the rate of increase of membership of satisfaction, the DM must determine his/her membership function $\mu_i(z_i(x, c_i, d_i))$ in a subjective manner.

It is significant to note here that, in the fuzzy approaches, we can treat two types of fuzzy goals; namely, fuzzy goals expressed in words such as " $z_i(x, c_i, d_i)$ should be in the vicinity of C_i " (called fuzzy equal) as well as " $z_i(x, c_i, d_i)$ should be substantially less than A_i or greater than B_i " (called fuzzy min or fuzzy max).

For this reason, in the following, as a generalization of the minimization problem (3.1), we consider the generalized α-MOLFP (Gα-MOLFP) problem expressed by :

$$\begin{array}{llll} \text{fuzzy} & \text{min} & z_i(x, c_i, d_i) & (i \in I_1), \\ \text{fuzzy} & \text{max} & z_i(x, c_i, d_i) & (i \in I_2), \\ \text{fuzzy} & \text{equal} & z_i(x, c_i, d_i) & (i \in I_3), \\ \text{subject to} & & x \in X(a, b), & (a, b, c, d) \in L_\alpha(\tilde{a}, \tilde{b}, \tilde{c}, \tilde{d}), \end{array} \tag{3.10}$$

where $I_1 \cup I_2 \cup I_3 = \{1, 2, \ldots, k\}, I_i \cap I_j = \phi, \ \ i, j = 1, 2, 3, \ \ i \neq j$.

For the membership functions of the Gα-MOLFP, it is reasonable to assume that $\mu_i(z_i(x, c_i, d_i))$, $i \in I_1$ and the right side functions of $\mu_i(z_i(x, c_i, d_i))$, $i \in I_3$ are strictly monotone decreasing and continuous functions with respect to $z_i(x, c_i, d_i)$, and $\mu_i(z_i\ (x,\ c_i,\ d_i))$, $i \in I_2$ and the left side of $\mu_i(z_i(x, c_i, d_i))$, $i \in I_3$ are strictly monotone increasing and continuous functions with respect to $z_i(x, c_i, d_i)$.

To be more specific, each membership function $\mu_i(z_i(x, c_i, d_i))$ of the Gα-MOLFP for $i \in I_1$, $i \in I_2$ or $i \in I_3$ is defined as follows :

(1) $i \in I_1$:

$$\mu_i(z_i(x, c_i, d_i)) = \begin{cases} 1 \quad \text{or} \to 1 & \text{if} \quad z_i(x, c_i, d_i) \leq z_{iR}^1, \\ D_{iR}(z_i(x, c_i, d_i)) & \text{if} \quad z_{iR}^1 \leq z_i(x, c_i, d_i) \leq z_{iR}^0, \\ 0 \quad \text{or} \to 0 & \text{if} \quad z_{iR}^0 \leq z_i(x, c_i, d_i). \end{cases} \tag{3.11}$$

(2) $i \in I_2$:

$$\mu_i(z_i(x, c_i, d_i)) = \begin{cases} 0 \quad \text{or} \to 0 & \text{if} \quad z_i(x, c_i, d_i) \leq z_{iL}^0, \\ D_{iL}(z_i(x, c_i, d_i)) & \text{if} \quad z_{iL}^0 \leq z_i(x, c_i, d_i) \leq z_{iL}^1, \\ 1 \quad \text{or} \to 1 & \text{if} \quad z_{iL}^1 \leq z_i(x, c_i, d_i), \end{cases} \tag{3.12}$$

(3) $i \in I_3$:

$$\mu_i(z_i(x,c_i,d_i)) = \begin{cases} 0 \quad \text{or} \to 0 & \text{if} \quad z_i(x,c_i,d_i) \le z_{iL}^0, \\ D_{iL}(z_i(x,c_i,d_i)) & \text{if} \quad z_{iL}^0 \le z_i(x,c_i,d_i) \le z_{iL}^1, \\ 1 & \text{if} \quad z_{iL}^1 \le z_i(x,c_i,d_i) \le z_{iR}^1, \\ D_{iR}(z_i(x,c_i,d_i)) & \text{if} \quad z_{iR}^1 \le z_i(x,c_i,d_i) \le z_{iR}^0, \\ 0 \quad \text{or} \to 0 & \text{if} \quad z_{iR}^0 \le z_i(x,c_i,d_i), \end{cases} \tag{3.13}$$

where $D_{iR}(z_i(x,c_i,d_i))$ or $D_{iL}(z_i(x,c_i,d_i))$ is respectively a strictly monotone decreasing or increasing continuous function with respect to $z_i(x,c_i,d_i)$ and may be linear or nonlinear, and z_{iL}^0 and z_{iR}^0 are unacceptable levels for $z_i(x,c_i,d_i)$ and z_{iL}^1 and z_{iR}^1 are totally desirable levels for $z_i(x,c_i,d_i)$.

When fuzzy equal is included in the fuzzy goals of the DM, it is desirable that $z_i(x,c_i,d_i)$ should be as close to C_i as possible. Consequently, the notion of α-Pareto optimal solutions defined in terms of objective functions cannot be applied. For this reason, we introduce the concept of M-α-Pareto optimal solutions [21,35] which is defined in terms of membership functions instead of objective functions, where M refers to membership.

Definition 3.4 (M-α-Pareto optimal solution)
$x^* \in X(a^*,b^*)$ is said to be an M-α-Pareto optimal solution to the Gα-MOLFP, if and only if there does not exist another $x \in X(a,b), (a,b,c,d) \in L_\alpha(\tilde{a},\tilde{b},\tilde{c},\tilde{d})$ such that $\mu_i(z_i(x,c_i,d_i) \ge \mu_i(z_i(x^*,c_i^*,d_i^*)), i = 1,\ldots,k$, with strict inequality holding for at least one i, where the corresponding values of parameters (a^*,b^*,c^*,d^*) are called α-level optimal parameters.

After eliciting the membership functions $\mu_i(z_i(x,c_i,d_i)), i = 1,...,k$ from the DM for each of the objective functions $z_i(x,c_i,d_i), i = 1,\ldots,k$, if we introduce a general aggregation function

$$\mu_D(\mu_1(z_1(x,c_1,d_1)),\ldots,\mu_k(z_k(x,c_k,d_k)),\alpha), \tag{3.14}$$

a general fuzzy α-multiobjective decision problem (Fα-MODP) can be defined by:

$$\begin{array}{ll} \max & \mu_D(\mu_1(z_1(x,c_1,d_1)),\ldots,\mu_k(z_k(x,c_k,d_k)),\alpha), \\ \text{subject to} & (x,a,b,c,d) \in P(\alpha), \quad \alpha \in [0,1], \end{array} \tag{3.15}$$

where $P(\alpha)$ is the set of M-α-Pareto optimal solutions and the corresponding α-level optimal parameters to the Gα-MOLFP.

Probably the most crucial problem in the Fα-MODP is the identification of an appropriate aggregation function which well represents the human decision makers' fuzzy preferences. If $\mu_D(\cdot)$ can be explicitly identified, then the Fα-MODP reduces to a standard mathematical programming problem. However, this rarely happens and as an alternative, it becomes evident that an interaction with the DM is necessary.

Throughout this section, we make the following assumptions.

Assumption 3.1
The fuzzy goals of the DM can be quantified by eliciting the corresponding membership functions through the interaction with the DM.

Assumption 3.2
$\mu_D(\cdot)$ exists and is known only implicitly to the DM, which means the DM cannot specify the entire form of $\mu_D(\cdot)$, but he/she can provide local information concerning his/her preference. Moreover, it is strictly increasing and continuous with respect to $\mu_i(\cdot)$ and α.

3.2 Minimax problems for MOLFP-FP

Having determined the membership functions for each of the objective functions, in order to generate a candidate for the satisficing solution which is also M-α-Pareto optimal, the DM is asked to specify the degree α of the α-level set and the reference levels of achievement of the membership functions, called reference membership values [15]. For the DM's degree α and reference membership values $\bar{\mu}_i, i = 1, \ldots, k$, the corresponding M-$\alpha$-Pareto optimal solution, which is in a sense close to his/her requirement or better than that if the reference levels are attainable, is obtained by solving the following minimax problem.

$$\min_{\substack{x \in X(a,b) \\ (a,b,c,d) \in L_\alpha(\tilde{a},\tilde{b},\tilde{c},\tilde{d})}} \max_{1 \le i \le k} (\bar{\mu}_i - \mu_i(z_i(x, c_i, d_i))), \tag{3.16}$$

or equivalently:

$$\begin{array}{ll} \min & v \\ \text{subject to} & \bar{\mu}_i - \mu_i(z_i(x, c_i, d_i)) \le v, i = 1, \ldots, k, \\ & a_j x \le b_j, j = 1, \ldots, m, \quad x \ge 0, \\ & (a, b, c, d) \in L_\alpha(\tilde{a}, \tilde{b}, \tilde{c}, \tilde{d}). \end{array} \tag{3.17}$$

In order to solve the formulated problem on the basis of the linear programming method, we first convert each constraint $\bar{\mu}_i - \mu_i(z_i(x, c_i, d_i)) \le v, i = 1, \ldots, k$, of the minimax problem (3.17) into the following form using the strictly monotone property of $D_{iL}(\cdot)$ and $D_{iR}(\cdot)$.

$$z_i(x, c_i, d_i) \le D_{iR}^{-1}(\bar{\mu}_i - v) \qquad i \in I_1 \cup I_3. \tag{3.18}$$

$$z_i(x, c_i, d_i) \ge D_{iL}^{-1}(\bar{\mu}_i - v) \qquad i \in I_2 \cup I_3. \tag{3.19}$$

Since $q_i(x, d_i) > 0$ for all $x \in X(a, b)$ and all i (by assumption), each constraint (3.18) and (3.19) can be converted as follows:

$$p_i(x, c_i) \leq D_{iR}^{-1}(\bar{\mu}_i - v)q_i(x, d_i) \qquad i \in I_1 \cup I_3. \tag{3.20}$$
$$p_i(x, c_i) \geq D_{iL}^{-1}(\bar{\mu}_i - v)q_i(x, d_i) \qquad i \in I_2 \cup I_3. \tag{3.21}$$

Then we introduce the following set-valued functions $R_{iL}(c_i, d_i)$, $R_{iR}(c_i, d_i)$, $i = 1, \ldots, k$, and $T_j(a_j, b_j)$, for each $(a, b, c, d) \in L_\alpha(\tilde{a}, \tilde{b}, \tilde{c}, \tilde{d})$.

$$R_{iL}(c_i, d_i) = \{(x, v) \mid p_i(x, c_i) \geq D_{iL}^{-1}(\bar{\mu}_i - v)q_i(x, d_i)\} \quad i \in I_2 \cup I_3. \tag{3.22}$$
$$R_{iR}(c_i, d_i) = \{(x, v) \mid p_i(x, c_i) \leq D_{iR}^{-1}(\bar{\mu}_i - v)q_i(x, d_i)\} \quad i \in I_1 \cup I_3. \tag{3.23}$$
$$T_j(a_j, b_j) = \{x \mid a_j x \leq b_j\}, \quad j = 1, \ldots, m. \tag{3.24}$$

Then, it can be verified that the following relations hold for $R_{iL}(\cdot, \cdot)$, $R_{iR}(\cdot, \cdot)$ and $T_j(\cdot, \cdot)$, when $x \geq 0$.

Proposition 3.1

(1) If $c_i^1 \leq c_i^2$, then $R_{iL}(c_i^1, d_i) \subset R_{iL}(c_i^2, d_i)$ and $R_{iR}(c_i^1, d_i) \supset R_{iR}(c_i^2, d_i)$.
(2) If $d_i^1 \leq d_i^2$, then $R_{iL}(c_i, d_i^1) \supset R_{iL}(c_i, d_i^2)$ and $R_{iR}(c_i, d_i^1) \subset R_{iR}(c_i, d_i^2)$.
(3) If $a_j^1 \leq a_j^2$, then $T_j(a_j^1, b_j) \supset T_j(a_j^2, b_j)$.
(4) If $b_j^1 \leq b_j^2$, then $T_j(a_j, b_j^1) \subset T_j(a_j, b_j^2)$.

It should be noted here that the feasible regions for c_i, d_i, a_j and b_j can be denoted respectively by the closed intervals $[c_{i\alpha}^L, c_{i\alpha}^R]$, $[d_{i\alpha}^L, d_{i\alpha}^R]$, $[a_{j\alpha}^L, a_{j\alpha}^R]$ and $[b_{j\alpha}^L, b_{j\alpha}^R]$.

Therefore, by making use of the results in Proposition 3.1, we can obtain an optimal solution to (3.17) by solving the following problem.

$$\begin{array}{ll} \min & v \\ \text{subject to} & p_i(x, c_{i\alpha}^L) \leq D_{iR}^{-1}(\bar{\mu}_i - v)q_i(x, d_{i\alpha}^R) \quad i \in I_1 \cup I_3, \\ & p_i(x, c_{i\alpha}^R) \geq D_{iL}^{-1}(\bar{\mu}_i - v)q_i(x, d_{i\alpha}^L) \quad i \in I_2 \cup I_3, \\ & a_{j\alpha}^L x \leq b_{j\alpha}^R, j = 1, \ldots, m, \quad x \geq 0. \end{array} \tag{3.25}$$

It is important to note here that in this formulation, if the value of v is fixed, it can be reduced to a set of linear inequalities. Obtaining the optimal solution v^* to the above problem is equivalent to determining the minimum value of v so that there exists an admissible set satisfying the constraints of (3.25). Since v satisfies $\bar{\mu}_{max} - 1 \leq v \leq \bar{\mu}_{max}$, where $\bar{\mu}_{max}$ denotes the maximum value of $\bar{\mu}_i$, $i = 1, \ldots, k$, we have the following method for solving this problem by combined use of the bisection method and the simplex method of linear programming.

Step 1: Set $v = \bar{\mu}_{max}$ and test whether an admissible set satisfying the constraints of (3.25) exists or not by making use of the phase one of the simplex method. If an admissible set exists, proceed. Otherwise, the DM must reassess his/her membership function.

Step 2: Set $v = \bar{\mu}_{max} - 1$ and test whether an admissible set satisfying the constraints of (3.25) exists or not using the phase one of the simplex method. If an admissible set exists, set $v^* = \bar{\mu}_{max} - 1$. Otherwise go to the next step, since the minimum v which satisfies the constraints of (3.25) exists between $\bar{\mu}_{max} - 1$ and $\bar{\mu}_{max}$.

Step 3: For the initial value of $v = \bar{\mu}_{max} - 0.5$, update the value of v using the bisection method as follows :
(a) $v_{n+1} = v_n - 1/2^{n+1}$ if admissible set exists for v_n,
(b) $v_{n+1} = v_n + 1/2^{n+1}$ if no admissible set exists for v_n.

In this way, we can determine the optimal solution v^*. Then the DM selects an appropriate standing objective from among the objectives $z_i(x, c_i, d_i), i = 1, \ldots, k$. For notational convenience, in the following, without loss of generality, let it be $z_1(x, c_1, d_1)$ and $1 \in I_1$. Then the following linear programming problem is solved for $v = v^*$.

$$\begin{aligned}
\min \quad & z_1(x, c^L_{1\alpha}, d^R_{1\alpha}) \\
\text{subject to} \quad & p_i(x, c^L_{i\alpha}) \leq D^{-1}_{iR}(\bar{\mu}_i - v^*) q_i(x, d^R_{i\alpha}) \quad i \in I_1 \cup I_3, \\
& p_i(x, c^R_{i\alpha}) \geq D^{-1}_{iL}(\bar{\mu}_i - v^*) q_i(x, d^L_{i\alpha}) \quad i \in I_2 \cup I_3, \\
& a^L_{j\alpha} x \leq b^R_{j\alpha}, j = 1, \ldots, m, \quad x \geq 0.
\end{aligned} \tag{3.26}$$

In order to solve this linear fractional programming problems, we can use Charnes-Cooper's variable transformation [2]:

$$t = 1/q_1(x, d^R_{1\alpha}), \qquad y = (x, 1)^T t, \tag{3.27}$$

and formulate the following standard linear programming problems :

$$\begin{aligned}
\min \quad & c^L_{1\alpha} y \\
\text{subject to} \quad & c^L_{i\alpha} y \leq D^{-1}_{iR}(\bar{\mu}_i - v^*) d^R_{i\alpha} y \quad i \in I_1 \cup I_3, \\
& c^R_{i\alpha} y \geq D^{-1}_{iL}(\bar{\mu}_i - v^*) d^L_{i\alpha} y \quad i \in I_2 \cup I_3, \\
& d^R_{1\alpha} y = 1, \\
& (a^L_{j\alpha}, -b^R_{j\alpha}) y \leq 0, j = 1, \ldots, m, \quad y \geq 0.
\end{aligned} \tag{3.28}$$

For convenience in our subsequent discussion, assume that the optimal solution y^* to (3.28) satisfies the following conditions :

$$\begin{aligned}
c^L_{i\alpha} y^* &= D^{-1}_{iR}(\bar{\mu}_i - v^*) d^R_{i\alpha} y^* \quad i \in I_1 \cup I_{3R}, \\
c^R_{i\alpha} y^* &= D^{-1}_{iL}(\bar{\mu}_i - v^*) d^L_{i\alpha} y^* \quad i \in I_2 \cup I_{3L}.
\end{aligned}$$

where $I_3 = I_{3L} \cup I_{3R}$ and $I_{3L} \cap I_{3R} = \phi$.

The relationships between the optimal solutions to (3.28) and the M-α-Pareto optimal concept of the Gα-MOLFP can be characterized by the following theorem.

Theorem 3.1

(1) If x^* is a unique optimal solution to (3.25), then x^* is an M-α-Pareto optimal solution to the Gα-MOLFP.

(2) If x^* is an M-α-Pareto optimal solution to the Gα-MOLFP, then x^* is an optimal solution to (3.25) for some $\bar{\mu} = (\bar{\mu}_1 \ldots, \bar{\mu}_k)$.

It should be noted here that for generating M-α-Pareto optimal solutions using Theorem 3.1, uniqueness of solution must be verified. In general, however, it is not easy to check numerically whether an optimal solution to (3.25) is unique or not.

Consequently, in order to test the M-α-Pareto optimality of a current optimal solution x^*, we formulate and solve the following linear programming problem :

$$\begin{array}{lll} \max & \bar{w} = \sum_{i=1}^{k} \varepsilon_i & \\ \text{subject to} & p_i(x, c_{i\alpha}^L) + \varepsilon_i = z_i(x^*, c_{i\alpha}^L, d_{i\alpha}^R) q_i(x, d_{i\alpha}^R) & i \in I_1 \cup I_{3R}, \\ & p_i(x, c_{i\alpha}^R) - \varepsilon_i = z_i(x^*, c_{i\alpha}^R, d_{i\alpha}^L) q_i(x, d_{i\alpha}^L) & i \in I_2 \cup I_{3L}, \\ & a_{j\alpha}^L x \leq b_{j\alpha}^R, j = 1, \ldots, m, \quad x \geq 0, & \\ & \varepsilon_i \geq 0, i = 1, \ldots, k. & \end{array} \tag{3.29}$$

Let $\bar{x}$ and $\bar{\varepsilon}$ be an optimal solution to (3.29). If all $\bar{w} = 0$, then x^* is an M-α-Pareto optimal solution. In case of $\bar{w} > 0$ and consequently at least one $\bar{\varepsilon}_\ell > 0$, we perform the following operations.

Step 1: Solve the following problem for any ℓ such that $\bar{\varepsilon}_\ell > 0$.

$$\begin{array}{lll} \min & \bar{z}_\ell(x, c_{\ell\alpha}, d_{\ell\alpha}) & \\ \text{subject to} & z_i(x, c_{i\alpha}^L, d_{i\alpha}^R) = z_i(\bar{x}, c_{i\alpha}^L, d_{i\alpha}^R) & \{i \mid \bar{\varepsilon}_i = 0\} \cap \{I_1 \cup I_{3R}\}, \\ & z_i(x, c_{i\alpha}^R, d_{i\alpha}^L) = z_i(\bar{x}, c_{i\alpha}^R, d_{i\alpha}^L) & \{i \mid \bar{\varepsilon}_i = 0\} \cap \{I_2 \cup I_{3L}\}, \\ & z_i(x, c_{i\alpha}^L, d_{i\alpha}^R) \leq z_i(\bar{x}, c_{i\alpha}^L, d_{i\alpha}^R) & \{i \mid \bar{\varepsilon}_i > 0\} \cap \{I_1 \cup I_{3R}\}, \\ & z_i(x, c_{i\alpha}^R, d_{i\alpha}^L) \geq z_i(\bar{x}, c_{i\alpha}^R, d_{i\alpha}^L) & \{i \mid \bar{\varepsilon}_i > 0\} \cap \{I_2 \cup I_{3L}\}, \\ & a_{j\alpha}^L x \leq b_{j\alpha}^R, \ j = 1, \ldots, m, \quad x \geq 0, & \end{array} \tag{3.30}$$

where $\bar{z}_\ell(x, c_{\ell\alpha}, d_{\ell\alpha})$ is defined as :

$$\bar{z}_\ell(x, c_{\ell\alpha}, d_{\ell\alpha}) = \begin{cases} z_\ell(x, c_{\ell\alpha}^L, d_{\ell\alpha}^R) & \ell \in I_1 \cup I_{3R}, \\ -z_\ell(x, c_{\ell\alpha}^R, d_{\ell\alpha}^L) & \ell \in I_2 \cup I_{3L}. \end{cases}$$

Step 2: Test the M-α-Pareto optimality for the solution to (3.30).
Step 3: If $\bar{w} = 0$, stop. Otherwise, return to Step 1.

Repeating this process at least $k-1$ iterations, an M-α-Pareto optimal solution can be obtained.

3.3. Interactive algorithm for MOLFP-FP

Now given the M-α-Pareto optimal solution for the degree α and the reference membership values specified by the DM by solving the corresponding minimax problem, the DM must either be satisfied with the current M-α-Pareto optimal solution and α, or update his/her reference membership values and/or the degree α. In order to help the DM express his/her degree of preference, trade-off information between a standing membership function and each of the other membership functions as well as between the degree α and the membership functions is very useful. Such a trade-off information is easily obtainable since it is closely related to the simplex multipliers of the problem (3.28).

To derive the trade-off information, define the following Lagrangian function L corresponding to the problem (3.28).

$$L = c_{1\alpha}^L y + \eta(d_{1\alpha}^R y - 1) + \sum_{i \in I_1 \cup I_{3R}} \Pi_{iR}\{c_{i\alpha}^L y - D_{iR}^{-1}(\bar{\mu}_i - v^*) d_{i\alpha}^R y\}$$

$$+ \sum_{i \in I_2 \cup I_{3L}} \Pi_{iL}\{D_{iL}^{-1}(\bar{\mu}_i - v^*) d_{i\alpha}^L y - c_{i\alpha}^R y\} + \sum_{j=1}^{m} \lambda_j (a_{j\alpha}^L, -b_{j\alpha}^R) y. \quad (3.31)$$

where η, Π_{iL}, Π_{iR} and λ_j are simplex multipliers corresponding to the constraints of (3.28).

Here, we assume that the problem (3.28) has a unique and nondegenerate optimal solution satisfying the following conditions :

(1) $\Pi_{iR} > 0 \quad i \in I_1 \cup I_{3R}, i \neq 1,$

(2) $\Pi_{iL} > 0 \quad i \in I_2 \cup I_{3L}.$

Then, by using the results in Haimes and Chankong [7], the following expression holds.

$$-\frac{\partial(z_1(x, c_{1\alpha}^L, d_{1\alpha}^R))}{\partial(z_i(x, c_{i\alpha}^L, d_{i\alpha}^R))} = \Pi_{iR} d_{i\alpha}^R y \qquad i \in I_1 \cup I_{3R},\ i \neq 1, \quad (3.32)$$

$$-\frac{\partial(z_1(x, c_{1\alpha}^L, d_{1\alpha}^R))}{\partial(z_i(x, c_{i\alpha}^R, d_{i\alpha}^L))} = -\Pi_{iL} d_{i\alpha}^L y \qquad i \in I_2 \cup I_{3L}. \quad (3.33)$$

Furthermore, using the strictly monotone decreasing or increasing property of $D_{iR}(\cdot)$ or $D_{iL}(\cdot)$ together with the chain rule, if $D_{iR}(\cdot)$ and $D_{iL}(\cdot)$ are differentiable at the optimal solution to (3.28), it holds that

$$-\frac{\partial \mu_1(z_1(x, c_{1\alpha}^L, d_{1\alpha}^R))}{\partial \mu_i(z_i(x, c_{i\alpha}^L, d_{i\alpha}^R))} = \Pi_{iR} d_{i\alpha}^R y \frac{D_{1R}'(z_1(x, c_{1\alpha}^L, d_{1\alpha}^R))}{D_{iR}'(z_i(x, c_{i\alpha}^L, d_{i\alpha}^R))} \quad i \in I_1 \cup I_{3R}, \; i \neq 1 \tag{3.34}$$

$$-\frac{\partial \mu_1(z_1(x, c_{1\alpha}^L, d_{1\alpha}^R))}{\partial \mu_i(z_i(x, c_{i\alpha}^R, d_{i\alpha}^L))} = -\Pi_{iL} d_{i\alpha}^L y \frac{D_{1R}'(z_1(x, c_{1\alpha}^L, d_{1\alpha}^R))}{D_{iL}'(z_i(x, c_{i\alpha}^R, d_{i\alpha}^L))} \quad i \in I_2 \cup I_{3L}. \tag{3.35}$$

Regarding a trade-off rate between $\mu_1(\cdot)$ and α, the following relation holds based on the sensitivity theorem [6].

$$\begin{aligned}\frac{\partial \mu_1(z_1(x, c_{1\alpha}^L, d_{1\alpha}^R))}{\partial \alpha} &= D_{1R}'(z_1(x, c_{1\alpha}^L, d_{1\alpha}^R))\left\{\frac{\partial(c_{1\alpha}^L)}{\partial \alpha} y + \eta \frac{\partial(d_{1\alpha}^R)}{\partial \alpha} y\right.\\ &+ \sum_{i \in I_1 \cup I_{3R}} \Pi_{iR}\left\{\frac{\partial(c_{i\alpha}^L)}{\partial \alpha} y - D_{iR}^{-1}(\bar{\mu}_i - v^*)\frac{\partial(d_{i\alpha}^R)}{\partial \alpha} y\right\}\\ &+ \sum_{i \in I_2 \cup I_{3L}} \Pi_{iL}\left\{D_{iL}^{-1}(\bar{\mu}_i - v^*)\frac{\partial(d_{i\alpha}^L)}{\partial \alpha} y - \frac{\partial(c_{i\alpha}^R)}{\partial \alpha} y\right\}\\ &\left.+ \sum_{j=1}^{m} \lambda_j \left(\frac{\partial(a_{j\alpha}^L)}{\partial \alpha}, -\frac{\partial(b_{j\alpha}^R)}{\partial \alpha}\right) y\right\}\end{aligned} \tag{3.36}$$

It should be noted here that in order to obtain the trade-off rate information from (3.34) and (3.35), all the constraints of the problem (3.28) must be active for the current optimal solution y^*, and y^* must satisfy the M-α-Pareto optimality test. Therefore, if there are inactive constraints, it is necessary to replace $\bar{\mu}_i$ for inactive constraints by $\mu_i(z_i(x^*, c_{i\alpha}^L, d_{i\alpha}^R)) + v^*$ or $\mu_i(z_i(x^*, c_{i\alpha}^R, d_{i\alpha}^L)) + v^*$, and solve the corresponding problem (3.28) for obtaining the simplex multipliers.

Following the above discussions, we can now construct an interactive algorithm in order to derive the satisficing solution for the DM from among the M-α-Pareto optimal solution set. The steps marked with an asterisk involve interaction with the DM.

Step 0 : Calculate the individual minimum and maximum of each objective function under the given constraints for $\alpha = 0$ and $\alpha = 1$.

Step 1* : Elicit a membership function $\mu_i(z_i(x, c_i, d_i))$ from the DM for each of the objective functions.

Step 2* : Ask the DM to select the initial value of α ($0 \leq \alpha \leq 1$) and set the initial reference membership values $\bar{\mu}_i, i = 1, \ldots, k$.

Step 3 : For the degree α and the reference membership values specified by the DM, solve the minimax problem and perform the M-α-Pareto optimality test.

Step 4* : The DM is supplied with the corresponding M-α-Pareto optimal solution and the trade-off rates between the membership functions and the degree α. If the DM is satisfied with the current membership function values of the M-α-Pareto optimal nsolution and α, stop. Otherwise, the DM must update the reference membership values and/or the degree α by considering the current values of the membership functions and α together with the trade-off rates between the membership functions and the degree α, and return to Step 3.

4. INTERACTIVE DECISION MAKING FOR MULTIOBJECTIVE NONLINEAR PROGRAMMING PROBLEMS WITH FUZZY PARAMETERS

4.1. Problem statement and solution concept

In general, the multiobjective nonlinear programming (MONLP) problem is represented as the following vector-minimization problem:

$$\begin{array}{ll} \min & (f_1(x), f_2(x), \cdots, f_k(x)) \\ \text{subject to} & x \in X = \{x \in E^n \mid g_j(x) \leq 0, j = 1, \cdots, m\}, \end{array} \qquad (4.1)$$

where x is an n-dimensional vector of decision variables, $f_1(x), \cdots, f_k(x)$ are k distinct objective functions of the decision vector x, $g_1(x), \ldots, g_m(x)$ are m inequality constraints, and X is the feasible set of constrained decisions.

Fundamental to the MONLP is the Pareto optimal concept, also known as a noninferior solution. Qualitatively, a Pareto optimal solution of the MONLP is one where any improvement of one objective function can be achieved only at the expense of another. Mathematically, a formal definition of a Pareto optimal solution to the MONLP is given below :

Definition 4.1 (Pareto optimal solution)
$x* \in X$ is said to be a Pareto optimal solution to the MONLP, if and only if there does not exist another $x \in X$ such that $f_i(x) \leq f_i(x^*), i = 1, \ldots, k$, with strict inequality holding for at least one i.

In practice, however, it would certainly be appropriate to consider that the possible values of the parameters in the description of the objective functions and the constraints usually involve the ambiguity of the experts' understanding of the real system. For this reason, in this paper, we consider the following multiobjective nonlinear programming problem with fuzzy parameters (MONLP-FP) [19,24,26]:

$$\begin{array}{ll} \min & f(x, \tilde{a}) = (f_1(x, \tilde{a}_1), \cdots, f_k(x, \tilde{a}_k)) \\ \text{subject to} & x \in X(\tilde{b}) = \{x \in E^n \mid g_j(x, \tilde{b}_j) \leq 0, j = 1, \ldots, m\}, \end{array} \qquad (4.2)$$

where $\tilde{a}_i = (\tilde{a}_{i1}, \ldots, \tilde{a}_{i,p_i})$, and $\tilde{b}_j = (\tilde{b}_{j1}, \ldots, \tilde{b}_{j,q_j})$ represent respectively fuzzy parameters involved in the objective function $f_i(x, \tilde{a}_i)$ and the constraint function $g_j(x, \tilde{b}_j)$.

These fuzzy parameters are assumed to be characterized as the fuzzy numbers by Dubois and Prade [5].

We now assume that $\tilde{a}_{ir}$ and $\tilde{b}_{js}$ in the MONLP-FP are fuzzy numbers whose membership functions are $\mu_{\tilde{a}_{ir}}(a_{ir})$ and $\mu_{\tilde{b}_{js}}(b_{js})$ respectively. For simplicity in the notation, define the following vectors :

$$a_i = (a_{i1}, \ldots, a_{i,p_i}), \qquad b_j = (b_{j1}, \ldots, b_{j,q_j}),$$
$$a = (a_1, \ldots, a_k), \qquad \tilde{a} = (\tilde{a}_1, \ldots, \tilde{a}_k), \qquad b = (b_1, \ldots, b_m), \qquad \tilde{b} = (\tilde{b}_1, \ldots, \tilde{b}_m).$$

Then we can introduce the following α-level set or α-cut [5] of the fuzzy numbers $\tilde{a}_{ir}$ and $\tilde{b}_{js}$.

Definition 4.2 (α-level set)
The α-level set of the fuzzy numbers $\tilde{a}_{ir}, (i = 1, \ldots, k,\ \ r = 1, \ldots, p_i)$ and $\tilde{b}_{js}(j = 1, \ldots, m,\ \ s = 1, \ldots, q_j)$ is defined as the ordinary set $L_\alpha(\tilde{a}, \tilde{b})$ for which the degree of their membership functions exceeds the level α :

$$\begin{aligned} L_\alpha(\tilde{a}, \tilde{b}) = \{(a, b) \mid \ & \mu_{\tilde{a}_{ir}}(a_{ir}) \geq \alpha \quad (i = 1, \ldots, k,\ \ r = 1, \ldots, p_i), \\ & \mu_{\tilde{b}_j}(b_j) \geq \alpha \quad (j = 1, \ldots, m,\ \ s = 1, \ldots, q_j)\}. \end{aligned} \tag{4.3}$$

For a certain degree α, the MONLP-FP can be understood as the following nonfuzzy α-multiobjective nonlinear programming (α-MONLP) problem [19,24, 26].

$$\begin{aligned} \min \quad & f(x, a) = \quad (f_1(x, a_1), f_2(x, a_2), \ldots, f_k(x, a_k)) \\ \text{subject to} \quad & x \in X(a, b) = \{x \in E^n \mid g_j(x, b_j) \leq 0, j = 1, \ldots, m\}, \\ & (a, b) \in L_\alpha(\tilde{a}, \tilde{b}). \end{aligned} \tag{4.4}$$

It should be emphasized here that in the α-MONLP the parameters a_{ir}, $i = 1, \ldots k$, $r = 1, \ldots, p_i$, b_{js}, $j = 1, \ldots m$, $s = 1, \ldots, q_j$, are treated as decision variables rather than constants.

On the basis of the α-level sets of the fuzzy numbers, we introduce the concept of α-Pareto optimal solutions to the α-MONLP [19 , 24 , 26].

Definition 4.3 (α-Pareto optimal solution)
$x^* \in X(b^*)$ is said to be an α-Pareto optimal solution to the α-MONLP, if and only if there does not exist another $(x, (a, b)) \in X(b) \times L_\alpha(\tilde{a}, \tilde{b})$ such that $f_i(x, a_i) \leq f_i(x^*, a_i^*), i = 1, \ldots, k$, with strict inequality holding for at least one i, where the

corresponding values of parameters (a^*, b^*) are called α-level optimal parameters and $\times$ denotes the Cartesian product.

For practical computational purposes, however, since only local solutions are guaranteed in solving a scalar optimization problem by any standard optimization technique, unless the problem is convex, we deal with local α-Pareto optimal solutions [19,24,26] instead of global α-Pareto optimal solutions.

Definition 4.4 (local α-Pareto optimal solution)
$x^* \in X(b^*)$ is said to be a local α-Pareto optimal solution to the α-MONLP, if and only if there does not exist another $(x, (a, b)) \in X(b) \times L_\alpha(\tilde{a}, \tilde{b}) \cap N(x^*, a^*, b^*; \delta)$ such that $f_i(x, a_i) \le f_i(x^*, a_i^*), i = 1, \ldots, k$, with strict inequality holding for at least one i, where the corresponding values of parameters (a^*, b^*) are called α-level optimal parameters and $N(x^*, a^*, b^*; \delta)$ denotes the δ-neighbourhood of (x^*, a^*, b^*).

As can be immediately understood from the definition, usually, (local) α-Pareto optimal solutions consist of an infinite number of points, and some kinds of subjective judgement should be added to the quantitative analyses by the DM. The DM must select his/her (local) satisficing or compromise solution from among (local) α-Pareto optimal solutions.

However, considering the imprecise nature of the DM's judgement, it is natural to assume that the DM may have imprecise or fuzzy goals for each of the objective functions in the α-MONLP. In a minimization problem, a fuzzy goal stated by the DM may be to achieve "substantially less than A". This type of statement can be quantified by eliciting a corresponding membership function.

In order to elicit a membership function $\mu_{f_i}(x, a_i)$ from the DM for each of the objective functions $f_i(x, a_i)$ in the α-MONLP, we first calculate the (local) individual minimum f_i^{min} and maximum f_i^{max} of each objective function $f_i(x, a_i)$ under the given constraints for $\alpha = 0$ and $\alpha = 1$. By taking account of the calculated (local) individual minimum and maximum of each objective function for $\alpha = 0$ and $\alpha = 1$, the DM must determine his/her subjective membership function $\mu_{f_i}(x, a_i)$ which is a strictly monotone decreasing function with respect to $f_i(x, a_i)$.

So far we have considered a minimization problem and consequently assumed that the DM has a fuzzy goal such as "$f_i(x, a_i)$ should be substantially less than A_i ". In the fuzzy approaches, we can further treat a more general case where the DM has two types of fuzzy goals, namely fuzzy goals expressed in words such as "$f_i(x, a_i)$ should be in the vicinity of C_i" (called fuzzy equal) as well as "$f_i(x, a_i)$ should be substantially less than A_i or greater than B_i " (called fuzzy min or fuzzy max). Such a generalized α-MONLP (Gα-MONLP) problem may now be expressed as:

$$\begin{array}{llll} \text{fuzzy} & \text{min} & f_i(x, a_i) & (i \in I_1), \\ \text{fuzzy} & \text{max} & f_i(x, a_i) & (i \in I_2), \\ \text{fuzzy} & \text{equal} & f_i(x, a_i) & (i \in I_3), \\ \text{subject to} & & x \in X(b), & (a, b) \in L_\alpha(\tilde{a}, \tilde{b}), \end{array} \tag{4.5}$$

where $I_1 \cup I_2 \cup I_3 = \{1, 2, \ldots, k\}, I_i \cap I_j = \phi, \quad i, j = 1, 2, 3, \quad i \neq j$.

In order to elicit a membership function from the DM for a fuzzy goal like "$f_i(x, a_i)$ should be in the vicinity of C_i ", it is obvious that we can use different functions to the left and right sides of C_i.

When fuzzy equal is included in the fuzzy goals of the DM, it is desirable that $f_i(x, a_i)$ should be as close to C_i as possible. Hence we introduce the concept of (local) M-α-Pareto optimal solutions [17,23,27] which are defined in terms of membership functions instead of objective functions, where M refers to membership.

Definition 4.5 ((local) M-α-Pareto optimal solution)
$x^* \in X(b^*)$ is said to be a local M-α-Pareto optimal solution to the Gα-MONLP, if and only if there does not exist another $(x, (a, b)) \in X(b) \times L_\alpha(\tilde{a}, \tilde{b}) \cap N(x^*, a^*, b^*; \delta)$ such that $\mu_{f_i}(x, a_i) \geq \mu_{f_i}(x^*, a_i^*), i = 1, \ldots, k$, with strict inequality holding for at least one i, where the corresponding values of parameters (a^*, b^*) are called α-level optimal parameters.

Having elicited the membership functions $\mu_{f_i}(x, a_i), i = 1, ..., k$ from the DM for each of the objective functions $f_i(x, a_i), i = 1, \ldots, k$, if we introduce a general aggregation function

$$\mu_D(\mu_f(x, a), \alpha) = \mu_D(\mu_{f_1}(x, a_1), \ldots, \mu_{f_k}(x, a_k), \alpha), \tag{4.6}$$

a general fuzzy α-multiobjective decision problem (Fα-MODP) can be defined by:

$$\begin{array}{ll} \max & \mu_D(\mu_f(x, a), \alpha), \\ \text{subject to} & (x, a, b) \in P(\alpha), \quad \alpha \in [0, 1], \end{array} \tag{4.7}$$

where $P(\alpha)$ is the set of M-α-Pareto optimal solutions and corresponding α-level optimal parameters to the Gα-MONLP.

Probably the most crucial problem in the Fα-MODP is the identification of an appropriate aggregation function which well represents the human decision makers' fuzzy preferences. If $\mu_D(\cdot)$ can be explicitly identified, then the Fα-MODP reduces to a standard mathematical programming problem. However, this rarely happens and as an alternative, it becomes evident that an interaction with the DM is necessary.

Throughout this section, we make the following assumptions.

Assumption 4.1
The fuzzy goals of the DM can be quantified by eliciting the corresponding membership functions through the interaction with the DM.

Assumption 4.2
$\mu_D(\cdot)$ exists and is known only implicitly to the DM, which means the DM cannot specify the entire form of $\mu_D(\cdot)$, but he/she can provide local information concerning his/her preference. Moreover, it is strictly increasing and continuous with respect to $\mu_i(\cdot)$ and α.

Assumption 4.3
All $f_i(x, a_i)$, $i = 1, ..., k$ and all $g_j(x, b_j)$, $j = 1, ..., m$ are continuously differentiable in their respective domains.

4.2. Augmented minimax problems for MONLP-FP

Having determined the membership functions for each of the objective functions, in order to generate a candidate for the (local) satisficing solution which is also (local) M-α-Pareto optimal, the DM is asked to specify the degree α of the α-level set and the reference levels of achievement of the membership functions, called the reference membership values [15].

For linear and linear fractional cases, in order to develop the linear programming based methods, the minimax problems together with the M-α-Pareto optimality tests have been used for generating the M-α-Pareto optimal solutions. For nonlinear case, however, in order to circumvent the necessity to perform the (local) M-α-Pareto optimality test in the minimax problems, it is reasonable to use augmented minimax problems instead of minimax problems.

Therefore, for the DM's degree α and the reference membership values $\bar{\mu}_{f_i}$, $i = 1, ..., k$, the following augmented minimax problem is recommended for generating the corresponding (local) M-α-Pareto optimal solution, which is in a sense close to his/her requirement or better than that if the reference membership values are attainable, where it is assumed that the difference $(\bar{\mu}_{f_i} - \mu_{f_i}(x, a_i))$ is of equal importance to the DM.

Thus

$$\min_{\substack{x \in X(b) \\ (a,b) \in L_\alpha(\tilde{a},\tilde{b})}} \left\{ \max_{1 \le i \le k} (\bar{\mu}_{f_i} - \mu_{f_i}(x, a_i)) + \rho \sum_{i=1}^{k} (\bar{\mu}_{f_i} - \mu_{f_i}(x, a_i)) \right\}, \tag{4.8}$$

or equivalently:

$$\begin{array}{ll} \min & v + \rho \sum_{i=1}^{k} (\bar{\mu}_{f_i} - \mu_{f_i}(x, a_i)), \\ \text{subject to} & \bar{\mu}_{f_i} - \mu_{f_i}(x, a_i) \le v, i = 1, \ldots, k, \\ & x \in X(b), \quad (a, b) \in L_\alpha(\tilde{a}, \tilde{b}). \end{array} \tag{4.9}$$

The term augmented is adopted because the term $\rho\sum_{i=1}^{k}(\bar{\mu}_{f_i} - \mu_{f_i}(x, a_i))$ is added to the usual minimax problems, where ρ is a sufficiently small positive scalar. Naturally, ρ should be a sufficiently small, but computationally significant, positive scalar. In most case, a computationally significant value of $\rho = 10^{-3} \sim 10^{-5}$ should suffice. Such an augmented minimax problem can be viewed as a modified fuzzy version of the augmented Tchebycheff norm problem of Steuer and Choo [30] or Choo and Atkins [4].

It should be emphasized here that the augmented minimax problem is simply used as a means of generating a (local) M-α-Pareto optimal solution, and if the DM is not satisfied with the current (local) M-α- Pareto optimal solution, it is possible for him/her to improve the solution by updating his/his reference membership values.

The relationships between the (local) optimal solutions of the augmented minimax problem and the (local) M-α-Pareto optimal concept of the Gα-MONLP can be characterized by the following theorem.

Theorem 4.1
(1) If (x^*, a^*, b^*) is a (local) optimal solution to the augmented minimax problem for some $\bar{\mu}_{f_i}$, $i = 1, ..., k$, then x^* is a (local) M-α-Pareto optimal solution and a^*, b^* are α-level (local) optimal parameters to the Gα-MONLP.
(2) If x^* is a (local) M-α-Pareto optimal solution and a^*, b^* are α- level (local) optimal parameters to the Gα-MONLP, then there exist $\bar{\mu}_{f_i}$, $i = 1, ..., k$ such that (x^*, a^*, b^*) is a (local) optimal solution to the augmented minimax problem with sufficiently small positive ρ.

4.3. Interactive algorithm for MONLP-FP

Now given the (local) M-α-Pareto optimal solution for the degree α and the reference membership values specified by the DM by solving the corresponding augmented minimax problem, the DM must either be satisfied with the current (local) M-α-Pareto optimal solution and α, or update the reference membership values and/or the degree α. In order to help the DM express his/her degree of preference, trade-off information between a standing membership function and each of the other membership functions as well as between the degree α and the membership functions is very useful. Fortunately, such a trade-off information is easily obtainable since it is closely related to the strict positive Lagrange multipliers of the augmented minimax problem.

To derive the trade-off information, define the Lagrangian function L for the augmented minimax problem (4.9) as follows :

$$L = v + \rho\sum_{i=1}^{k}(\bar{\mu}_{f_i} - \mu_{f_i}(x, a_i))$$

$$+\sum_{i=1}^{k}\lambda_i^{\mu}(\bar{\mu}_{f_i}-\mu_{f_i}(x,a_i)-v)+\sum_{j=1}^{m}\lambda_j^{g}g_j(x,b_j)$$

$$+\sum_{i=1}^{k}\sum_{r=1}^{p_i}\lambda_{ir}^{a}(\alpha-\mu_{\tilde{a}_{ir}}(a_{ir}))+\sum_{j=1}^{m}\sum_{s=1}^{q_j}\lambda_{js}^{b}(\alpha-\mu_{\tilde{b}_{js}}(b_{js})) \tag{4.10}$$

In the following , for notational convenience , we denote the decision variable in the augmented minimax problem (4.9) by $y=(x,v,a,b)$ and let us assume that the augmented minimax problem has a unique local optimal solution y^* satisfying the following three assumptions.

Assumption 4.4
y^* is a regular point of the constraints of the augmented minimax problem, i.e., the gradients of the active constraints are linearly independent.

Assumption 4.5
The second-order sufficiency conditions are satisfied at y^*.

Assumption 4.6
There are no degenerate constraints at y^*, i.e., all active constraints have strictly positive corresponding Lagrange multipliers.

Then the following existence theorem, which is based on the implicit function theorem [6] , holds.

Theorem 4.2
Let $y^*=(x^*,a^*,b^*)$ be a unique local solution of the augmented minimax problem (4.9) satisfying Assumptions 4.4, 4.5 and 4.6. Let $\lambda^*=(\lambda^{\mu^*},\lambda^{a^*},\lambda^{b^*},\lambda^{g^*})$ denote the Lagrange multipliers corresponding to the constraints of (3.9). Then there exist a continuously differentiable vector valued function $y(\cdot)$ and $\lambda(\cdot)$ defined on some neighborhood $N(\alpha)$ so that $y(\alpha^*)=y^*$, $\lambda(\alpha^*)=\lambda^*$, where y(α) is a unique local solution of the augmented minimax problem (4.9) for any $\alpha\in N(\alpha^*)$ satisfying Assumptions 4.4, 4.5 and 4.6, and $\lambda(\alpha)$ is the Lagrange multiplier corresponding to the constraints of (4.9).

In Theorem 4.2,

$$\inf_{x,v,a,b}\Big\{v+\rho\sum_{i=1}^{k}(\bar{\mu}_{f_i}-\mu_{f_i}(x,a_i))\mid\bar{\mu}_{f_i}-\mu_{f_i}(x,a_i)\leq v,$$
$$i=1,\ldots,k,x\in X(b),\ \ (a,b)\in L_\alpha(\tilde{a},\tilde{b})\Big\}$$

can be viewed as the optimal value function of the augmented minimax problem (4.9) for any $\alpha\in N(\alpha^*)$. Therefore, the following theorem holds under the same assumptions in Theorem 4.2.

Theorem 4.3
If all the assumptions in Theorem 4.2 are satisfied, then the following relations hold on some neighborhood $N(\alpha^*)$ of α.

$$\frac{\partial(v+\rho\sum_{i=1}^{k}(\bar{\mu}_{f_i}-\mu_{f_i}(x,a_i)))}{\partial\alpha}=\frac{\partial L}{\partial\alpha}=\sum_{i=1}^{k}\sum_{r=1}^{p_i}\lambda_{ir}^{a}+\sum_{j=1}^{m}\sum_{s=1}^{q_j}\lambda_{js}^{b} \tag{4.11}$$

If all the constraints of the augmented minimax problem are active, namely if $v(\alpha^*)=\bar{\mu}_{f_i}-\mu_{f_i}(x(\alpha^*),a_i(\alpha^*))$, then the following theorem holds.

Theorem 4.4
Let all the assumptions in Theorem 4.2 be satisfied. Also assume that all the constraints of the augmented minimax problem are active. Then it holds that

$$\left.\frac{\partial\mu_{f_i}(x,a_i)}{\partial\alpha}\right|_{\alpha=\alpha^*}=-\frac{1}{1+\rho k}\left\{\sum_{i=1}^{k}\sum_{r=1}^{p_i}\lambda_{ir}^{a^*}+\sum_{j=1}^{m}\sum_{s=1}^{q_j}\lambda_{js}^{b^*}\right\}\quad i=1,\ldots,k. \tag{4.12}$$

Regarding a trade-off rate between $\mu_{f_1}(x)$ and $\mu_{f_i}(x)$ for each $i=2,...,k$, by extending the results in Haimes and Chankong [7], it can be proved that the following theorem holds [34].

Theorem 4.5
Let all the assumptions in Theorem 4.2 be satisfied. Also assume that the constraints of the augmented minimax problem are active. Then it holds that

$$\left.\frac{\partial\mu_{f_i}(x,a_i)}{\partial\mu_{f_1}(x,a_1)}\right|_{\alpha=\alpha^*}=-\frac{\lambda_1^{\mu^*}+\rho}{\lambda_i^{\mu^*}+\rho}\quad i=1,\ldots,k. \tag{4.13}$$

It should be noted here that in order to obtain the trade-off rate information from (4.12) and (4.13), all the constraints of the augmented minimax problem must be active. Therefore, if there are inactive constraints, it is necessary to replace $\bar{\mu}_{f_i}$ for inactive constraints by $\mu_{f_i}(x^*,a_i^*)+v^*$ and solve the corresponding augmented minimax problem for obtaining the Lagrange multipliers.

Following the above discussions, we can now construct an interactive algorithm in order to derive the (local) satisficing solution for the DM from among the (local) M-α-Pareto optimal solution set. The steps marked with an asterisk involve interaction with the DM.

Step 0 : Calculate the (local) individual minimum f_i^{min} and maximum f_i^{max} of each objective function $f_i(x)$ under the given constraints for $\alpha=0$ and $\alpha=1$.

Step 1* : Elicit a membership function $\mu_{f_i}(x, a_i)$ from the DM for each of the objective functions.

Step 2* : Ask the DM to select the initial values of α $(0 \leq \alpha \leq 1)$ and set the initial reference membership values $\bar{\mu}_{f_i} = 1$, $i = 1, \ldots, k$.

Step 3 : For the degree α and the reference membership values specified by the DM, solve the corresponding augmented minimax problem.

Step 4* : The DM is supplied with the corresponding M-α-Pareto optimal solution and the trade-off rates between membership functions and the degree α. If the DM is satisfied with the current membership function values of the M-α-Pareto optimal solution and α, stop. Otherwise, the DM must update his/her reference membership values and/or the degree α by considering the current values of the membership functions and α together with the trade-off rates between the membership functions and the degree α, and return to Step 3.

5. INTERACTIVE COMPUTER PROGRAMS AND AN ILLUSTRATIVE EXAMPLE

Interactive decision making processes for multiobjective linear, linear fractional and nonlinear programming problems with fuzzy parameters discussed thus far include eliciting a membership function for each of the objective functions together with reference membership values and degree α from the DM. Thus, interactive utilization of computer facilities is highly recommended. Based on the methods described in sections 2, 3 and 4, we have developed the corresponding interactive computer programs for solving multiobjective linear, linear fractional and nonlinear programming problems with fuzzy parameters. The entire programs are written in FORTRAN, because FORTRAN language is very popular among many scientists and also almost all computer facilities are available for loading the programs with minor changes.

Our programs include graphical representations by which the DM can visualize the shapes of his/her membership functions, and he/she can find incorrect assessments or inconsistent evaluations promptly, revise them immediately and proceed to the next stage more easily.

Each of our computer programs is composed of one main program and several subroutines. The main program calls in and runs the subprograms with commands indicated by the user (DM). The major commands prepared in our programs are summarized as follows.

(1) MINMAX : Displays the calculated (local) individual minimum and maximum of each of the objective functions under the given constraints for $\alpha = 0$ and $\alpha = 1$.

(2) MF : Elicit a membership function from the DM for each of the objective functions.
(3) GRAPH : Depicts graphically the shape of the membership function for each of the objective functions.
(4) GO : Derives the (local) satisficing solution for the DM from among the (local) M-α-Pareto optimal solution set by updating the reference membership values and/or the degree α.
(5) STOP : Exists from the program.
(6) SAVE : Saves all the necessary information, which has been put in, in a file.
(7) READ : Restores the information which was saved in the file.

In our computer program, the DM can select his/her membership function in a subjective manner from among the following five types of functions; linear [38], exponential, hyperbolic [13], hyperbolic inverse and piecewise linear [8] functions. Then the parameter values are determined through the interaction with the DM. Here, it is assumed that $\mu_{f_i}(x) = 0$ if $f_i(x) \geq f_i^0$ and $\mu_{f_i}(x) = 1$ if $f_i(x) \leq f_i^1$, where f_i^0 is an unacceptable level for $f_i(x)$ and f_i^1 is a totally desirable level for $f_i(x)$. In the following description, for notational convenience, we deal with the fuzzy min membership functions in the multiobjective nonlinear programming problem.

(1) Linear membership function :

$$\mu_{f_i}(x) = (f_i(x) - f_i^0)/(f_i^1 - f_i^0).$$

The linear membership function can be determined by asking the DM to specify the two points f_i^0 and f_i^1 within f_i^{min} and f_i^{max}, where f_i^a represents the value of $f_i(x)$ such that the degree of membership function $\mu_{f_i}(x)$ is a.

(2) Exponential membership function :

$$\mu_{f_i}(x) = \alpha_i(1 - \exp(-\beta_i(f_i(x) - f_i^0)/(f_i^1 - f_i^0))).$$

The exponential membership function can be determined by asking the DM to specify the three points f_i^0, $f_i^{0.5}$ and f_i^1 within f_i^{max} and f_i^{min}, where β_i is a shape parameter.

(3) Hyperbolic membership function :

$$\mu_{f_i}(x) = (1/2)\tanh(\beta_i(f_i(x) - \gamma_i)) + \delta_i.$$

The hyperbolic membership function can be determined by asking the DM to specify the four points $f_i^0, f_i^{0.5}, f_i^1$ and γ_i within f_i^{max} and f_i^{min}, where β_i is a shape parameter and γ_i is associated with the point of inflection.

(4) Hyperbolic inverse membership function :

$$\mu_{f_i}(x) = \alpha_i \tanh^{-1}(\beta_i(f_i(x) - \gamma_i)) + \delta_i.$$

The hyperbolic inverse membership function can be determined by asking the DM to specify the four points $f_i^0, f_i^{0.5}, f_i^1$ and γ_i within f_i^{max} and f_i^{min}, where β_i is a shape parameter and γ_i is associated with the point of inflection.

(5) Piecewise linear membership function :

$$\mu_{f_i}(x) = \sum_{j=1}^{N_i} \alpha_{ij} \mid f_i(x) - g_{ij} \mid + \beta_i f_i(x) + \gamma_i.$$

Here, it is assumed that $\mu_{f_i}(x) = t_{ir} f_i(x) + s_{ir}$ for each segment $g_{i,r-1} \leq f_i(x) \leq g_{ir}$, where t_{ir} is the slope and s_{ir} is the y-intercept for the section of the curve initiated at $g_{i,r-1}$ and terminated at g_{ir}. The piecewise linear membership function can be determined by asking the DM to specify the degree of membership in each of several values of objective functions within $f_i^{\max}$ and $f_i^{\min}$.

We now demonstrate the interaction processes using our computer program by means of an illustrative example which is designed to test the program. Consider the following three objective nonlinear programming problem with fuzzy parameters.

$$\begin{array}{lll} \text{fuzzy} & \text{min} & f_1(x, \tilde{a}_1) = (x_1 + 5)^2 + \tilde{a}_{11} x_2^2 + 2(x_3 - \tilde{a}_{12})^2 \\ \text{fuzzy} & \text{min} & f_2(x, \tilde{a}_2) = \tilde{a}_{21}(x_1 - 45)^2 + (x_2 + 15)^2 + 3(x_3 + \tilde{a}_{22})^2 \\ \text{fuzzy} & \text{equal} & f_3(x, \tilde{a}_3) = \tilde{a}_{31}(x_1 + 20)^2 + \tilde{a}_{32}(x_2 - 45)^2 + (x_3 + 15)^2 \\ \text{subject to} & & g_1(x, \tilde{b}_1) = \tilde{b}_{11} x_1^2 + \tilde{b}_{12} x_2^2 + \tilde{b}_{13} x_3^2 \leq 100, \\ & & 0 \leq x_i \leq 10, \;\; i = 1, 2, 3. \end{array}$$

The membership functions for the fuzzy numbers $\tilde{a}_{11}, \ldots, \tilde{a}_{32}$ and $\tilde{b}_{11}, \ldots, \tilde{b}_{13}$ in this example are explained in Table 1. where L and E represent respectively linear and exponential membership functions.

Table 1. Fuzzy numbers for numerical example

$\tilde{p}$	(p_1,	p_2,	p_3,	p_4)	left	right
$\tilde{a}_{11}$	(3.8,	4.0,	4.0,	4.3)	L	E
$\tilde{a}_{12}$	(48.5,	50.0,	50.0,	52.0)	E	E
$\tilde{a}_{21}$	(1.85,	2.0,	2.0,	2.2)	E	L
$\tilde{a}_{22}$	(18.2,	20.0,	20.0,	22.5)	L	E
$\tilde{a}_{31}$	(2.9,	3.0,	3.0,	3.15)	E	L
$\tilde{a}_{32}$	(4.7,	5.0,	5.0,	5.35)	L	L
$\tilde{b}_{11}$	(0.9,	1.0,	1.0,	1.1)	E	E
$\tilde{b}_{12}$	(0.8,	1.0,	1.0,	1.2)	E	E
$\tilde{b}_{13}$	(0.85,	1.0,	1.0,	1.15)	E	L

In applying our computer program to this problem, suppose that the interaction with the hypothetical DM establishes the following membership functions and corresponding assessment values for the three objective functions.

f_1 : exponential, $(f_1^0, f_1^{0.5}, f_1^1) = (5400, 5000, 3300)$

f_2 : hyperbolic, $(f_2^0, f_2^{0.5}, f_2^1, b_2) = (6900, 4600, 3900, 4400)$

f_3 : left : exponential, $(f_3^0, f_3^{0.5}, f_3^1) = (7800, 8200, 10000)$
right : hyperbolic inverse, $(f_3^0, f_3^{0.5}, f_3^1, b_3) = (13300, 11000, 10000, 12000)$

(Local) M-α-Pareto optimal solutions are obtained by solving the augmented minimax problem using the revised version of the generalized reduced gradient (GRG) [11] program called GRG2 [12].

In this example, at the $5th$ iteration, the (local) satisficing solution of the DM is derived. The whole interactive processes are summarized in Table 2.

Table 2. Interactive processes

Iteration	1	2	3	4	5
$\bar{\mu}_1$	1	0.5	0.55	0.55	0.65
$\bar{\mu}_2$	1	0.6	0.50	0.50	0.55
$\bar{\mu}_3$	1	0.8	0.75	0.75	0.75
α	0.9	0.9	0.9	0.7	0.7
μ_1	0.5678	0.4558	0.5543	0.5979	0.6392
μ_2	0.5678	0.5558	0.5043	0.5479	0.5392
μ_3	0.5678	0.7558	0.7543	0.7979	0.7392
f_1	4917.71	5048.17	4934.94	4877.16	4816.81
f_2	4475.23	4496.28	4591.57	4510.35	4526.11
f_3	10828.78	10422.53	10425.39	10342.49	10455.02
x_1	8.4567	8.0461	7.9090	8.1024	8.1775
x_2	5.2592	6.0275	6.0594	6.0905	5.8785
x_3	1.8210	1.2689	1.8437	1.9837	2.2585
$-\partial\mu_2/\partial\mu_1$	0.6433	0.4845	0.5757	0.6824	0.7638
$-\partial\mu_3/\partial\mu_2$	1.5303	1.2998	1.6415	1.9187	2.1046
$-\partial\mu_i/\partial\alpha$	0.2143	0.2390	0.2164	0.2193	0.2074

6. CONCLUSIONS

In this paper, we have proposed the interactive decision making methods for multiobjective linear, linear fractional and nonlinear programming problems with fuzzy parameters respectively to cope with the imprecise nature of human judgements. As the conclusions of this paper, the desirable features of our proposed methods will be summarized as follows.

(1) The experts' ambiguous understanding of the nature of the parameters in the problem-formulation process can be incorporated.
(2) The fuzzy goals of the DM can be quantified by eliciting the corresponding membership functions, which may be nonlinear.
(3) The concept of α-multiobjective linear, linear fractional and nonlinear programming together with (local) M-α-Pareto optimality have been introduced on the basis of the α-level sets of the fuzzy numbers.
(4) For the degree α and the reference membership values specified by the DM, the corresponding (local) M-α-Pareto optimal solution can be obtained by solving the (augmented) minimax problems.
(5) Concerning multiobjective linear and linear fractional programming problems, for generating a candidate for the satisficing solution which is also M-α-Pareto optimal, minimax problems were adopted, and consequently, it was shown that the formulated minimax problems can be solved based mainly on the well-known linear programming method.
(6) In the nonlinear case, however, since some nonlinear programming codes are necessary, it was recommended to adopt augmented minimax problems instead of minimax problems for circumventing the necessity to perform the (local) M-α-Pareto optimality tests for the current solution to the minimax problems.
(7) On the basis of the proposed methods, the time-sharing computer programs for all the proposed methods have been written in FORTRAN to implement man-machine interactive procedures.
(8) An illustrative numerical example demonstrated the feasibility and efficiency of both the proposed method and its interactive computer program by simulating the responses of the hypothetical DM.

However, applications must be carried out in cooperation with a person actually involved in decision making. From such experiences the proposed method and its computer program must be revised.

REFERENCES

[1] Bellman, R.E. and Zadeh, L.A. (1970) "Decision making in a fuzzy environment", Management Science 17, 141-164.
[2] Charnes, A. and Cooper, W.W. (1962) "Programming with linear fractional criteria", Naval Research Logistics Quarterly 9, 181- 186.
[3] Chankong, V. and Haimes, Y.Y. (1983) Multiobjective Decision Making : Theory and Methodology, North-Holland, New York.
[4] Choo, E.U. and Atkins, D.R. (1983) "Proper efficiency in nonconvex multicriteria programming", Mathematics of Operations Research 8, 467-470.
[5] Dubois, D. and Prade, H. (1980) Fuzzy Sets and Systems : Theory and Applications, Academic Press.
[6] Fiacco, A.V. (1983) Introduction to Sensitivity and Stability Analysis in Nonlinear Programming, Academic Press.

[7] Haimes, Y.Y. and Chankong, V. (1979) "Kuhn-Tucker multipliers as trade-offs in multiobjective decision-making analysis", Automatica 15, 59-72.

[8] Hannan, E.L. (1981) "Linear programming with multiple fuzzy goals", Fuzzy Sets and Systems 6, 235-248.

[9] Kornbluth, J.S.H. and Steuer, R.E. (1981) "Multiple objective linear fractional programming", Management Science 27, 1024-1039.

[10] Kornbluth, J.S.H. and Steuer, R.E. (1981) "Goal programming with linear fractional criteria", European Journal of Operational Research, 8, 58-65.

[11] Lasdon, L.S., Fox, R.L. and Ratner, M.W. (1974) "Nonlinear optimization using the generalized reduced gradient method", Revue Francaise d'Automatique, Informatique et Research Opérationnelle, 3, 73-103.

[12] Lasdon, L.S., Waren, A.D. and Ratner, M.W. (1980) "GRG2 user's guide", Technical Memorandum, University of Texas.

[13] Leberling, H. (1981) "On finding compromise solution in multicriteria problems using the fuzzy min-operator", Fuzzy Sets and Systems, 6, 105-118.

[14] Orlovski, S.A. (1984) "Multiobjective programming problems with fuzzy parameters", Control and Cybernetics 13, 175-183.

[15] Sakawa, M. and Yano, H. (1985) "An interactive fuzzy satisficing method using augmented minimax problems and its application to environmental systems", IEEE Transactions on Systems, Man, and Cybernetics, SMC-15, 720-729.

[16] Sakawa, M. and Yano, H. (1985) "Interactive decision making for multiobjective linear fractional programming problems with fuzzy parameters", Cybernetics and Systems : An International Journal, 16, 377-394.

[17] Sakawa, M. and Yano, H. (1986) "An interactive fuzzy satisficing method for multiobjective nonlinear programming problems with fuzzy parameters", IIASA (International Institute for Applied Systems Analysis), Laxenburg, Austria, Collaborative paper CP-86-15.

[18] Sakawa, M. and Yano, H. (1986) "An interactive fuzzy satisficing method for multiobjective linear programming problems with fuzzy parameters", Proc. of the 4-th IFAC/IFORS Symposium, Large Scale Systems : Theory and Applications, Zurich, Switzerland, 497-502.

[19] Sakawa, M. and Yano, H. (1986) "An interactive method for multiobjective nonlinear programming problems with fuzzy parameters" in R. Trappl (ed.), Cybernetics and Systems'86, D.Reidel Publishing Company, 607-614.

[20] Sakawa, M. and Yano, H. (1986) "Interactive decision making for multiobjective linear programming problems with fuzzy parameters", in G. Fandel, M.Grauer, A. Kurzhanski and A.P.Wierzbicki (eds.), Large-Scale Modelling and Interactive Decision Analysis, Springer- Verlag, 88-96.

[21] Sakawa, M. and Yano, H. (1987) "An interactive fuzzy satisficing method for multiobjective linear fractional programming problems with fuzzy parameters", in Y. Sawaragi, K. Inoue and H. Nakayama (eds.), Towards Interactive and Intelligent Decision Support Systems , Vol.2 , Proceedings, Kyoto , Japan

, Springer-Verlag, 338-347.

[22] Sakawa, M. and Yano, H. (1986) "Personal computer-aided interactive decision making for multiobjective linear fractional programming problems with fuzzy parameters", Proc. of International Computer Symposium 1986 (ICS'86), National Cheng-Kung University, Taiwan , 1868-1876.

[23] Sakawa, M. and Yano, H. (1987) "Interactive fuzzy decisionmaking for multiobjective nonlinear programming problems with fuzzy parameters", Proc. of 10th World Congress of IFAC (IFAC'87), Vol.7, Munich, FGR, 408-413.

[24] Sakawa, M. and Yano, H. (1987) "An interactive satisficing method for multiobjective nonlinear programming problems with fuzzy parameters", in J. Kacprzyk and S.A.Orlovski (eds.), Optimization Models Using Fuzzy Sets and Possibility Theory, D.Reidel Publishing Company, 258-271.

[25] Sakawa, M. and Yano, H. (1988) "An interactive algorithm for multiobjective linear programming problems with fuzzy coefficients", in R.Trappl (ed.), Cybernetics and Systems'88, Kluwer Academic Publishers, 693-700.

[26] Sakawa, M. and Yano, H. (1989) "Interactive decision making for multiobjective nonlinear programming problems with fuzzy parameters", Fuzzy Sets and Systems, 29, 315-326.

[27] Sakawa, M. and Yano, H. (1989) "An interactive fuzzy satisficing method for multiobjective nonlinear programming problems with fuzzy parameters", Fuzzy Sets and Systems, 30, 221-238.

[28] Sakawa, M. and Yano, H. (1989) "Interactive computer package for multiobjective programming problems with fuzzy parameters", Proc. of Third IFSA Congress, Seattle, 307-310.

[29] Seo, F. and Sakawa, M. (1988) Multiple Criteria Decision Analysis in Regional Planning : Concepts, Methods and Applications, D. Reidel Publishing Company.

[30] Steuer, R.E. and Choo, E.U. (1983) "An interactive weighted Tchebycheff procedure for multiple objective programming", Mathematical programming 26, 326-344.

[31] Tanaka, H. and Asai, K. (1981) "A formulation of linear programming problems by fuzzy function", Systems and Control, 25, 351-357 (in Japanese).

[32] Tanaka, H. and Asai, K. (1984) "Fuzzy linear programming problems with fuzzy numbers", Fuzzy Sets and Systems 13, 1-10.

[33] Wierzbicki, A.P. (1979) "The Use of Reference Objectives in Multiobjective Optimization - Theoretical Implications and Practical Experiences, IIASA (International Institute for Applied System Analysis), Laxenburg, Austria, Working Paper WP-79-66.

[34] Yano, H. and Sakawa, M. (1987) "Trade-off rates in the weighted Tchebycheff norm method", Large Scale Systems, 13, 167-177.

[35] Yano, H. and Sakawa, M. "Interactive fuzzy decision making for generalized multiobjective linear fractional programming problems with fuzzy parame-

ters", Fuzzy Sets and Systems, 32, 245-261.
[36] Zadeh, L.A. (1975) "The concept of a linguistic variables and its application to approximate reasoning-1", Information Science 8, 199-249.
[37] Zeleny, M. (1982) Multiple Criteria Decision Making, McGraw-Hill, New York.
[38] Zimmermann, H.J. (1978) "Fuzzy programming and linear programming with several objective functions", Fuzzy Sets and Systems, 1, 45-55.
[39] Zimmermann, H.J. (1983) "Fuzzy mathematical programming", Computers and Operations Research, 10, 291-298.

Part III
Chapter 2

A POSSIBILISTIC APPROACH FOR MULTIOBJECTIVE PROGRAMMING PROBLEMS. EFFICIENCY OF SOLUTIONS

M. Delgado, J.L. Verdegay, M.A. Vila

Departamento de Ciencias de la Computacion
e Inteligencia Artificial

Facultad de Ciencias

Universidad de Granada

18071 Granada, Spain

ABSTRACT. Fuzzy Sets and Possibility Theory have been extensively used in designing a wide class of resolution methods for Multicriteria Decision Problems and more particularly in Multiobjective Programming. Such methods appear without common relation as each of them arises from a particular and suitable set of hypotheses about factors such that the framework of the problem, its nature, decision-maker's behavior, etc. In this paper we present a general formulation for using possibility theory, or fuzzy sets, to solve multiobjective problems, which contains all of aforementioned methods as particular cases because all of them may be obtained from a common auxiliary problem. First we analyze how this auxiliary problem is constructed, its main properties as well as different solution methods. After that, the efficiency of the corresponding solutions is dealt with. The paper finishes with some remarks and examples.

1. Introduction.

Consider a multiobjective programming problem

$$\max_{x \in X} \{f_1(x), f_2(x), \ldots, f_q(x)\} \qquad (1)$$

X being a subset of $\mathbb{R}^n$ and $f_i : \mathbb{R}^n \longrightarrow \mathbb{R}$, $i \in Q = \{1,2,\ldots,q\}$, without any additional condition for the moment.

Although we consider $X \subset \mathbb{R}^n$, most results of this paper can be directly extended to more general spaces.

It is well known there exists no concept of "optimal solution" universally accepted and valid for any multiobjective problem. On contrary, several approaches and points of view are used to produce a lot of solution methods, (Multiattribute Utility, Goal Programming, Vector Maximum, etc.) each providing

R. Slowinski and J. Teghem (eds.),
Stochastic versus Fuzzy Approaches to Multiobjective Mathematical Programming under Uncertainty, 229–248.

a kind of solution with a set of specific features and properties. Choosing a particular method depends on several factors such as the own nature of the problem, decision-maker's requests, the ranking of the objectives and its evaluation, etc. (see Cohon (1978), Zeleny (1982) or Chankong and Haimes (1983)).

Anyway, the main difference between a single objective problem and a multi-objective one is that the first has "optimal solution" (if it exists) whereas for the second one can only look for what is usually called a "good compromise solution". In fact, each of the aforementioned solution methods provides a kind of compromise solution being "good" according to the point of view underlying the method and the subsequent hypotheses.

Let us observe the term "good compromise solution" is very vague (a fuzzy statement) and a lot of subjective, imprecisely stated or vague factors (such that "decision-maker's satisfaction", "goals achievement", etc.) must be handled and evaluated to characterize what will be considered as "good compromise solution" in any particular case. For these reasons, Fuzzy Sets and Possibility Theory have had a very big impact in Multicriteria Decision and Multiobjective Programming, and several solution methods have been developed by using fuzzy subsets or possibility distributions to model imprecise or vague concepts.

Narasimhan (1980), Hannan (1981a, 1981b), Ignizio (1982), Rubin et al. (1984), Sakawa (1984), and several other, have introduced the called "Fuzzy Goal Programming". In this approach a set of goals about objectives is assumed to be imposed by the decision maker. Moreover these goals, their interdependences and/or their ranking (priorities among them) are not precisely established but with vague statements (linguistic labels) being represented by fuzzy subsets of corresponding spaces, usually the real line. Tiwari et al. (1985), (1986), (1987) and Driankov (1987), among others, have studied decision models with priorities and interdependent goals. Sakawa and Yano (1986) have developed an interesting interactive method using augmented minimax problems for multi-objective nonlinear programming with fuzzy goals. Werners (1987) introduced a different possibilistic interactive approach to obtain compromise solutions for multiobjective problems with crisp or fuzzy goals.

On the other hand, Leung (1982,1983) has introduced the concept of "Fuzzy Ideal" originating what is called "Fuzzy Compromise Programming".

From a different point of view, several authors, e.g. Zimmermann (1978,1985), Delgado (1983), Delgado et al. (1985), LLena (1985), etc. have proposed to model decision-maker's "satisfaction" (satisfaction in general sense), by means of fuzzy sets or possibility distributions.

The main aim of this paper is to present a unified framework to use Fuzzy Sets or Possibility Theory in Multicriteria Decision or Multiobjective Programming. Moreover its general properties are explored, focusing on efficiency of the solutions.

2. A Unified Approach

Suppose that an application

$$h_i : \mathbb{R} \longrightarrow [0,1] \qquad i\in Q$$

may be established for any objective of (1), such that $h_i(r)$ measures how much the value $r\in\mathbb{R}$ fulfill decision-maker's requirements about the i-th objective (satisfaction, achievement of goals, nearness to a fuzzy ideal, etc.).

It is clear that we can consider $h_i:\mathbb{R}^n \longrightarrow [0,1]$ just by taking $h_i(x) = h_i[f_i(x)]$, $i\in Q$. Now $h_i(x)$ measures for each x how good is such alternative with respect to the i-th objective. In other words, $h_i(x)$ may be considered the membership degree of x to the (fuzzy) subset of "good solutions" for the ith objective, i.e. the possibility of x being a good solution for that objective.

A "good compromise solution" for (1) may be defined as a $x\in X$ being "as good as possible" for the whole set of objectives. By the nature of the $h_i(.)$, looking for such a kind of solution by means of the auxiliary problem

$$\operatorname*{Max}_{x\in X} \; [h_1(x),h_2(x),\ldots,h_q(x)] \qquad (2)$$

it seems quite reasonable. Observe (2) is a new multiobjective problem like (1) with the functions $h_i(.)$ instead of the originals $f_i(.)$. It can be solved by

(i) Some general method such that Goal Programming, Multi attribute Utility. Vector Maximum Techniques, etc. or

(ii) Some specific method being designed having regard to $h_i(x)$ $i\in Q$ are membership functions or possibility distributions.

In the following we will explore this way more deeply. Because Max $[h_1(x),h_2(x),\ldots,h_q(x)]$ may be interpreted as a synthetical notation of a conjunction statement (maximize jointly all objectives), it is reasonable to think of using a t-norm N to represent the connective AND (see Schwaizer et al (1969)) as $h_i(.)$, $i\in Q$, are membership, fulfillment or performance degrees with values in [0,1]. In this way (2) becomes the single-objective problem

$$\operatorname*{Max}_{x\in X} \; N[h_1(x),h_2(x),\ldots,h_q(x)] \qquad (3)$$

T-norms have been extensively used in Fuzzy Sets and Possibility Theory to describe the connective AND (disjunction) in different cases and applications.

Choosing a concrete one depends on the problem, its environment and the behavior of the decision-maker, whose reasoning rules are attempted to model (see Alsina et al. (1980), Thole et al. (1981), Dubois (1980, 1983) and Zimmermann et al. (1980)).

Let us remark that Zadeh originally considered min or prod to define the intersection of fuzzy sets and in fact these t-norms have been the most used to represent the connective and. R. Bellman and L.A. Zadeh introduced in 1974 the concept of "fuzzy decision" as intersection of fuzzy goals and fuzzy constraints by means of the minimum too.

Zimmermann and Zysno (1980) claimed that most of decisions taken in the real world are neither non-compensatory nor fully compensatory and dealt with compensatory operators for decision making and J.R. Rao et al. (1980) have developed a method for finding numerical compensation in multicriteria problem. We do not consider such a kind of operators here focusing the use of t-norms only.

On the other hand, we can solve (2) by means of the so called "parametric approach" (Verdegay (1982), Delgado (1983), Chanas (1983), Delgado et al. (1985a, 1985b)). The basic idea is that an ill-defined problem ought have an ill-defined solution like a well-defined problem has a well-defined solution, and thus a fuzzy-solution of (1) is looked for. It is obtained by solving a parametric programming problem which is constructed by means of the α-cuts of $h_i(.)$, $i \in Q$.

In the following we shall study how the problem (1) may be solved through (2), particularly the efficiency of the solutions. We pay a special attention to the problem (3), emphasizing the cases of Minimum and Product (the most usual t-norms).

3. Some definitions and concepts

3.1. Efficient and M-efficient solutions

It is well known that a key concept in Multicriteria Decision is that of "efficient solution" (non-dominated or Pareto-optimal), which substitutes the concept of optimal solution used in single objective models.

Definition 1. An $x^{*}\in X$ is said to be efficient for (1) iff there exits no x^{**} such that $f_i(x^{**}) \geq f_i(x^{*})$ $\forall i$ with strict inequality for some i.

It is clear any decision-maker must look for a compromise solution being efficient, because a dominated action is worse than a non-dominated one.

Let us observe (2) is a multiobjective problem too, and thus it makes sense to give

Definition 2. An $\bar{x}\in X$ is said to be M-efficient for (1) iff it is efficient for (2), i.e. there exists no $\bar{\bar{x}}$ such that $h_i(\bar{\bar{x}}) \geq h_i(\bar{x})$, with strict inequality for some i.

Sakawa (1984) and Zimmermann (1985), have used a stronger version of this definition according to which a point in X is called F-efficient iff it is efficient for both problems (1) and (2). On the other hand, Sakawa and Yano (1986) introduced a local definition of M-efficiency for nonlinear programming problems. We have introduced the definition 2 because we think of dealing with global solutions for problems (1) and (2) separately is preferable.

Obtaining a dominated solution in a certain multicriteria problem is quite possible, depending on the used solution method (see Chankong-Haimes (1973)). Furthermore, it is clear that only M-efficient solutions may be looked for when (1) is solved trough (2). From this reasons, it makes sense to question:

1.- When is obtained an M-efficient solution by solving (2)?,

2.- When is efficient an M-efficient solution?.

In later sections we shall attempt to answer these two questions.

3.2. Assessing decision-maker's request about objectives

As the model is to maximize the objectives, one can expect the decision-maker to have goals in the form,

"to obtain on the ith objective a gain Q_i-greater than F_i"

Q_i being a linguistic (fuzzy) quantifier like than "a bit", "rather", "much", etc.

That kind of goals are linguistic properties (elastic constraints) about real numbers and thus they induce possibility distributions on $\mathbb{R}$ (fuzzy sets on the real line) in the form:

$$h_i(r) = \begin{cases} 0 & \text{if } r \le g_i \\ v_i(r) & \text{if } r \in [g_i, G_i], \\ 1 & \text{if } r \ge G_i \end{cases} \qquad (4)$$

being continuous, with $v_i(r)$ strictly increasing in $[g_i, G_i]$. This interval and the concrete $v_i(.)$ depend on g_i, Q_i and decision-maker's evaluation of such quantifier (see Kacprzyk et al. (1984)). Graphically:

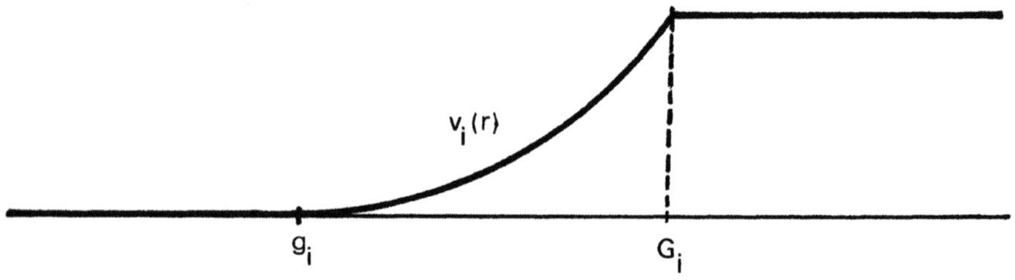

For the sake of convenience, this kind of possibility distributions will be called increasing in the remain of the paper.

A different approach arises when the decision-maker evaluates satisfaction instead of fixing goals. When the assessments of this satisfaction for any objective are given by linguistic statements, possibility distributions on $\mathbb{R}$ are generated again and one can expect the functions to be like (4) (with different meaning, of course) because

a) the greater the gain is the greater the satisfaction must be,

b) a lower bound g_i must exist such that there is no satisfaction on the ith objective when its value is smaller than g_i,

c) an upper bound G_i must exist such that decision-maker's requests are quite full when the gain is greater than G_i.

Zimmermann (1978), suggests to take g_i and G_i equal to the minimum and the maximum of f_i on X respectively and $v_i(r) = (r-g_i)/(r-G_i)$. The idea is to have a unified way of fixing $h_i(.)$ for all objectives. However we prefer the decision maker to be free to choose g_i, G_i and $v_i(.)$ (shape and values) as the model is more general and a sizeable diminution of the computational effort is obtained (see Delgado (1983)).

Remark 1. If we consider a model with the objectives to be minimized, it is clear we can reformulate it like (1) by changing the sign all the objectives. However, one can choose to preserve the original problem and then only the functions $h_i(.)$ ought be changed to measure satisfaction or performance. The problem (2) remains unchanged because this satisfaction or performance always is to be maximized. It is obvious these $h_i(.)$ must be

$$h_i(r) = \begin{cases} 1 & \text{if } r \leq g_i \\ w_i(r) & \text{if } r \in [g_i, G_i] \\ 0 & \text{if } r \geq G_i \end{cases}$$

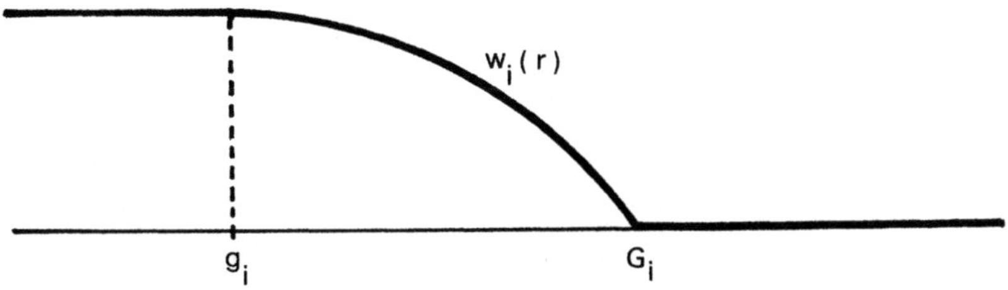

because now the decision-maker prefers a smaller gain on the objectives. For the sake of convenience this kind of possibility distributions will be called decreasing in the remain of the paper.

4. Obtaining M-efficient solutions.

In this section we are going to analyze the methods which may be used to solve (2) and the conditions to obtain M-efficient solutions for (1). First we consider some general and classical Multiobjective Programming methods. Technics for problems with special structure or properties will not be considered here as we have assumed no hypotheses about the functions or the constraint set. In the second part of this section we deal with using t-norms, that is, solving (2) through (3). We focus on this method and its properties.

4.1. General methods of Multiobjective Programming

For more detailed explanations about next and other methods see Cohon (1978), Zeleny (1982) and Chankong and Haimes (1983).

4.1.1. Weighting objectives

The problem (2) is replaced by

$$\underset{x \in X}{\text{Max}} \sum_{i=1,q} w_i h_i(x)$$

with $w_i \geq 0$, $\forall i$. Usually the weights are taken normalized and such that $\sum w_i = 1$. Using weights allows to handle non-preemptive priorities. When $w_i > 0$ $\forall i$, the optimal solution of this problem is M-efficient for (1).

4.1.2. Constraints method

Now the problem (2) is changed into

$$\begin{aligned} &\text{Max } h_k(x) \\ &\quad h_i(x) \geq L_i \qquad i \in Q-\{k\} \\ &\quad x \in X \end{aligned} \tag{5}$$

k being an index either arbitrarily chosen or selected according to some preemptive priorities. The values L_i $i \in Q-\{k\}$ are determined by the decision maker according to his requests about the objectives. Observe they are security levels.

The problem (5) provides M-efficient solutions for (1) under Lin's conditions (see Chankong and Haimes op. cit.).

There exists the possibility of stating (5) with equalities,

$$\begin{aligned} &\text{Max } h_k(x) \\ &\text{s.t:} \\ &\quad h_i(x) = L_i, \; i \in Q-\{k\} \\ &\quad x \in X \end{aligned} \tag{6}$$

having conditions for M-efficiency like (5).

Constraints method is closely related with solving (3) by using the minimum as we shall show later.

4.1.3. Hybrid method

The problem (2) is replaced by

$$\begin{aligned} &\operatorname*{Max}_{\text{s.t:}} \sum_{i=1,q} w_i h_i(x) \\ &\qquad h_j(x) \geq L_j \quad j\in Q \qquad (7) \\ &\qquad x\in X \end{aligned}$$

with $w_i \geq 0\ \forall i$. If $w_i>0\ \forall i$, then any optimal solution of (7) is M-efficient for (1).

Obviously, for (5), (6) or (7) values L_i, $i\in Q$, such that the corresponding problem has feasible solutions, must be chosen.

4.1.4. Goal Programming

The decision maker establishes goals η_i for the possibility values $h_i(.)$, $i\in Q$, and then the following problem is solved

$$\begin{aligned} &\text{Min} \sum_{i=1,q} F_i(d_i^-,d_i^+) \\ &\text{s.t:} \qquad h_i(x) - d_i^+ + d_i^- = \eta_i \\ &\qquad\qquad x\in X \\ &\qquad\qquad d_i^-,\ d_i^+ \geq 0,\ i\in Q \end{aligned}$$

where d_i^- and d_i^+ are variables measuring deviation of the ith objective with respect to the corresponding goal and $F_i(d_i^-,d_i^+)$ a penalty of such deviations, $i\in Q$.

Goal Programming may provide either M-efficient or dominated solutions depending on the values of the goals.

Narasimhan (1980), Hannan (1981a, 1981b), Rubin et. al. (1984) and Sakawa (1984) have suggested this approach to solve Fuzzy Goal Programming, where the functions $h_i(.)$ $i\in Q$, represent the possibility distributions induced by the fuzzy goals.

4.2. Using t-norms

As we have said the idea is to solve (2) by means of the auxiliary problem

$$\operatorname*{Max}_{x\in X} N[h_1(x),h_2(x),\ldots,h_q(x)] \qquad (8)$$

where the choice of a particular t-norm depends upon several factors such as the nature of the problem, the environment or decision maker's knowledge representation model.

Proposition 1. Let N be an arbitrary t-norm. If $\bar{x}$ is the only optimal solution of (8), then $\bar{x}$ is M-efficient.

Proof. If $\bar{x}$ is not efficient, it must exist an $\bar{\bar{x}}\in X$ such that $h_i(\bar{x}) \le h_i(\bar{\bar{x}})$. Thus $N[h_1(\bar{x}),\dots.h_q(\bar{x})] \le N[h_1(\bar{\bar{x}}),\dots,h_q(\bar{\bar{x}})]$, and $\bar{\bar{x}}$ must also be an optimal solution for (8). That contradicts the uniqueness of $\bar{x}$ and therefore $\bar{\bar{x}}$ does not exist and $\bar{x}$ must be M-efficient.#

Proposition 2. If $\bar{x}$ is an optimal solution for (8) such that $h_i(\bar{x})\neq 0$ $\forall i$, and N is strictly monotonous, then $\bar{x}$ is efficient for (1).

Proof. If $\bar{x}$ were not M-efficient, then a $\bar{\bar{x}}$ such that $h_i(\bar{\bar{x}}) \ge h_i(\bar{x}) > 0$ $\forall i$ with strict inequality for some i must exist and thus

$$N[h_1(\bar{\bar{x}}),\dots,h_q(\bar{\bar{x}})] < N[h_1(\bar{x}),\dots,h_q(\bar{x})]$$

because N is strictly monotonous, but this inequality is impossible as $\bar{x}$ is an optimal solution of (8). Therefore we conclude $\bar{\bar{x}}$ does not exist, that is $\bar{x}$ is efficient.#

Minimum and Product are the only t-norms being used in the literature to solve (2) trough (8). The associated problems are, respectively

$$\underset{x\in X}{\text{Max}} \bigwedge_i h_i(x) \qquad (9)$$

$$\underset{x\in X}{\text{Max}} \prod_i h_i(x) \qquad (10)$$

On the other hand, (9) can be equivalently converted into

$$\begin{aligned} &\text{Max } \rho \\ &h_i(x)\ge\rho \quad i\in Q \qquad (11) \\ &x\in X,\ \rho\in[0,1] \end{aligned}$$

which can be seen as directly obtained from (2) by using a modified version of the constraints method. The problem (9) (or (11)), has been extensively used because the minimum is the most usual t-norm in Fuzzy Sets and Possibility Theory. In fact it was the one originally considered by Zadeh. Moreover R.

Bellman and L.A. Zadeh introduced in 1974 the concept of "fuzzy decision" in a somewhat different context but using the minimum too.

From the propositions 1 and 2 two conclusions arise at once:

1.- An optimal solution of (9) is M-efficient when it is the only one because the minimum is not a strictly monotonous t-norm. This property may be obtained by applying Lin's conditions to (11).

2.- An optimal solution of (10), say $\bar{x}$, will be M-efficient for (1) if either it is the only one or $h_i(\bar{x})\neq 0$ $\forall i$, because the product is a strictly monotonous t-norm and thus propositions 1 and 2 hold.

In the general case, when (8) has several optimal solutions, their M-efficiency need to be investigated. For that we may use the following adaptation of the classical test of efficiency (see Chankong and Haimes, op. cit.)

4.3. Test of M-efficiency

Let $\bar{x}$ be a feasible solution of (1).

(A) Solve the problem $T(\bar{x})$

$$\begin{aligned} &\text{Max} \sum_{i=1,q} s_i \\ &\text{s.t:} \\ &\qquad h_i(x)-s_i=h_i(\bar{x}) \quad i\in Q \\ &\qquad x\in X,\ s_i\geq 0 \quad i\in Q \end{aligned}$$

(B) Let $(\bar{\bar{x}},\bar{\bar{s}})$ be an optimal solution of $T(\bar{x})$. If $\bar{\bar{s}}_i = 0$ $\forall i$, then $\bar{\bar{x}}$ is M-efficient for (1). Otherwise $\bar{x}$ is M-efficient.

Observe $T(\bar{x})$ is always feasible as $\bar{x}$ is a feasible solution anyway.

For the existence of M-efficient solutions we can state

Proposition 3. In the set of the optimal solutions to (8) (if any) there exists one, at least, being M-efficient for (1).

Proof. Let $\bar{x}$ be an optimal solution of (8) and x any feasible solution in $T(\bar{x})$. Obviously $h_i(x) \geq h_i(\bar{x})$ $\forall i$, and thus $N[h_1(x),\dots,h_q(x)] \geq N[h_1(\bar{x}),\dots,h_q(\bar{x})]$, that is, x must be optimal for (8) too. Therefore, by the properties of the M-efficiency test, we conclude either, $\bar{x}$ is M-efficient or there exists $\bar{\bar{x}}$ being also optimal and M-efficient.

Summarizing the results of this section, we propose the following two steps procedure to obtain M-efficient solutions:

S.1) Solve (2) either by a direct method or changing it into (8). If the M-efficiency of the solution may be guaranteed (propositions 1 and 2) then stop. Otherwise go to S.2.

S.2) Applying the previous test to either establish that the current solution is actually M-efficient or find another which certainly was.

5. Efficiency of the M-efficient solutions

Any decision maker will look for efficient solutions of (1) because they are preferable to dominated ones. But (2) may be used to solve (1), and thus, it is clear only an M-efficient solution (if any) can be guaranteed. Observe this fact is reasonable in our approach as we suppose the satisfaction or the performance is to be maximized. In this section the conditions for the efficiency of a M-efficient solutions are dealt with.

Proposition 4. Let x^* be M-efficient for (1) with $h_i(x^*) \in (0,1)$ $\forall i$. If all the functions $h_i(.)$ are increasing, then x^* is efficient for (1).

Proof. If x^* is dominated, then $x^{**} \in X$ and $i_o \in Q$ such that $f_i(x^*) \le f_i(x^{**})$ $\forall i$, $f_{i_o}(x^*) < f_{i_o}(x^{**})$ must exist. Since $h_i(x^*) \in (0,1)$ and $h_i(x)$ are strictly increasing on $h_i^{-1}[(0,1)]=v_i^{-1}[(0,1)]$, $i \in Q$, then $h_i(x^*) \le h_i(x^{**})$, $\forall i$, $h_{i_o}(x^*) <$ $< h_{i_o}(x^{**})$ what contradicts the M-efficiency of x^*.

Corollary. If x^* is the only M-efficient solution of (1), then it is efficient whichever the values $h_i(x^*)$, $i \in Q$, may be.

Proof. It is immediate from proposition 4.

To investigate the efficiency of any $x^* \in X$, the following test (like the one for M-efficiency) may be used.

A) solve the problem $T'(x^*)$

$$\mathrm{Max}\{\textstyle\sum_{i=1,q} u_i \ / \ f_i(x) - u_i = f_i(x^*),\ x \in X,\ u_i \ge 0,\ i \in Q\}$$

B) Let (x^{**}, u^{**}) be the optimal solution of $T'(x^*)$. If $u_i^{**}=0$ $\forall i$, then x^{**} is efficient for (1). Otherwise x^* is efficient.

When one applies this test all $h_i(.)$ being increasing, the following points are to be borne in mind:

(i) any efficient solution is M-efficient,

(ii) any feasible solution of $T'(x^*)$ must be M-efficient,

(iii) when x^* is M-efficient but non efficient, that is there exists i_o such that $u_{i_o}^{**} = 0$, then $h_{i_o}(x^*) \in \{0,1\}$

From the results of the two last sections we can state

Proposition 5. Let x^* an optimal solution of (8) with all the h_i being increasing. x^* is efficient for (1) if either a) x^* is the only optimal solution, or b) N is strictly monotonous and $h_i(x^*) \in (0,1)$ $\forall i$.

Proof. a) follows from proposition 1 and the properties of the efficiency test. b) follows from propositions 2 and 3.

Zimmermann (1978), Hannan (1979), Leberling (1981), Buckley (1983) and LLena (1985), have directly studied the efficiency of the solutions obtained from (9) or (10). They use increasing possibility functions (like most authors) and their results can be summarized as follows

1) An optimal solution of (9) is efficient when it is the only one.
2) An optimal solution of (10) is efficient when either it is the only one or the corresponding values of all functions $h_i(.)$ are in (0,1).

Let observe these facts are particular cases of the proposition (5), as the product is a strictly monotonous t-norm.

6. The parametric approach. Obtaining a fuzzy solution

In this work, like most authors in the quoted references, we have assumed that the solution of (2) is a point or points in X, but anyway a crisp subset of the feasible space X.

Several authors (Verdegay (1982), Chanas (1983), Delgado (1983) and Delgado et al. (1985a, 1985b)), have investigated giving a fuzzy solution to ill defined problems like a well defined (crisp) solution is always looked for in well defined (non-vague) problems. All of these authors have used the Theorem of Representation for Fuzzy Sets (see Negoita and Ralescu (1975)) to transform the original problem into an indexed family of "α-cut problems" which generates an indexed family of "α-cut solutions" from which the fuzzy solution is obtained.

Delgado et al. (1985a, 1985b) have specifically dealt with multiobjective problems. In the following we summarize their approach.

By using the Representation Theorem, (2) may be changed into

$$\begin{aligned} &\text{Max } h_k(x) \\ &h_i(x) \geq \alpha \quad i \in Q-\{k\} \qquad (12) \\ &x \in X, \ \alpha \in (0,1] \end{aligned}$$

where the constraints $h_i(x) \geq \alpha$ arise from the α-cuts of that fuzzy sets whose membership functions were h_i. On the other hand, the index k for the objective may be arbitrarily chosen but it may be also selected according to subjective criteria and/or decision's maker requests.

In general (12) is a parametric programming problem on $\alpha \in (0,1]$, whose solution may be represented as the indexed family $\{x^*(\alpha), \alpha \in (0,1]\}$ which generates (via Representation Theorem again) a fuzzy subset of X being considered the fuzzy solution of (1).

Four points deserve to be remarked:

(i) for any $\alpha \in (0,1]$, (12) represent an application of the constraint method to (2). So, for a certain α we obtain M-efficient solutions under Lin's conditions for h_i $\forall i$ and X,

(ii) when all $h_i(.)$ are increasing, (12) may be changed into

$$\begin{aligned} \text{Max:} \quad & f_k(x) \\ \text{s.t:} \quad & \\ & f_i(x) \geq v_i^{-1}(\alpha), \ i \in Q-\{k\} \\ & \alpha \in (0,1], \ x \in X \end{aligned}$$

which is an application of the constraints method to (1) for any α. Thus, efficient solutions are obtained when Lin's conditions hold for f_i $\forall i$ and X,

(iii) the crisp (point) solutions which arise applying the non-fuzzy approaches for fuzzy Multiobjective problems lie in the set $\{x^{*}(\alpha),\ \alpha\in(0,1]\}$ associated to some particular values of α. Specifically, the solution of (8) for the minimum is $x^{*}(\bar{\alpha})$, $\bar{\alpha}$ being the solution of the equation $h_k(x^{*}(\alpha)) = \alpha$.

(iv) By using this α-cuts decomposition, an interactive solution process may be carried out through an increasing sequence of satisfaction levels:The decision maker is requested to fix a value of α, say α^{*}, according to his aspirations and preferences. If he accepts the solution obtained from (12) with α equal to α^{*},then the process stops. Otherwise a new α^{*} greater than the previous must be established and the process restarts. To establish this new value the information provided by the solution associated to the previous level may be used.

7. Example

To illustrate the previous developments, we shall consider the following biobjective linear programming problem:

$$\text{Max } \{2x_1+x_2;\ -x_1-2x_2\ /\ x_1+x_2 \le 4,\ 3x_1+x_2 \ge 6,\ x_1,x_2 \ge 0\} \qquad (13)$$

Let us observe the objectives are almost opposing and thus no global solution can exist. In fact, the first objective attains its maximum at $O_1=(4,0)$ whereas the second does at $O_2 = (2,0)$. Thus, the decision maker only can look for compromise solutions. Looking at the geometry of this problem is easy to see the efficient solutions lie on the segment $\{(4-2\lambda,\ 0),\ \lambda \in [0,1]\}$.

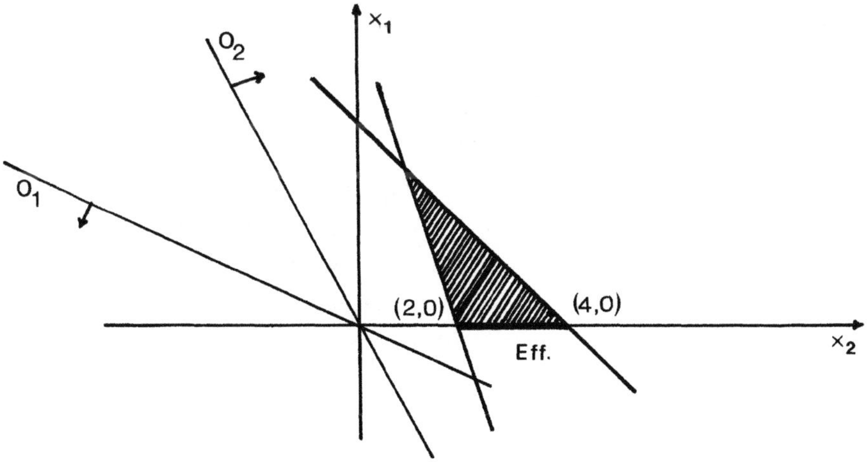

Let us consider the decision maker establish:

G_1) On the first objective I wish to gain more than -5, but I am satisfied when its value is greater than -3,

G_2) On the second objective I wish to gain more than 4 but I am satisfied when the value is greater than 7.

By assuming, for the sake of convenience, both $v_1(.)$ and $v_2(.)$ are linear, the possibility distributions generated by the above linguistic statements about the satisfying goal values, are

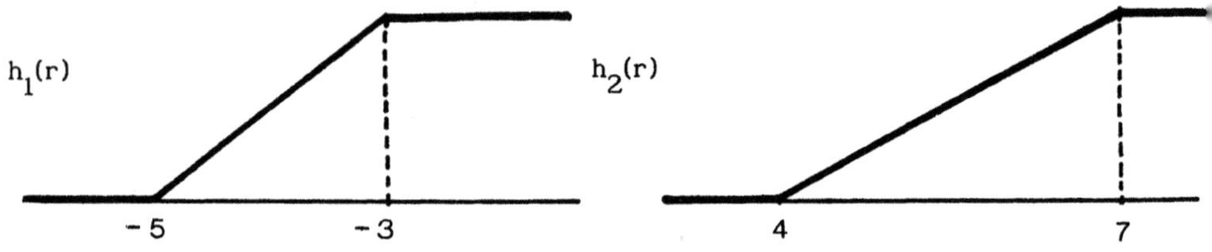

Let us suppose decision maker chooses the minimum operator to represent his evaluation of the connective AND. Therefore he changes the original problem into:

$$\begin{aligned} &\text{Max Min } [h_1(2x_1+x_2), h_2(-x_1-2x_2)] \\ &\text{s.t:} \\ &x_1+x_2 \le 4 \\ &3x_1+x_2 \ge 6 \\ &x_1, x_2 \ge 0 \end{aligned} \quad (14)$$

which in its turn may be solved through

$$\begin{aligned} &\text{Max } y \\ &\text{s.t:} \\ &h_1(2x_1+x_2) \ge y \\ &h_2(-x_1-2x_2) \ge y \\ &x_1+x_2 \le 4 \\ &3x_1+x_2 \ge 6 \\ &y \in [0,1],\ x_1, x_2 \ge 0 \end{aligned} \quad (15)$$

and it is easy to obtain the optimal solution

$$x^{*} = (23/7,0) \qquad (16)$$

which is efficient as it lies in the segment $\{(4-2\lambda,0),\lambda\in[0,1]\}$. This fact agrees with proposition 5 because x^{*} is the only optimal solution.

Let us suppose the decision maker looks for a fuzzy solution instead of a crisp one. Applying the parametric approach the problem to be solved is

$$\begin{aligned} \text{Max } & h_1(2x_1+x_2) \\ & h_2(-x_1-2x_2) \geq \alpha \\ & \alpha\in[0,1] \\ & x_1+x_2 \leq 4 \\ & 3x_1+x_2 \geq 6 \\ & x_1,x_2 \geq 0 \end{aligned} \qquad (17)$$

whose optimal solution is

$$x^{*}(\alpha) = \begin{cases} (4,0) & \alpha \in [0,1/2] \\ (-2\alpha+5,0) & \alpha \in [1/2,1] \end{cases}$$

that defines a fuzzy solution for our original problem. Let us observe $x^{*}(\alpha)$ is efficient for any α and thus we might speak on efficient fuzzy solution.

The above optimal solution x^{*} is just $x^{*}(6/7)$ and it is easy to prove 6/7 is the only solution of the fixed point equation $h_1(2x_1^{*}(\alpha) + x_2^{*}(\alpha)) = \alpha$.

Now let us consider decision's maker evaluation of the connective AND is by the product operator. Thus the original problem is changed into

$$\begin{aligned} \text{Max } & \{h_1(2x_1+x_2)\cdot h_2(-x_1-2x_2)\} \\ \text{s.t: } & \\ & x_1+x_2 \leq 4 \\ & 3x_1+x_2 \geq 6 \\ & x_1,x_2 \geq 0 \end{aligned}$$

By using the results developed in Delgado et al. (1985c), it is easy to obtain the optimal solution

$$x^{**} = (3.5,0)$$

that is $x^{*}(6/8)$. Agreeing with proposition 5, x^{**} is efficient again (belongs to segment $\{(4-2\lambda), \lambda\in[0,1]\}$).

References

Alsina, C., E. Trillas and L.Valverde (1980), On non distributive logical connectives for Fuzzy Sets Theory. Busefal, 3, 18-28.

Bellman, R. and M.A. Girschick (1970), Decision making in a fuzzy environment. Management Science, 17, B-141-164.Buckley, J.J. (1983) Fuzzy Programming and the Pareto optimal set. Fuzzy Sets and Systems, 10, 57-64.

Chanas, S. (1983), The use of Parametric Programming in Fuzzy Linear Programming. Fuzzy Sets and Systems, 11, 243-251.

Chankong, V. and Y.Y. Haimes (1983), Multiobjective Decision Making, North Holland.

Cohon, J.L (1978), Multiobjective Programming and Planning, Academic Press, (1978).

Delgado, M. (1983), A resolution method for multiobjective problems. European Journal of Operations Research, 13, 165-172.

Delgado, M., J.L. Verdegay and M.A. Vila (1985a), Resolucion por Programacion Lineal Parametrica del problema multiobjetivo lineal difuso. Trabajos de Estadistica e Investigacion Operativa, 36, 126-137.

Delgado, M., J.L. Verdegay and M.A. Vila (1985b), Un metodo de resolucion para problemas multiobjetivo con metas difusas. Proceedings of the XV Reunion Nacional de la Sociedad Española de Estadistica e Investigacion Operativa, Oviedo (Spain), 150-160.

Delgado, M., J.L. Verdegay and M.A. Vila (1985c), Solving the biobjective linear programming problem: a fuzzy approach. In M.M. Gupta, A. Kandel, W. Bandler, J.B. Kiszka (eds.), Approximate Reasoning in Expert Systems, Elsevier Science Publishers, Bv, North Holland, 317-322.

Driankov, D. (1987) An outline of a fuzzy sets approach to decision making with interdependent goals. Fuzzy Sets and Systems, 21, 3.

Dubois, D. (1980), Triangular norms for fuzzy sets. Proceedings of the Second Seminar on Fuzzy Sets Theory (E. Klement ed.), 39-68.

Dubois, D. (1983), Modeles mathematiques de l'imprecis et de l'incertain en vue d'applications aux techniques d'aide a la decision. Tesis d'Etat,Universite of Grenoble.

Dubois, D. and H. Prade (1980)E, Fuzzy sets and Systems. Theory and Applications. Academic Press, New York.

Hannan, E.L. (1981a) Linear Programming with multiple fuzzy goals. Fuzzy Sets and Systems, 6, 235-248.

Hannan, E.L. (1979), On the efficiency of the product operator in Fuzzy Programming with multiple objectives. Fuzzy Sets and Systems, 2, 259-262.

Ignizio, J.P. (1982), On the (re)discovery of fuzzy goal programming, Decision Sci. 13, 331-336.

Kacprzyk, J. and R.R. Yager (1984), Linguistic quantifiers and belief qualification in fuzzy multicriteria and multistage Decision Making. Control and Cybernetics, 13, 155-174.

Leberling, H. (1981), On finding compromise solutions in multicriteria problems using the fuzzy min-operator. Fuzzy Sets and Systems, 6, 105-118.

Leung, Y. (1982), Multicriteria conflict solution through a theory of displaced Fuzzy Ideal. In M.M. Gupta and E. Sanchez (eds), Approximate Reasoning in Decision analysis. North Holland, 381-390.

Leung, Y. (1983), A concept of a Fuzzy Ideal for multicriteria conflict solution. In P.P. Wang (ed.) Advances in Fuzzy Sets Possibility Theory and Applications, Plenum Press, 387-403.

Llena, J. (1985), On Fuzzy Linear Programming. European Journal of Operations Research, 22, 216-223.

Narasimhan, R. (1980), Goal Programming in a fuzzy environment. Decision Sciences, 11, 325-336.

Narasimhan, R. (1981), On Fuzzy Goal Programming. Some comments. Decision Sciences, 12, 532-538.

Negoita, C.V. and D. Ralescu (1975), Applications of Fuzzy Sets to Systems Analysis. Birkhauser Verlag.

Rao, J.J., R.N. Tiwari and B.K. Mohanty (1988), A method for finding numerical compensation for fuzzy multicriteria decision problem. Fuzzy Sets and Systems, 25, 33-42,

Rubin, P.A. and R. Narasimhan (1984), Fuzzy Goal Programming with nested priorities. Fuzzy Sets and Systems, 14, 115-130.

Sakawa, M. (1984), Interactive Fuzzy Goal Programming for multiobjective non linear problems and its applications to water quality management, Control and Cybernetics, 13, 217-228.

Sakawa, M. and H. Yano (1986), Interactive fuzzy decision making for multi-objective nonlinear programming using augmented minimax problems. Fuzzy Sets and Systems, 20, 31-44.

Schweizer, B. and A. Sklar (1963), Associative functions and abstract semi groups. Publi. Math Debrecen, 10, 69-81.

Tanaka, H., T. Okuda and H. Asai (1974), On Fuzzy Mathematical Programming. Journal of Cybernetics, 3, 37-46.

Thole, U., H.-J. Zimmermann and P. Zysno (1981), On the suitability of minimum and product operators for the intersection of Fuzzy Sets. Fuzzy Sets and Systems, 2, 167-180.

Tiwari, R.N., S. Dharmar and J.R. Rao (1986), Priority structure in fuzzy goal programming. "Fuzzy Sets and Systems, 19, 251-259.

Tiwari, R.N., J.R. Rao and S. Dharmar (1985), Some aspects of fuzzy goal Programming. International Symposium on Mathematical Modelling of Ecological Environmental and Biological Systems, Kanpur, India.

Tiwari, R.N., S. Dharmar and J.R. Rao (1987), Fuzzy goal programming-An additive model. Fuzzy Sets and Systems, 24, 27-34.

Verdegay, J.L. (1982), Fuzzy Mathematical Programming. In M.M. Gupta and E. Sanchez (eds.) Fuzzy Information and Decision Processes, North Holland, 257-273

Werners, B. (1987), An interactive fuzzy programming system. Fuzzy Sets and Systems, 23, 131-148.

Zadeh, L.A. (1978), Fuzzy Sets as a basis for a Theory of Possibility. Fuzzy Sets and Systems, 1, 3-29.

Zeleny, M. (1982), Multiple Criteria Decision Making. Mc Graw Hill.

Zimmermann, H.J. (1978), Fuzzy Programming and Linear Programming with several objective functions, Fuzzy Sets and Systems, 1, 45-55.

Zimmermann, H.J. (1985), Applications of Fuzzy Sets Theory to Mathematical Programming. Information Sciences, 367, 29-58.

Zimmermann, H.J. and P. Zysno (1980), Latent Connectives in Human Decision Making. Fuzzy Sets and Systems, 4, 37-51.

Part III
Chapter 3

'FLIP': AN INTERACTIVE METHOD FOR MULTIOBJECTIVE LINEAR PROGRAMMING WITH FUZZY COEFFICIENTS

R. SŁOWIŃSKI
Institute of Computer Science
Technical University of Poznań
60-965 Poznań, Poland

ABSTRACT. FLIP is an interactive method for solving multiobjective linear programming problems with fuzzy coefficients in the objective functions and on the both sides of the constraints. Taking into account that fuzzy aspiration levels are defined on all the objectives, the key question is how to compare left- and right-hand-sides in objectives and constraints which are fuzzy numbers for given $\underline{x}$. Assuming an L-R representation of fuzzy numbers, a comparison principle has been proposed enabling transformation of the fuzzy MOLP problem into a multiobjective linear fractional program. It is then solved with an interactive method using a linear programming procedure as the only optimiser. In FLIP, interaction with the user takes place at two levels: first, when safety parameters have to be defined in the transformation phase, and second, when the associate deterministic problem is solved. At the latter level, the user is assisted by a graphical display of solutions.

1. INTRODUCTION

FLIP solves the following multiobjective linear programming (MOLP) problem with fuzzy coefficients:

$$(\text{P1}) \qquad \widetilde{\text{minimize}} \begin{bmatrix} \tilde{z}_1 = \underline{\tilde{c}}_1\underline{x} \\ \cdots\cdots \\ \tilde{z}_k = \underline{\tilde{c}}_k\underline{x} \end{bmatrix}$$

$$\text{s.t.} \quad \underline{\tilde{a}}_i\underline{x} \lesssim \tilde{b}_i \qquad i = 1,\ldots,m_1 \qquad \text{(A)}$$

$$\underline{\tilde{a}}_i\underline{x} \gtrsim \tilde{b}_i \qquad i = m_1+1,\ldots,m \qquad \text{(B)}$$

$$\underline{x} \geq \underline{0}$$

R. Slowinski and J. Teghem (eds.),
Stochastic versus Fuzzy Approaches to Multiobjective Mathematical Programming under Uncertainty, 249–262.

where $\underline{x}$ is a columm vector of n decision variables, $\underline{\tilde{c}}_1,\ldots,\underline{\tilde{c}}_k$ are row vectors of fuzzy cost coefficients corresponding to the objective functions $\tilde{z}_1,\ldots,\tilde{z}_k$, $\underline{\tilde{a}}_i$ is the i-th row of the matrix of fuzzy coefficients, and $\tilde{b}_i$ is its corresponding fuzzy right-hand-side. The equality constraints are excluded from the above formulation since they can be obviously represented by pairs of inequality constraints. It is assumed, moreover, that for the particular objectives, the decision maker (DM) is in a position to define fuzzy aspiration levels, thought of as goals, denoted by $\tilde{g}_1,\ldots,\tilde{g}_k$. All fuzzy coefficients are given as fuzzy numbers, i.e. normal and convex fuzzy subsets of the real line.

FLIP has been proposed to solve a water supply planning problem under uncertainty. Its presentation in (Słowiński, 1986) has been preceded by a brief survey of approaches to fuzzy mathematical programmming proposed before 1984. The survey has shown that there was no method which would deal with a multiobjective linear programming problem with fuzzy coefficients in the objective functions and on the both sides of the constraints. Since then, a lot of new work has been done; a short updated review has been made in (Słowiński, 1987), together with some refinements of FLIP. More comprehensive surveys have been done recently by Dubois (1987), Luhandjula (1989), and Inuiguchi, Ichihashi and Tanaka (1990). Others, like Delgado, Verdegay and Vila (1989), Rommelfanger (1989), and Sakawa and Yano (1989), presented their new methods on the background of existing ones.

The idea of FLIP relies on an observation that for a given $\underline{x}$, the main question to be answered consists in the comparison of the left- and right-hand-sides in objectives and constraints which are fuzzy numbers. Assuming an L-R representation of fuzzy coefficients, a comparison principle has been proposed (Słowiński, 1986) which allows a transformation of the fuzzy MOLP problem into a multiobjective linear fractional programming problem. The best compromise solution of the latter problem ensures the "best consistency" between the goals and the objective functions, and satisfies the constraints with a given "safety".

In the next section, we present the comparison principle of fuzzy numbers which is used in the transformation of the fuzzy MOLP problem into a deterministic mathematical program. The transformation is described in section 3, and in section 4, the way of solving the associate deterministic problem is outlined. In section 5, the steps of FLIP are summarized with a special emphasis on interactive steps.

2. COMPARISON OF L-R FLAT FUZZY NUMBERS

A convenient representation of a fuzzy number $\tilde{a}$ is a quadruple of parameters $(a^L, a^U, \alpha, \beta)$ of its membership function

$$\mu_{\tilde a}(x)=\begin{cases} L((a^L-x)/\alpha) & \text{if } x < a^L \\ 1 & \text{if } a^L \le x \le a^U \\ R((x-a^U)/\beta) & \text{if } x > a^U \end{cases}$$

where $[a^L,a^U]$ is an interval of the "most possible" values, α and β are nonnegative left and right "spreads" of $\tilde a$, respectively, and L, R are symmetric bell-shaped reference functions that are decreasing in $(-\infty, \infty)$ and $L(0)=R(0)=1$, $L(1)=R(1)=0$; $\tilde a$ is said to be an L-R flat fuzzy number. If $a^L=a^U=a$, then a is called the "middle"* value of $\tilde a$. When the spreads are zero and $a^L=a^U=a$, then $\tilde a$ is a nonfuzzy (crisp) number equal to a.

Let $\tilde a=(a^L,a^U,\alpha,\beta)_{LR}$ and $b=(b^L,b^U,\gamma,\delta)_{L'R'}$ be flat fuzzy numbers. In order to evaluate the possibility for $\tilde b$ to be greater than $\tilde a$, we use two different indices: δ and π, called optimistic and pessimistic, respectively.

(a) Optimistic index σ

Let us suppose a situation presented in Fig.1. Optimistic index $\sigma(\tilde b>\tilde a)$ is defined as an ordinate of the intersection point of R' and L, i.e. the right slope of $\tilde b$ with the left slope of $\tilde a$:

$$\sigma(b>a) = R'((d-b^U)/\delta) = L((a^L-d)/\alpha) = \omega \tag{1}$$

where d is an abscissa of the intersection point. The smaller $\sigma(\tilde b>\tilde a)$ is, the less true is the assertion that $\tilde b$ is greater than $\tilde a$. Let us observe that (1) is equivalent to

$$b^U + \delta R'^{-1}(\omega) = a^L - \alpha L^{-1}(\omega) \tag{2}$$

and

$$\delta R'^{-1}(\omega) + \alpha L^{-1}(\omega) = a^L - b^U \tag{3}$$

*For a flat fuzzy number, middle value means $(a^L+a^U)/2$.

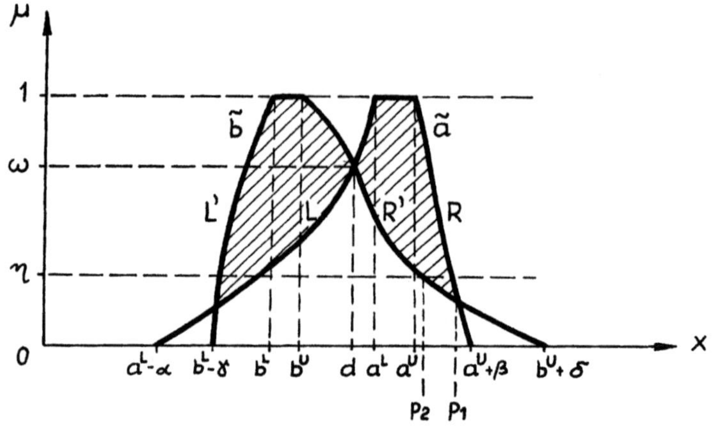

Fig. 1. Comparison of flat fuzzy numbers $\tilde{a}$ and $\tilde{b}$

For $F(\omega) = (\delta R'^{-1}(\omega) + \alpha L^{-1}(\omega))^{-1}$, we have

$$F^{-1}(\omega) = a^L - b^U \tag{4}$$

which implies

$$\sigma(\tilde{b}>\tilde{a}) = \omega = F(a^L - b^U) \tag{5}$$

Similarly, $\sigma(\tilde{a}>\tilde{b})$ is defined as an ordinate of the intersection point of R and L', i.e. the right slope of $\tilde{a}$ with the left slope of $\tilde{b}$:

$$\sigma(\tilde{a}>\tilde{b}) = \psi = G(b^L - a^U) \tag{6}$$

where $G(\psi)=(\gamma L'^{-1}(\psi) + \beta R^{-1}(\psi))^{-1}$.

If L=R' and L'=R then (5) and (6) take the form

$$\sigma(\tilde{b}>\tilde{a}) = L((a^L - b^U)/(\alpha+\delta)) \tag{7}$$

$$\sigma(\tilde{a}>\tilde{b}) = R((b^L - a^U)/(\beta+\gamma)) \tag{8}$$

In order to say that $\tilde{b}>\tilde{a}$, we need both $\sigma(\tilde{a}>\tilde{b})$ and $\sigma(\tilde{b}>\tilde{a})$. If for instance $\sigma(\tilde{b}>\tilde{a})\geq 1$, we know that either $\tilde{b}$ is greater than $\tilde{a}$, or both fuzzy numbers are too close to be separated. Then we may choose a threshold $0\leq\tau\leq 1$ and admit that $\tilde{b}$ is greater than $\tilde{a}$ at level τ as soon as $\sigma(\tilde{a}>\tilde{b})\leq\tau$. If $\min\{\sigma(\tilde{a}>\tilde{b}),\sigma(\tilde{b}>\tilde{a})\}\geq\tau$, we say that $\tilde{a}$ and $\tilde{b}$ are approximately equal.

In the case of weak inequality $\tilde{b}\geq\tilde{a}$, we only need $\sigma(\tilde{b}>\tilde{a})$. Indeed, for $\sigma(\tilde{b}>\tilde{a})\geq 1$ the weak inequality is satisfied for any value of $\sigma(\tilde{a}>\tilde{b})$. Then we may choose a credibility constant $0\leq\tau\leq 1$ and admit that $\tilde{b}\geq\tilde{a}$ at credibility level τ as soon as $\sigma(\tilde{b}>\tilde{a})\geq\tau$.

The comparison index σ can be seen as optimistic because even for a high value of $\sigma(\tilde{b}>\tilde{a})$, the possibility for $x\in\mathbb{R}$, fuzzily restricted to belong to $\tilde{a}$, to be greater than $y\in\mathbb{R}$, fuzzily restricted to belong to $\tilde{b}$, may be quite big. As a measure of this possibility one can consider the hatched area marked in Fig. 1. Thus, in order to make the comparison more credible, we should use σ conjointly with the pessimistic index π.

(b) Pessimistic index π

The pessimistic index follows from the comparison of the right slopes of $\tilde{a}$ and $\tilde{b}$ at some level $0\leq\eta\leq 1$:

$$\pi(\tilde{b}>_{\eta}\tilde{a}) = b^U + \delta R'^{-1}(\eta) - a^U - \beta R^{-1}(\eta) = p_2 - p_1 \tag{9}$$

$$\pi(\tilde{a}>_{\eta}\tilde{b}) = a^U + \beta R^{-1}(\eta) - b^U - \delta R'^{-1}(\eta) = p_1 - p_2 \tag{10}$$

For $\eta=0$ we have

$$\pi(\tilde{b}>_{0}\tilde{a}) = b^U + \delta - a^U - \beta \tag{11}$$

$$\pi(\tilde{a}>_{0}\tilde{b}) = a^U + \beta - b^U - \delta \tag{12}$$

Thus, a joint use of $\sigma(\tilde{b}>\tilde{a})$ and $\pi(\tilde{b}>_{\eta}\tilde{a})$ consists in the comparison of both slopes of $\tilde{a}$ which is supposed to be smaller with the right slope of $\tilde{b}$ which is supposed to be greater.

Now, we can admit that $\tilde{b} \geq \tilde{a}$ at credibility levels τ and η if and only if

$$\sigma(\tilde{b} > \tilde{a}) \geq \tau \tag{13}$$

and
$$\pi(\tilde{b} >_{\eta} \tilde{a}) \geq \theta \tag{14}$$

where $\tau, \eta \in [0,1]$ and $\theta \in (-\infty, \infty)$. $\theta \geq 0$ means that for any pair (x,y) such that $x \geq a^U$, $y \geq b^U$ and $0 \leq \mu_a(x) = \mu_b(y) \leq \eta$, inequality $y \geq x$ is true. A negative value of θ makes possible that inequality $y \geq x$ is not true.

The constants τ, η and θ are called "safety parameters" because they are responsible for the safety of the assertion that $\tilde{b}$ is greater than $\tilde{a}$. Instead of safety, we could speak of course about risk which is a complementary term. Let us remark that using τ, η and θ one can control the surface of the hatched area marked in Fig. 1 which corresponds to the risk of the above assertion.

3. THE ASSOCIATE DETERMINISTIC PROBLEM

Conditions (13) and (14) can be used directly to transform the fuzzy MOLP problem into an associate deterministic one. Let us separately analyse the constraints and the objective functions of the fuzzy MOLP problem.

Let us suppose that fuzzy coefficients of the constraints are given as L-R flat fuzzy numbers:

$$\tilde{\underline{a}}_i = (\underline{a}_i^L, \underline{a}_i^U, \underline{\alpha}_i, \underline{\beta}_i)_{LR}, \qquad \tilde{b}_i = (b_i^L, b_i^U, \gamma_i, \delta_i)_{RL}, \qquad i=1,\ldots m.$$

For the sake of clarity, we assume that the reference functions of all fuzzy coefficients are of two kinds only: L and R. It can be seen from the preceding considerations that this assumption is not necessary for the calculation of σ and π (cf. (5), (6) and (9), (10)).

It should be specified that all arithmetic operations on fuzzy numbers taking place in (P1) are extended operations in the sense of Zadeh's principle. For any $\underline{x} \geq \underline{0}$, the left side of the i-th constraint can be summarized to the flat fuzzy number:

$$\tilde{\underline{a}}_i\underline{x} = (\underline{a}_i^L\underline{x},\ \underline{a}_i^U\underline{x},\ \underline{\alpha}_i\underline{x},\ \underline{\beta}_i\underline{x})_{LR}, \qquad i=1,\dots,m.$$

It is easy to verify that for $i=1,\dots,m_1$

$$\sigma(\tilde{b}_i > \tilde{\underline{a}}_i\underline{x}) = L((\underline{a}_i^L\underline{x} - b_i^U)/(\underline{\alpha}_i\underline{x} + \delta_i))$$

$$\pi(\tilde{b}_i >_{\eta_i} \tilde{\underline{a}}_i\underline{x}) = b_i^U + \delta_i L^{-1}(\eta_i) - \underline{a}_i^U\underline{x} - \underline{\beta}_i\underline{x}R^{-1}(\eta_i)$$

and for $i=m_1+1,\dots,m$

$$\sigma(\tilde{\underline{a}}_i\underline{x} > \tilde{b}_i) = R((b_i^L - \underline{a}_i^U\underline{x})/(\underline{\beta}_i\underline{x} + \gamma_i))$$

$$\pi(\tilde{\underline{a}}_i\underline{x} >_{\eta_i} \tilde{b}_i) = \underline{a}_i^U\underline{x} + \underline{\beta}_i\underline{x}R^{-1}(\eta_i) - b_i^U - \delta_i L^{-1}(\eta_i)$$

For given values of safety parameters τ_i, η_i and θ_i, $i=1,\dots,m$, the constraints (A) and (B) of (P1) may be transformed to:

$$L((\underline{a}_i^L\underline{x} - b_i^U)/(\underline{\alpha}_i\underline{x} + \delta_i)) \geq \tau_i \qquad i=1,\dots,m_1 \qquad (15)$$

$$R((b_i^L - \underline{a}_i^U\underline{x})/(\underline{\beta}_i\underline{x} + \gamma_i)) \geq \tau_i \qquad i=m_1+1,\dots,m \qquad (16)$$

$$b_i^U + \delta_i L^{-1}(\eta_i) - \underline{a}_i^U\underline{x} - \underline{\beta}_i\underline{x}R^{-1}(\eta_i) \geq \theta_i \qquad i=1,\dots,m_1 \qquad (17)$$

$$\underline{a}_i^U\underline{x} + \underline{\beta}_i\underline{x}R^{-1}(\eta_i) - b_i^U - \delta_i L^{-1}(\eta_i) \geq \theta_i \qquad i=m_1+1,\dots,m \qquad (18)$$

Basing on the property of strict monotonicity of reference functions, one can transform (15) and (16) into the following equivalent conditions:

$$\underline{a}_i^L\underline{x} - b_i^U \leq L^{-1}(\tau_i)(\underline{\alpha}_i\underline{x} + \delta_i) \qquad i=1,\dots,m_1 \qquad (19)$$

$$b_i^L - \underline{a}_i^u \underline{x} \leq R^{-1}(\tau_i)(\underline{\beta}_i \underline{x} + \gamma_i) \qquad i=m_1+1,\ldots,m \qquad (20)$$

In the case of fuzzy objective functions, σ can be used to evaluate the degree of consistency between fuzzy objectives and fuzzy goals. Let the fuzzy cost coefficients and fuzzy goals be:

$$\tilde{\underline{c}}_l = (\underline{c}_l^L, \underline{c}_l^U, \underline{\varepsilon}_l, \underline{\kappa}_l)_{LR}, \qquad \tilde{g}_l = (g_l, 0, \nu_l)_{LL}, \qquad l=1,\ldots,k$$

where $g_l^L = g_l^U = g_l$ and, moreover, the left spread of $\tilde{g}_l$, $l=1,\ldots,k$, is equal to zero because it is immaterial for the evaluation of consistency between goals and objectives. Here again, the equality of the reference functions is not a necessary assumption. For any $\underline{x}$, the components of the l-th objective function can be summarized with the extension principle to the flat fuzzy number:

$$\tilde{\underline{c}}_l \underline{x} = (\underline{c}_l^L \underline{x},\ \underline{c}_l^U \underline{x},\ \underline{\varepsilon}_l \underline{x},\ \underline{\kappa}_l \underline{x})_{LR}, \qquad l=1,\ldots k.$$

On the basis of (7), we can write for $l=1,\ldots,k$

$$\sigma(\tilde{g}_l > \tilde{\underline{c}}_l \underline{x}) = L((\underline{c}_l^L \underline{x} - g_l)/(\underline{\varepsilon}_l \underline{x} + \nu_l)) \qquad (21)$$

In order to ensure the best consistency between goals and objectives, $\sigma(\tilde{g}_l > \tilde{\underline{c}}_l \underline{x})$, $l=1,\ldots,k$ should be maximized. In consequence, we orrive to the following deterministic mathematical programming problem equivalent to problem (P1):

$$\text{(P2) maximize} \begin{bmatrix} f_1(\underline{x}) = L((\underline{c}_1^L \underline{x} - g_1)/(\underline{\varepsilon}_1 \underline{x} + \nu_1)) \\ \cdots\cdots\cdots\cdots\cdots\cdots \\ f_k(\underline{x}) = L((\underline{c}_k^L \underline{x} - g_k)/(\underline{\varepsilon}_k \underline{x} + \nu_k)) \end{bmatrix}$$

$$\text{s.t. } \underline{a}_i^L \underline{x} - b_i^U \leq L^{-1}(\tau_i)(\underline{\alpha}_i \underline{x} + \delta_i) \qquad i=1,\ldots,m_1 \qquad (A1)$$

$$b_i^L - \underline{a}_i^u \underline{x} \leq R^{-1}(\tau_i)(\underline{\beta}_i \underline{x} + \gamma_i) \qquad i=m_1+1,\ldots,m \qquad (B1)$$

$$b_i^U + \delta_i L^{-1}(\eta_i) - \underline{a}_i^U \underline{x} - \underline{\beta}_i \underline{x} R^{-1}(\eta_i) \geq \theta_i \qquad i=1,\dots,m_1 \qquad \text{(A2)}$$

$$\underline{a}_i^U \underline{x} + \underline{\beta}_i \underline{x} R^{-1}(\eta_i) - b_i^U - \delta_i L^{-1}(\eta_i) \geq \theta_i \qquad i=m_1+1,\dots,m \qquad \text{(B2)}$$

$$\underline{x} \geq \underline{0} \qquad \text{(C)}$$

If reference function L of fuzzy cost coefficients is linear, i.e. $L(y)=1-y$, than problem (P2) becomes a multiobjective linear fractional programming (MLFP) problem:

$$\text{(P3) maximize} \begin{bmatrix} f_1(\underline{x}) = 1 - (\underline{c}_1^L \underline{x} - g_1)/(\underline{\varepsilon}_1 \underline{x} + \nu_1) \\ \dots\dots\dots\dots\dots\dots\dots\dots \\ f_k(\underline{x}) = 1 - (\underline{c}_k^L \underline{x} - g_k)/(\underline{\varepsilon}_k \underline{x} + \nu_k) \end{bmatrix}$$

s.t. (A1), (B1), (A2), (B2), (C).

In order to be sure that the denominatores of $f_1(\underline{x}),\dots,f_k(\underline{x})$ are greater than zero for any feasible $\underline{x}$, we may admit that $\nu_l>0$, $l=1,\dots,k$, which is quite natural.

4. INTERACTIVE STEPS

FLIP can be summarized in the following steps:

Step 1. Formulation of problem (P1) and definition of fuzzy coefficients.

Step 2. Definition of fuzzy aspiration levels $\tilde{g}_l$ $(l=1,\dots,k)$ on objectives.

Step 3. Definition of safety parameters τ_i, η_i and $\theta_i (i=1,\dots,m)$ for fuzzy constraints.

Step 4. Formulation of the associate multiobjective deterministic problem (P2) or (P3).

Step 5. Application of an interactive method for solving the associate deterministic problem formulated in step 4.

Step 6. If a best compromise solution has been found then stop, otherwise return to step 3 for revision of safety parameters. Retraction to steps 1 and 2 for redefinition of fuzzy coefficients and/or aspiration levels is also possible.

Our experience indicates that an interaction with the DM in searching of the best compromise solution is very beneficial for the final decision. So, in step 5, in order to solve the associate multiobjective deterministic problem, we propose to use an interactive method (cf. Vanderpooten, 1990). If the associate deterministic problem has the form of the MLFP problem (P3), a very convenient interactive procedure is that of Choo and Atkins (1980).
Let us recall its general idea informally.

In order to find a starting efficient point of the MLFP problem, the objective functions $f_1,\ldots,f_k$ are aggregated by the Chebyshev norm which is the maximum weighted deviation from the ideal point $\underline{u}^*$:

(P4) minimize $\max_l\{\varphi_l(u_l^* - f_l(\underline{x}))\}$

s.t. (A1), (B1), (A2), (B2), (C)

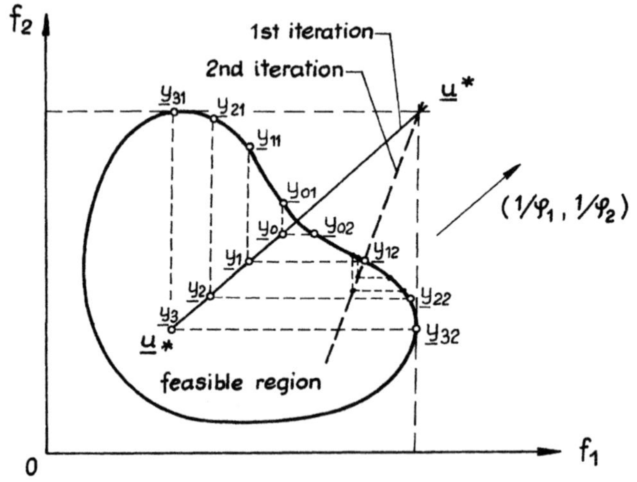

Fig. 2. An idea of the interactive procedure by Choo and Atkins (1980)

where $\underline{\phi} = (\phi_1,\ldots,\phi_k)$ is a weighting vector defining a "direction" of the Chebyshev norm, i.e. a line passing through an ideal and a nadir point, $\underline{u}^*$ and $\underline{u}_*$, respectively (cf. Fig.2).

The Chebyshev norm minimization chooses the "corner" closest to $\underline{u}^*$ and still in contact with the feasible region. This final point of contact ensures weak efficiency. In the case of the MLFP, (P4) can be transformed to a linear program with a single parameter. One can thus use any convenient univariate search method over this parameter to find point $\underline{y}_0$ as close to efficient as we like. The "closeness" here is not critical and even a rough approximation will be quite sufficient. This point is taken as starting point for the interactive part of the algorithm. This part crucially involves the DM. The search direction from $\underline{u}^*$ to $\underline{y}_0$ is then extended to include several more points $\underline{y}_1$, $\underline{y}_2,\ldots,\underline{y}_p$. Next, taking each criterion in turn, say f_1 first, f_1 is maximized subject to all other criteria being at least equal to their values at $\underline{y}_0$, then $\underline{y}_1$, all the way to $\underline{y}_p$. This gives a sequence of (weakly) efficient points $\underline{y}_{01}$, $\underline{y}_{11},\ldots,\underline{y}_{p1}$ for criterion 1 and $\underline{y}_{20}$, $\underline{y}_{21}$, etc. for criterion 2, and so on for all the criteria. The minimization of $f_l(\underline{x})$ for $l=1,\ldots,k$ is of a linear fractional subject to linear constraints, so the Charnes and Cooper transformation into a linear program can be used. Thus, at each step a single-objective linear programming problem has to be solved. The calculated sample of the efficient border is then presented to the DM who is asked to select the one that best fits his needs. The selected point becomes the new starting point and the procedure is repeated, but now the search space is focused on a part of the efficient border of most interest. The interactive process ceases when the most satisfactory efficient point is reached.

An important advantage of the above algorithm is that the only optimization procedure to be used is a linear programming one. Moreover, it has a simple scheme and allows retractions to the points which have been found uninteresting in previous iterations.

In a microcomputer implementation of FLIP, the efficient points proposed to the DM are presented both numerically, in terms of $\underline{x}$ and middle values of $\tilde{z}_l(\underline{x})$, $l=1,\ldots,k$, and graphically, in terms of mutual positions of fuzzy numbers corresponding to objectives and aspiration levels on the one hand, and to left- and right-hand-sides of constraints, on the other hand (cf. Czyżak, 1990; Czyżak and Słowiński, 1989; Słowiński and Teghem, 1990). In this way, the DM gets quite a complete idea about the quality of each proposed solution. The quality is evaluated taking into account:

- scores of fuzzy objectives in relation to the goals,
- dispersion of values of the fuzzy objectives due to uncertainty,
- safety of the solution or, using a complementary term, the risk of violation of the constraints (cf. the hatched areas in Fig.1).

So, the definition of the best compromise involves an analysis of the compromise among criteria and an evaluation of the safety of the corresponding solution.

It steems from the above considerations that the DM intervenes in two steps of FLIP. First, when fixing the safety parameters (step 3), and then in the course of the guided generation and evaluation of the efficient points of the associate deterministic problem (step 5). Thus, the interaction with the DM takes place at two levels. As to the first one (step 3), it is worth noting that there are two practical ways of controlling the safety of solutions using parameters τ_i, η_i and θ_i:

(a) fix $\eta_i= 0$, $i=1,\ldots,m$, and control the optimistic safety with τ_i, and the pessimistic safety with θ_i, $i=1,\ldots,m$, or

(b) fix $\theta_i= 0$, $i=1,\ldots,m$, and control the optimistic safety with τ_i, and the pessimistic safety with η_i, $i=1,\ldots,m$.

The safety parameters are defined taking into account their interval of variation and the knowledge acquired in preceding iterations about the dependency between safety and the quality of the compromise among criteria.

If way (a) is chosen, an (approximate) interval of variation of θ_i at level $\eta_i=0$, $i=1,\ldots,m$, can be calculated in the following way. Let the interval be denoted by $[\theta_i^L,\theta_i^U]$, and the set of feasible solutions by D.

It is clear that

$$\theta_i^U = b_i^U + \delta_i - b_i^L + \gamma_i \qquad \text{for } i=1,\ldots m_1 \qquad (22)$$

and $$\theta_i^U = \max_{\underline{x} \in D} \{\underline{a}_i^U\underline{x} + \underline{\beta}_i\underline{x} - \underline{a}_i^L\underline{x} + \underline{\alpha}_i\underline{x}\} \qquad \text{for } i=m_1+1,\ldots,m \qquad (23)$$

which corresponds to a perfect safety of solutions. The lower bound of the interval is defined as:

$$\theta_i^L = \min_{\underline{x} \in D} \{b_i^U + \delta_i - \underline{a}_i^U\underline{x} - \underline{\beta}_i\underline{x}\} \qquad \text{for } i=1,\ldots m_1 \qquad (24)$$

and $$\theta_i^L = \min_{\underline{x} \in D} \{\underline{a}_i^U\underline{x} + \underline{\beta}_i\underline{x} - b_i^U - \delta_i\} \qquad \text{for } i=m_1+1,\ldots,m \qquad (25)$$

In practice, an exact calculation of $[\theta_i^L, \theta_i^U]$, $i=1,\ldots,m$, may lead to too large intervals with very low utility. The following procedure permits "killing two birds with one stone". One substitutes all fuzzy coefficients of problem (P1) by their middle values and calculate the pay-off table of the multiobjective deterministic problem obtained from (P1). It needs solving k single-objective linear programming problems. In this way, a set S of k solutions is obtained. The result can be used in two ways. First, one can substitute D by S in formulae (23), (24), (25). Then $[\theta_i^L, \theta_i^U]$, $i=1,\ldots,m$, become approximate intervals having often a greater practical utility than exact ones. Second, the individual optima from the diagonal of the pay-off table may serve as suggestions for middle values of fuzzy goals. The columns of the pay-off table may also give an idea about the right spreads of the goals. This option is useful when the DM has no his own definition of the aspiration levels.

To conclude, let us remark that since its appearance, FLIP has been successfully applied to several real-world problems from the field of water supply systems (Słowiński et al., 1986; Słowiński, 1987) and agriculture (Czyżak, 1989; Czyżak and Słowiński, 1989; Czyżak, 1990). A simple didactic example is solved in Chapter IV-2 of this book.

REFERENCES

Choo, E.U., Atkins, D.R. (1980) 'An interactive algorithm for multicriteria programming', Computers and Operations Research 7, 81-87.

Czyżak, P. (1989) 'Multicriteria agricultural problem solving under uncertainty', Foundations of Control Engineering 14, no.2.

Czyżak, P. (1990) 'Application of the FLIP method to farm structure optimization problems', Chapter III-4 in this volume.

Czyżak, P. Słowiński, R. (1989) 'Multiobjective diet optimization problem under fuzziness', in J.L. Verdegay, M. Delgado (eds.), The Interface between Artificial Intelligence and Operations Research in Fuzzy Environment, Verlag TUV Rheinland, Koln, pp. 85-103.

Delgado, M., Verdegay, J.L., Vila, M.A. (1989) 'A general model for fuzzy linear programming', Fuzzy Sets and Systems 29, 21-30.

Dubois, D. (1987) 'Linear programming with fuzzy data', in J.C. Bezek (ed.), Analysis of Fuzzy Information, Vol.3: Applications in Engineering and Science, CRC Press, Boca Raton, pp. 241-263.

Inuiguchi, M., Ichihashi, H., Tanaka, H. (1990) 'Fuzzy programming: a survey of recent developments", Chapter I-4 in this volume.

Luhandjula, M.K. (1989) 'Fuzzy optimization: an appraisal', Fuzzy Sets and Systems 30, 257-282.

Rommelfanger, H. (1989) 'Interactive decision making in fuzzy linear optimization problems', European J. Operational Res. 41, 210-217.

Sakawa, M., Yano, H. (1989) 'Interactive fuzzy decision making for generalized multiobjective linear programming problems with fuzzy parameters', Fuzzy Sets and Systems 29, 315-326.

Słowiński, R. (1986) 'A multicriteria fuzzy linear programming method for water supply system development planning', Fuzzy Sets and Systems 19, 217-237.

Słowiński, R. (1987) 'An interactive method for multiobjective linear programming with fuzzy parameters and its application to water supply planning', in J. Kacprzyk, S.A. Orlovski (eds.), Optimization Models using Fuzzy Sets and Possibility Theory, D. Reidel, Dordrecht, pp. 396-414.

Słowiński, R., Teghem, J. (1990) 'Comparison study of STRANGE and FLIP', Chapter IV-2 in this volume.

Słowiński, R., Urbaniak, A., Węglarz, J. (1986) 'Probabilistic and fuzzy approaches to capacity expansion planning of a water supply system', in L. Valadares Tavares, J. Evaristo da Silva (eds.), Systems Analysis Applied to Water and Related Land Resources, Pergamon Press, Oxford, pp. 93-98.

Vanderpooten, D. (1990)' Multiobjective programming: basic concepts and approaches', Chapter I-2 in this volume.

Part III
Chapter 4

APPLICATION OF THE 'FLIP' METHOD TO FARM STRUCTURE OPTIMIZATION UNDER UNCERTAINTY.

P. CZYŻAK
Center of Education and Development in Agriculture
61-659 Poznań, ul.Winogrady 63, Poland

ABSTRACT. A farm structure optimization problem is considered which is formulated as a multicriteria linear programming problem with uncertain coefficients. The uncertain data are expressed in terms of L-R type fuzzy numbers. According to a principle of comparison of fuzzy numbers proposed in the 'FLIP' method, the orginal problem is transformed into a multicriteria linear fractional programming problem which is then solved using an interactive procedure. A real numerical example of a farm structure optimization problem is solved using the 'FLIP' software on micro.

1. INTRODUCTION

Agriculture is one of those domains in which attempts of using optimization methods have been made for many years. There were formulated many different models - from diet models, through models of a selected branch of agriculture, up to complex models that allow farm structure optimization. They usually had deterministic and single objective character because existing optimization methods didn't handled more complicated models of large size. So significant simplification of reality led, however, to low applicability of results.

In practice, every economic activity involves both a piece of uncertainty and a necessity of multicriteria evaluation of alternative decisions. In respect to agriculture, typical elements of economic play are influenced by such unpredictable factors as market fluctuations, climat and soil conditions. It is also typical for agricultural models that evaluation of alternative solutions involves multiple criteria, like cost or income, risk and ecological implications.

When handling uncertain data in an optimization model, we have to choose among two approaches: stochastic and fuzzy. First of them requires definition of probability distributions or probable scenarios for uncertain data. In the second approach, the decision maker defines the range of variability for uncertain data, points out the most possible value and a possibility distribution for other values within a given range.

R. Slowinski and J. Teghem (eds.),
Stochastic versus Fuzzy Approaches to Multiobjective Mathematical Programming under Uncertainty, 263–278.

In problems related to agriculture, one has to deal with a great subjectivity in estimation of uncertain data. So, it seems that the use of the second approach is more natural.

Both indicated aspects of agricultural problems, i.e. multicriteria evaluation of solutions and modelling of uncertain data using fuzzy approach, are taken into account in the 'FLIP' method proposed by R. Słowiński [6]. Until now, in the field of agriculture, the FLIP method was applied to optimization of a feeding dose (see Czyżak [2], Czyżak & Słowiński [3]).

In this chapter, we present an application of the 'FLIP' method to a farm structure optimization problem.

In section 2, we formulate the multicriteria farm structure optimization problem under uncertainty. The main steps of the FLIP software are presented in section 3. In the last section, we solve a real farm structure optimization problem using the FLIP software on an IBM compatible microcomputer.

2. FORMULATION OF THE FARM STRUCTURE OPTIMIZATION PROBLEM

To build a farm structure optimization model, detailed knowledge about plant raising, physiological needs of farm animal species and also about agricultural market mechanisms is necessary.

In general,we can say that such a model should include all available information concerning activities of a considered farm. By activities we understand here all operations taking place in the farm and requiring specified expenses and/or bringing certain profit, e.g. breeding some kind of animal, rising some kind of plant, buying or selling fodders, fertilizers, manpower and so on.

Making use of so defined model, the method of decision support should help the farmer in defining a desirable direction of farm development.

In this section, we formulate the model of farm structure optimization as multicriteria linear programming problem in which coefficients in criteria and on the both sides of constraints are, in general, L-R type fuzzy numbers. The model concerns a time horizon equal to 1 year.

A solution to the problem is defined as a vector of decision variables $\underline{x} = [x_1,\dots,x_n]$, where x_j denotes a size of activity j and n is a number of all activities taking place in the farm.

The region of variation of $\underline{x}$ is bounded by six types of constraints. Most of them are written in the form of balance conditions.

Balance of arable land and permanent grassland

$$\sum_{j\in A_t} x_j = AL \tag{1}$$

$$\sum_{j\in A_g} x_j = PG \tag{2}$$

where $A_t \subseteq \{1,\dots,n\}$ is a subset of activities j using arable land, AL is a size of arable land, $A_g \subseteq \{1,\dots,n\}$ is a subset of activities j using grassland, PG is a size of permanent grassland, $A_t \cap A_g = \emptyset$.

Balance of crop succession

$$\sum_{j \in A_f^p} x_j - \sum_{j \in A_p} x_j \geq 0 \tag{3}$$

where $p \in \{1,\dots,P\}$, P is a number of different kinds of plants, e.g. spring crops, winter crops; $A_p \subseteq \{1,\dots,n\}$ is a subset of plants of kind p, $A_f^p \subseteq \{1,\dots,n\}$ is a subset of plants being forecrop for plants of kind p.

Constraints on the area of some groups of plants

$$\sum_{j \in A_p} x_j \leq A_p^{max} AL \tag{4}$$

$$\sum_{j \in A_p} x_j \geq A_p^{min} AL \tag{5}$$

where $A_p \subseteq \{1,\dots,n\}$ is a subset of plants of kind p, $A_p^{max} \in [0,1]$ and $A_p^{min} \in [0,1]$ are, respectively, maximal and minimal contribution of plants of kind p in the arable land.

Manpower balance

$$\sum_{j \in A_l} \tilde{a}_{jl} x_j - x_l \leq \tilde{R}_l \tag{6}$$

where $l \in \{1,\dots L\}$, L is a number of periods of extended manpower demand, $A_l \subseteq \{1,\dots,n\}$ is a subset of activities j that require manpower in the l-th period, $\tilde{a}_{jl}$ denotes labour consumption of the j-th activity in the l-th period, x_l - hire of workers in the l-th period, and $\tilde{R}_l$ - available amount of manpower in the l-th period.

Balance of artificial fertilizers

$$\sum_{j \in A_f} \tilde{d}_{jf} x_j - x_f \leq 0 \tag{7}$$

where $f \in \{1,\dots,F\}$, F is a number of different fertilizers or components of fertilizers, e.g. phosphorus, nitrogen, potassium; $A_f \subseteq \{1,\dots,n\}$ is a subset of activities that require fertilizer f, $\tilde{d}_{jf}$ is a quantity of fertilizer f required by activity j; x is an amount of fertilizer f to be bought.

Balance of fodder
In our model, we take animals into consideration in a specific way. We have in Poland a lot of small farms in which the number of breeded animals is, usually, constant and small. So, in the present model, we consider the number of animals as fixed for a given farm, and set up a condition describing fodder requirement for these animals.

$$\sum_{j\in A_k} \tilde{h}_{jk} x_j + x_k \geq \sum_{s\in S_k} \tilde{T}_{sk} h_s \tag{8}$$

where $k \in \{1,\dots,K\}$, K is a number of different fodders, $A_k \subseteq \{1,\dots,n\}$ is a subset of activities that produce fodder k, $\tilde{h}_{jk}$ is an amount of fodder k produced by one unit of activity j; x_k denotes indispensable purchase of fodder k; S_k is a subset of animal species that require fodder k; h_s is a number of animals of kind s; $\tilde{T}_{sk}$ is an average amount of fodder k consumed by animal of kind s.

In order to evaluate a farm structure we shall use three following criteria:

-Gross profit

$$\sum_{j\in B_s} \tilde{c}^1_j x_j - \sum_{j\in B_b} \tilde{c}^2_j x_j + \sum_{s=1}^{S} \tilde{T}_s h_s \tag{9}$$

where $B_s \subseteq \{1,\dots,n\}$ is a subset of activities whose results can be selled; $\tilde{c}^1_j$ is a price of a unit of activity j; $B_b \subseteq \{1,\dots,n\}$ is a subset of activities that depend on purchasing or consuming something; $\tilde{c}^2_j$ is a cost of a unit of product j. The last part of the formula is a constant that expresses the value of gross profit from animal production; S is a number of kinds of animals in the farm, $\tilde{T}_s$ is an average gross profit for animal of kind s, h_s is a number of animals of kind s.

-Structure-forming plants area

$$\sum_{j\in B_p} \tilde{c}^3_j x_j \tag{10}$$

where B_p is a subset of plants that have a structure-forming character, $\tilde{c}_j^3$ expresses a structure-forming possibility of plant j per unit of arable land.

-Manpower hire

$$\sum_{l=1}^{L} x_l \tag{11}$$

where L is a number of periods of extended manpower demand, x_l denotes a deficiency of labour in the l-th period.

The criterion of gross profit and structure-forming plants area should be maximized while the citerion of manpower hire should be minimized.

Finnaly, we get a MOLP problem which consist in searching for vector $\underline{x}$ which satisfies constraints (1)-(8) and yields the best compromise between criteria (9)-(11).

The imprecision of some coefficients is denoted by ~. The specific problem (1)-(11), can be rewritten in the following general form:

$$\text{MAX}\begin{bmatrix} \tilde{\underline{c}}_1^T\underline{x} \\ \ldots \\ \tilde{\underline{c}}_k^T\underline{x} \end{bmatrix} \tag{12}$$

subject to

$$\tilde{\underline{a}}_i\underline{x} \le \tilde{b}_i \qquad i=1,\ldots,m' \tag{13}$$

$$\tilde{\underline{a}}_i\underline{x} \ge \tilde{b}_i \qquad i=m'+1,\ldots,m \tag{14}$$

$$x_j \ge 0 \qquad j=1,\ldots,n. \tag{15}$$

where $\tilde{\underline{c}}_l$ is a column vector of cost coefficients of criterion l ($l=1,\ldots,k$), $\tilde{\underline{a}}_i$ are constraint coefficients ($i=1,\ldots,m$, is a row number), while $\tilde{b}_i$ ($i=1,\ldots,m$) stand for right-hand-sides of constraints.

3. MAIN STEPS OF THE 'FLIP' SOFTWARE

To solve problem (12),(13)-(15) we use the 'FLIP' method described in [7]. In this section, we describe a microcomputer implementation of the most important steps of this method.

Step I. *Formulation of problem (12),(13)-(15) and definition of fuzzy coefficients.*

Entering of data, in particular fuzzy coefficients, is supported in a graphical way. Decision maker (DM) can observe both numbers defining a coefficient and its graphical representation being displayed on a screen (fig.1).

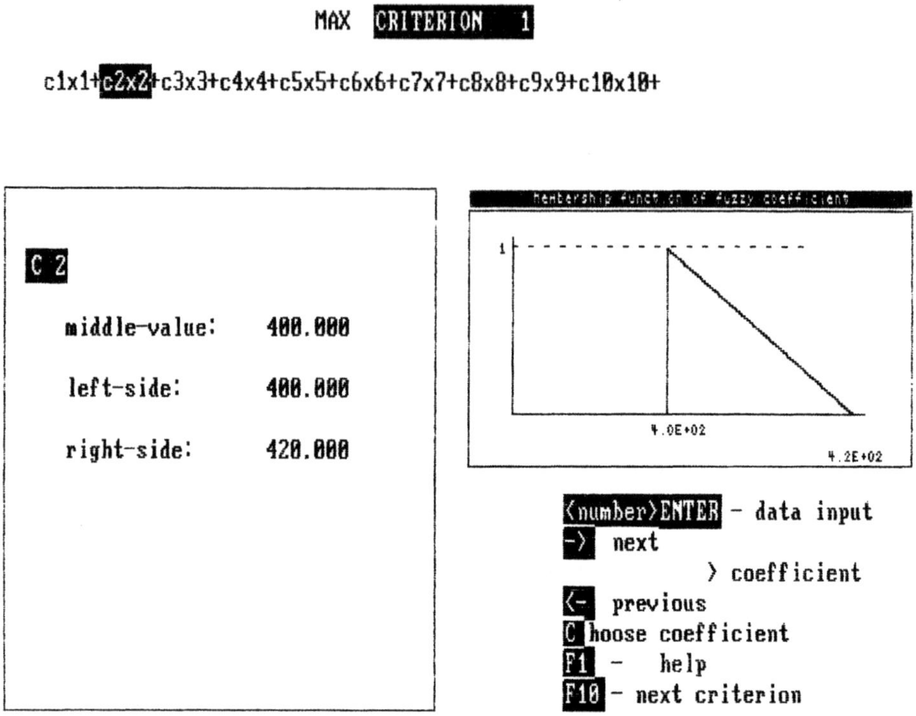

Figure 1. Entering of data.

The FLIP software makes possible relatively flexible defining of fuzzy coefficients. It allows using piecewise linear functions under the restriction that the number of linear pieces to the left and to the right of the middle value does not exceed 3 (fig.2).

Step II. *Definition of fuzzy aspiration levels* $\tilde{g}_l$ *(l=1,...,k) for criteria.*
To facilitate deciding about aspiration levels, individual optima for the criteria are calculated under the assumption that fuzzy coefficients of the problem take their middle values. The results are displayed suggesting attainable levels of criteria.

Step III. *Definition of safety coefficients* τ_i, η_i *and* Θ_i *(i=1,...,m) for constraints.*
To support DM, standard values are proposed for each coefficient: $\tau_i=0.6$ (i=1,...,m); $\eta_i=0$ (i=1,...,m) that is equivalent to the requirement that there is no risk of violation of fuzzy constraints at level 0; θ_i

(i=1,...,m) is equal to the middle value of an approximated range of variability. The DM can accept or change this values.

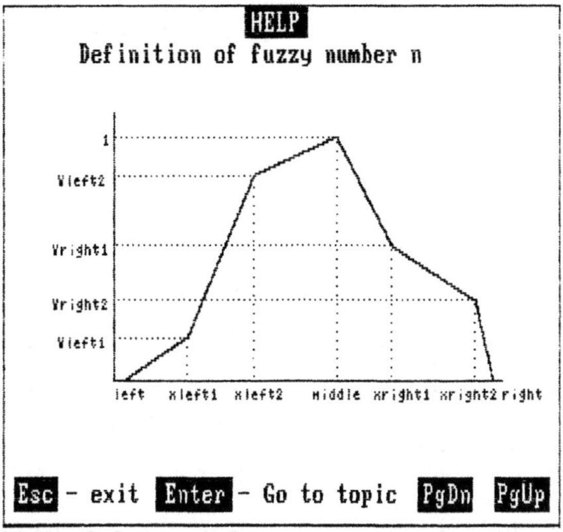

Figure 2. Definition of fuzzy number n.

Step IV. *Transformation of the fuzzy problem to an associate deterministic problem.*
Basing on the proposed principle of comparison of fuzzy numbers, the fuzzy problem is transformed to a multiobjective linear fractional problem. A starting point of the Choo-Atkin's method is then searched. The message about termination of this search is announced.

Step V. *Generation of efficient points.*
The DM must choose the number of compromise solutions he wants to compare and then the efficient border is sampled. The obtained solutions are presented to the DM both graphically and numerically.

Step VI. *Comparison of efficient points.*
If a best compromise has been obtained, then stop, otherwise the program goes back to step III, where safety coefficients are redefined, or to step V, where efficient points from the neighbourhood of the most interesting point are generated. This point is a new starting point for the next iteration of the Choo-Atkin's procedure.

4. SOLUTION OF A REAL FARM STRUCTURE OPTIMIZATION PROBLEM

Let us consider a real-life problem of searching for the best structure of a typical polish private farm. The considered farm has 20 hectars (ha) of arable land and 4 ha of permanent grassland. The farmer possesses 6 sows and 4 cows. We take into account 26 activities which can be divided into 5 groups:

- plant production for sale: winter wheat, winter barley, triticale, spring wheat, spring barley, rape, pea, potatos, sugar beet;
- plant production for fodders consumed in the farm: winter barley, spring barley, triticale, spring wheat, potatos, fodder beet, lucerne, clover, corn;
- permanent grassland cultivation;
- purchase of fertilizers: phosphorus, nitrogen, potassium;
- purchase of a nutritive fodder;
- manpower hire: in the spring, summer and autumnal period.

All these activities correspond to decision variables $x_1, x_2, \ldots, x_{26}$ defining their dimension, e.g. number of hectars of winter barley for sale, number of kilograms of phosphorus to be bought, number of hours of manpower hire in the spring period.

We take into consideration the following constraints:
- balance of arable land;
- 2 balances of crop succesion: for spring crops and for rape;
- 3 constraints on the area of group of plants: for crops, for winter crops and for root crops;
- 3 balances of manpower: for spring, summer and autumnal periods of extend manpower demand;
- 3 balances of artificial fertilizers: for phosphorus, nitrogen and potassium;
- 4 balances of fodders: for green fodder, for potatos, for fodder beet and for nutritive fodder;
- balance of permanent grassland.

The matrix of constraints is presented in Tab.1. Definition of fuzzy coefficients is given in Tab.2.

Three criteria are used to evaluate solutions:

- gross profit

$$f_1=\tilde{c}_1^1 x_1+\ldots+\tilde{c}_9^1 x_9-\tilde{c}_{10}^2 x_{10}-\ldots-\tilde{c}_{26}^2 x_{26}+\sum_{s=1}^{S} \tilde{T}_s h_s$$

where fuzzy cost coefficients are presented in Tab.3 and constant $\sum_{s=1}^{S} \tilde{T}_s h_s$ is equal to (22000,0,0).

- structure-forming plants area

$$f_2=c_6^3 x_6+c_7^3 x_7+c_{16}^3 x_{16}+c_{17}^3 x_{17}$$

where $c_6^3=0.5$, $c_7^3=c_{16}^3=c_{17}^3=1$.

TABLE 1. Set of constraints.

x_1	x_2	x_3	x_4	x_5	x_6	x_7	x_8	x_9	x_{10}	x_{11}	x_{12}	x_{13}	x_{14}
1	1	1	1	1	1	1	1	1	1	1	1	1	1
			-1	-1				1		-1		-1	
				1	-1					1			
1	1	1	1	1					1	1	1	1	1
								1	1				
1	1	1							1		1		
$\tilde{a}^7_1$	$\tilde{a}^7_2$	$\tilde{a}^7_3$	$\tilde{a}^7_4$	$\tilde{a}^7_5$	$\tilde{a}^7_6$	$\tilde{a}^7_7$	$\tilde{a}^7_8$	$\tilde{a}^7_9$	$\tilde{a}^7_{10}$	$\tilde{a}^7_{11}$	$\tilde{a}^7_{12}$	$\tilde{a}^7_{13}$	$\tilde{a}^7_{14}$
$\tilde{a}^8_1$	$\tilde{a}^8_2$	$\tilde{a}^8_3$	$\tilde{a}^8_4$	$\tilde{a}^8_5$	$\tilde{a}^8_6$	$\tilde{a}^8_7$	$\tilde{a}^8_8$	$\tilde{a}^8_9$	$\tilde{a}^8_{10}$	$\tilde{a}^8_{11}$	$\tilde{a}^8_{12}$	$\tilde{a}^8_{13}$	$\tilde{a}^8_{14}$
$\tilde{a}^9_1$	$\tilde{a}^9_2$	$\tilde{a}^9_3$	$\tilde{a}^9_4$	$\tilde{a}^9_5$	$\tilde{a}^9_6$	$\tilde{a}^9_7$	$\tilde{a}^9_8$	$\tilde{a}^9_9$	$\tilde{a}^9_{10}$	$\tilde{a}^9_{11}$	$\tilde{a}^9_{12}$	$\tilde{a}^9_{13}$	$\tilde{a}^9_{14}$
160	140	120	120	120	180	20	120	160	10	15	35	15	160
120	120	120	100	100	120	100	120	120	120	100	120	100	120
100	100	90	80	80	180	80	120	120	90	80	90	80	120
								300					
													250
									45	40	43	40	

x_{15}	x_{16}	x_{17}	x_{18}	x_{19}	x_{20}	x_{21}	x_{22}	x_{23}	x_{24}	x_{25}	x_{26}	sg.	RHS
1	1	1	1									=	20
1			1									>=	0
												>=	0
												<=	14
1												<=	5
												<=	10
$\tilde{a}^7_{15}$	$\tilde{a}^7_{16}$	$\tilde{a}^7_{17}$	$\tilde{a}^7_{18}$	$\tilde{a}^7_{19}$					-1			<=	$\tilde{b}_7$
$\tilde{a}^8_{15}$	$\tilde{a}^8_{16}$	$\tilde{a}^8_{17}$	$\tilde{a}^8_{18}$	$\tilde{a}^8_{19}$						-1		<=	$\tilde{b}_8$
$\tilde{a}^9_{15}$	$\tilde{a}^9_{16}$	$\tilde{a}^9_{17}$	$\tilde{a}^9_{18}$	$\tilde{a}^9_{19}$							-1	<=	$\tilde{b}_9$
160	30	30	200	200	-1							<=	0
120	120	120	140	120		-1						<=	0
120	100	100	120	100			-1					<=	0
	400	350	550	350								>=	3000
												>=	480
900												>=	450
								1				>=	360
				1								=	4

TABLE 2. Fuzzy coefficients of constraints.

$\tilde{a}^i_j = (a_{ij}, \alpha_{ij}, \beta_{ij})$									
j	a_{7j}	α_{7j}	β_{7j}	a_{8j}	α_{8j}	β_{8j}	a_{9j}	α_{9j}	β_{9j}
1	15	1	1	40	2	2	35	2	3
2	15	1	1	40	2	2	10	1	1
3	15	1	1	40	2	2	35	2	3
4	25	2	3	40	2	2	15	1	1
5	25	2	3	40	2	2	15	1	1
6	30	3	3	90	4	5	10	1	1
7	30	2	3	75	3	3	20	2	3
8	110	3	5	15	1	1	120	4	5
9	80	4	4	20	2	2	160	5	6
10	30	2	3	40	2	3	10	1	1
11	25	2	2	40	2	3	15	2	2
12	15	1	1	40	2	3	35	2	3
13	25	2	3	40	2	3	15	1	1
14	110	3	5	15	1	1	120	4	4
15	80	2	3	20	2	2	160	5	6
16	30	2	2	34	2	3	10	1	1
17	30	2	2	34	2	3	10	1	1
18	25	2	2	10	1	1	60	3	3
19	10	1	1	40	2	3	40	2	2
$\tilde{b}_i = (b_i, \gamma_i, \delta_i)$									
	b_7	γ_7	δ_7	b_8	γ_8	δ_8	b_9	γ_9	δ_9
	600	0	50	580	30	30	930	0	90

TABLE 3. Fuzzy cost coefficients.

$\tilde{c}^s_i = (c^s_i, \lambda^s_i, \rho^s_i)$, s=1 for i<10, s=2 for i≥10							
i	c^s_i	λ^s_i	ρ^s_i	i	c^s_i	λ^s_i	ρ^s_i
1	500	0	20	14	780	30	0
2	400	0	20	15	300	20	0
3	380	0	16	16	140	5	0
4	350	0	20	17	150	10	0
5	405	0	20	18	250	10	0
6	565	0	25	19	100	5	0
7	500	0	30	20	0.1	0.01	0
8	1400	0	50	21	0.75	0.25	0
9	600	0	30	22	0.04	0.02	0
10	180	20	0	23	13	2	0
11	170	10	0	24	0.8	0.1	0
12	180	10	0	25	0.9	0.2	0
13	175	10	0	26	1	0.2	0

- manpower hire

$$f_3=c^4_{24}x_{24}+c^4_{25}x_{25}+c^4_{26}x_{26}$$

where $c^4_{24}=c^4_{25}=c^4_{26}=1$.

After entering the data we get individual optima for the criteria that are calculated with fuzzy coefficients of the problem fixed on their middle values:

$$\text{Max } \hat{f}_1 = 20716,$$

$$\text{Max } \hat{f}_2 = 17.580,$$

$$\text{Min } \hat{f}_3 = 180.52.$$

According to the farmer suggestion, we take the following aspiration levels:

$$\text{for } f_1 - \tilde{g}_1 = (20716, 1716, 0),$$

$$\text{for } f_2 - \tilde{g}_2 = (4, 0.5, 0.5),$$

$$\text{for } f_3 - \tilde{g}_3 = (500, 50, 50).$$

Now we must define safety coefficients. Let's them take in conformity with computer suggestion $\tau_i=0.6$, $\eta_i=0$ and $\Theta_i=16.7$ for $i=1,\ldots,m$. Approximate ranges of variation of Θ_i are: $\Theta_7\in[-20.1, 50]$, $\Theta_8\in[-47.7, 20]$ and $\Theta_9\in[13.4, 90]$.

Then, the fuzzy problem is transformed into the deterministic one. For this new problem the pay-off table is calculated and a starting point of the Choo-Atkin's method is searched for. The first starting point is $y_0=(-2.64, 13.65, -0.72)$.

Now we start with examination of the efficient border of the transformed problem looking for the best compromise solution. In each iteration, the number of compromise solution S to be compared must be a multiple of the number of criteria. Let's take designation l_j for the l-th solution obtained in the j-th iteration.

Iteration 1.
S=15. We obtain 15 solutions listed in Tab. 4. In this table, only middle values of criteria for particular efficient points are shown. We have got a wide range of criteria values. After first short analysis we can point out solutions that are not attractive because of too high deviation of values of particular objectives from their aspiration levels, e.g for gross profit - solutions $1_1, 2_1, 3_1, 5_1, 6_1, 8_1, 9_1, 11_1, 12_1, 14_1, 15_1$.

From among solutions $4_1, 7_1, 10_1, 13_1$ solution 10_1 is chosen as a new starting point for the next iteration. Now we are looking for a solution with a lower value of criterion z_3.

TABLE 4.

	1_1	2_1	3_1	4_1	5_1	6_1	7_1	8_1	9_1
z_1	15618	15645	15644	18891	14022	14126	20088	12506	12539
z_2	10.8	10.8	10.8	8.5	14.9	8.5	6.1	17.6	6.1
z_3	536	536	538	648	648	373	760	690	278

	10_1	11_1	12_1	13_1	14_1	15_1
z_1	20568	11178	13339	20763	9940	14605
z_2	3.7	17.6	3.8	1.3	17.6	1.4
z_3	872	600	241	983	516	215

Iteration 2.

Let's take S=12. We have got 8 solutions (Tab.5) and 4 of generated problems are contradictory. When we are searching through the feasible region, we make successive reductions of it and it can lead, in some cases, to contradictory problems.

In this iteration we have got three very interesting solutions: 1_2, 7_2 and 10_2. For our DM, solution 7_2 is the best (Fig.3).

TABLE 5.

	1_2	3_2	4_2	6_2	7_2	9_2	10_2	12_2
z_1	19045	20567	19647	20557	19878	20547	19952	20537
z_2	3.7	3.7	3.1	1.9	4.0	0.2	4.0	0.1
z_3	503	937	539	905	575	897	611	891

As we can see, for solution 7_2, the value of the gross profit and the structure-forming plants area are within the range of the fuzzy aspiration levels. To get full evaluation of solution 7_2, let us also analyse the state of fuzzy constraints. In the upper windows of Fig.4, we have constraints corresponding to the demand of manpower in the spring and summer periods. As we can see, this demand has been satisfied with some surplus and no risk of violation of these constraints. For a constraint in the lower window that corresponds to the demand of manpower in the autumnal period, the risk of violation is very small. So, we can try to relax constraints 7, 8, 9 by changing safety coefficients. Let's assume τ_i=0.6, η_i=0 and Θ_i=0 for all i.

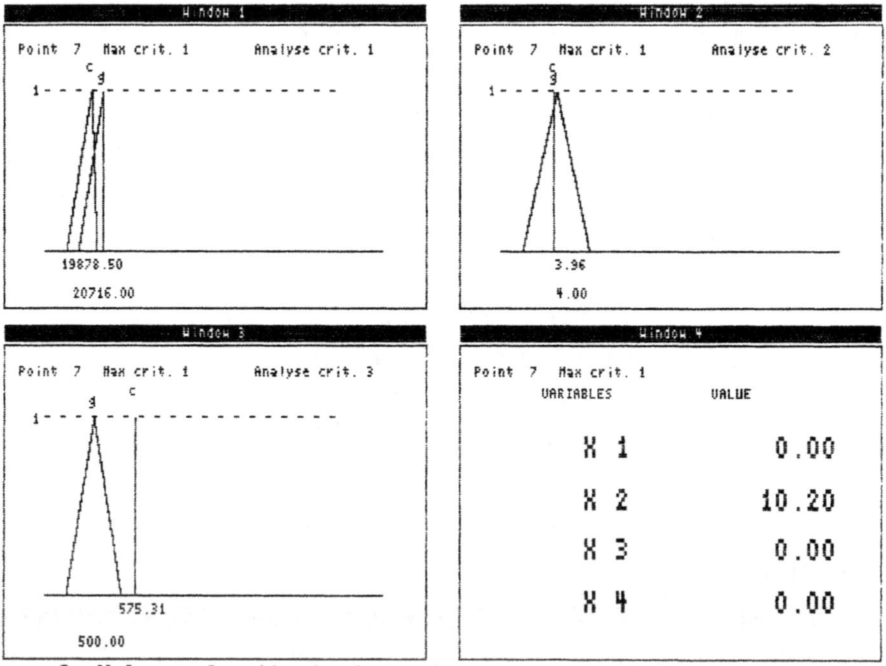

Figure 3. Values of criteria for solution 7_2.

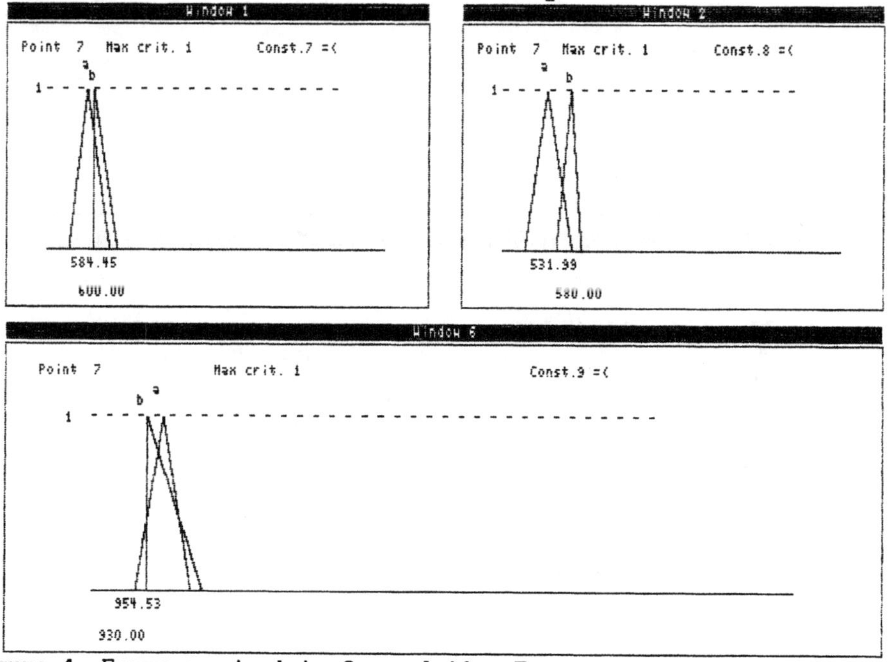

Figure 4. Fuzzy constraints for solution 7_2.

Iteration 1'.
For S=15 we obtain solutions listed in Tab.6.

TABLE 6.

	$1_{1'}$	$2_{1'}$	$3_{1'}$	$4_{1'}$	$5_{1'}$	$6_{1'}$	$7_{1'}$	$8_{1'}$	$9_{1'}$
z_1	15642	15652	15651	18969	14039	14133	20167	12524	12554
z_2	10.6	10.6	10.6	8.3	14.7	8.3	5.9	17.6	5.9
z_3	492	492	493	602	602	333	713	656	236

	$10_{1'}$	$11_{1'}$	$12_{1'}$	$13_{1'}$	$14_{1'}$	$15_{1'}$
z_1	20595	11202	13319	20692	9969	14596
z_2	3.4	17.6	3.5	2.1	17.6	1.1
z_3	823	566	198	933	483	171

Iteration 2'.
Taking solution $10_{1'}$ as a starting point and S=12, we get 8 new solutions (Tab.7).

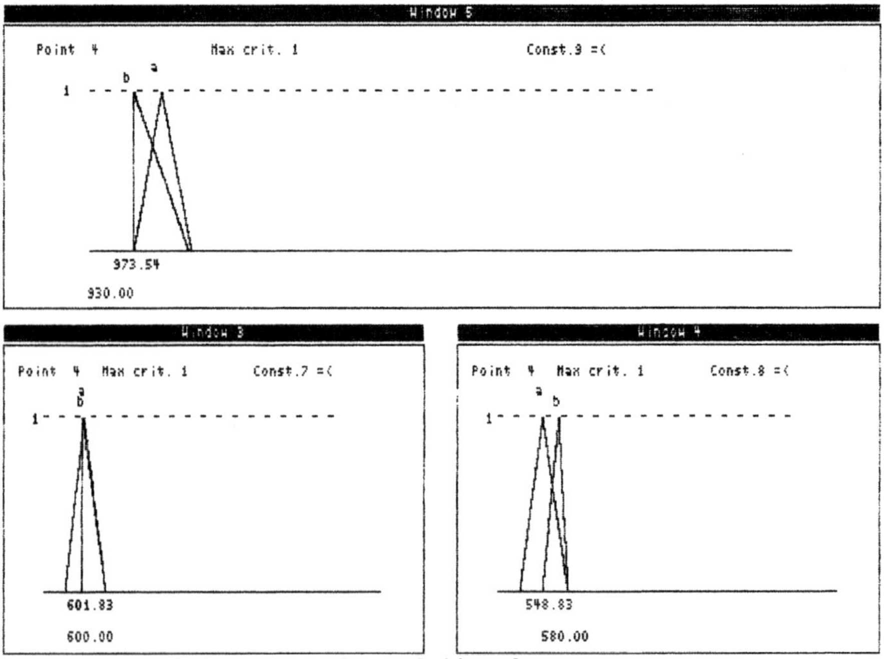

Figure 5. Fuzzy constraints for solution $4_{2'}$.

TABLE 7.

	$1_{2'}$	$3_{2'}$	$4_{2'}$	$6_{2'}$	$7_{2'}$	$9_{2'}$	$10_{2'}$	$12_{2'}$
z_1	19707	20594	19954	20582	20058	20570	20164	20558
z_2	3.2	3.4	4.0	1.7	4.0	0.0	4.0	0.3
z_3	503	862	541	841	584	833	624	826

For solution $4_{2'}$ the middle values of all objectives are within the range of their fuzzy aspiration levels. Fig.5 shows the state of fuzzy constraints for this solution. As we can see, there is no risk of violation of constraints 7 and 8 (lower windows) and a little bigger risk for constraint 9, then in solution 7_2.

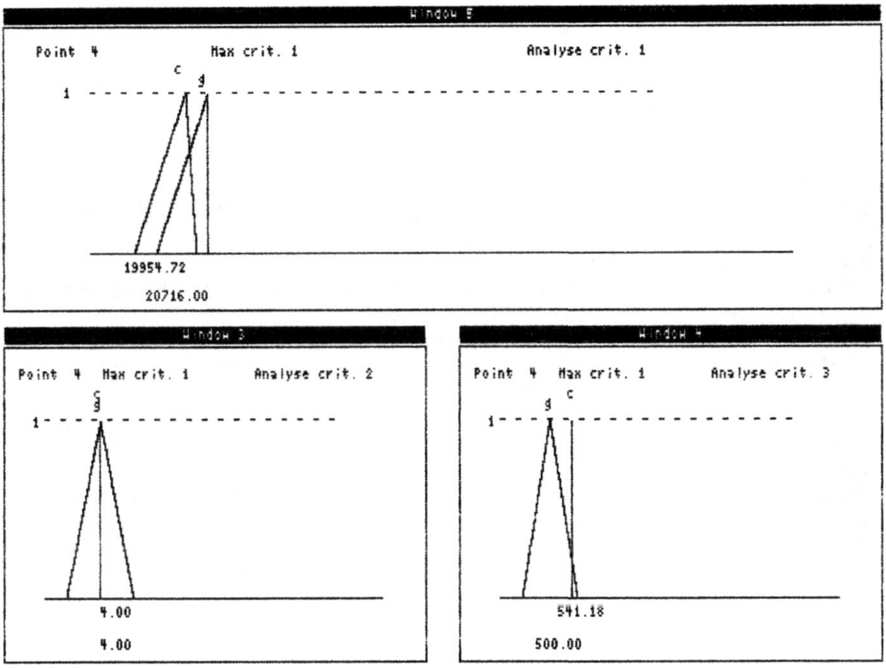

Figure 6. Values of criteria for solution $4_{2'}$.

Finally, the DM has accepted solution $4_{2'}$ as the best compromise. The values of corresponding criteria are presented in Fig.6.

So, we have got the following definition of the best-compromise farm structure:

a) plant production for sale:

- 0.80 ha of winter wheat,
- 9.20 ha of winter barley,
- 0.67 ha of spring barley,

- 2.75 ha of potatos;

b) plant production for fodder consumed in the farm:
   - 1.92 ha of potatos,
   - 0.5 ha of fooder beat,
   - 4 ha of lucerne;

c) 4 ha of permanent grassland under cultivation;

d) purchase of fertilizers:
   - 3267 kg of phosphorus,
   - 2866 kg of nitrogen,
   - 2397 kg of potassium;

e) manpower hire:
   - 281 hour's in the spring period,
   - 260 hour's in the summer period.

REFERENCES

[1] Choo,E.U. and Atkins,D.R. (1980) 'An interactive algorithm for multicriteria programming', Computers & Operations Res. **7**, 81-87.

[2] Czyżak,P. (1989) 'Multicriteria agricultural problem solving under uncertainty', Foundations of Control Engineering **14** (3).

[3] Czyżak,P. and Słowiński,R. (1989) 'Multiobjective diet optimization problem under fuzziness',in J.L.Verdegay & M.Delgado (eds.), The Interface Between Artificial Intelligence and Operations Research in Fuzzy Environment, Verlag TUV Rheinland, Cologne.

[4] Marszałkowicz,T. (1976) 'Methods of Optimal Programming in Agriculture' (in Polish), PWE, Warszawa.

[5] Munford,A.G. (1989) 'A microcomputer system for formulating animal diets which may involve liquid raw materials',European J. of Operational Research **41**(3).

[6] Slowiński,R. (1986) 'A multicriteria fuzzy linear programming method for water supply system development planning', Fuzzy Sets and Systems **19**,217-237.

[7] Słowiński,R. (1990) 'FLIP: an interactive method for multiobjective linear programming with fuzzy coefficients', in R.Słowiński & J.Teghem. (eds.), Stochastic versus Fuzzy Approaches to Multiobjective Mathematical Programming under Uncertainty, D.Reidel, Dordrecht, chapter III-3.

Part III
Chapter 5

FULPAL - AN INTERACTIVE METHOD FOR SOLVING (MULTIOBJECTIVE) FUZZY LINEAR PROGRAMMING PROBLEMS

H.ROMMELFANGER
Institute of Statistics and Mathematics
University of Frankfurt, Mertonstr.17
D-6000 Frankfurt am Main
Federal Republic of Germany

Abstract

This paper presents a method for solving (multicriteria) linear programs, where the right-hand sides as well as the coefficients in the constraints and/or the objective function(s) may be fuzzy. This approach is based on a new interpretation of the inequality-relation " $\lesseqqgtr$ ". Here a fuzzy constraint is replaced by a crisp inequality and a fuzzy objective function (utility function). This interpretation coincides with the well-known concepts in crisp inequalities and in soft constraints.
The decision procedure is modelled as an interactive man-machine process, called FULPAL (Fuzzy linear programming based on aspiration levels), which can be controlled by aspiration levels. This general method includes the procedure for solving LP-problems with soft constraints, proposed by ZIMMERMANN [1978],[ROMMELFANGER 1983], [WERNERS 1984] a.o.

1. Introduction

In literature many methods have been proposed for solving optimization models of the form

$$\left.\begin{array}{l} \tilde{Z}_1(x) = \tilde{C}_{11}x_1 +\dots+ \tilde{C}_{1n}x_n \\ \vdots \qquad\qquad \vdots \qquad\qquad \vdots \\ \tilde{Z}_K(x) = \tilde{C}_{K1}x_n +\dots+ \tilde{C}_{Kn}x_n \end{array}\right] \rightarrow \text{Max} \tag{1}$$

subject to

$$\tilde{A}_{i1}x_1 +\dots+ \tilde{A}_{in}x_n \lesssim \tilde{B}_i \quad , \quad i = 1,2,..,m$$
$$x_j \geq 0 \quad , \quad j = 1,\dots,n \; ,$$

where the coefficients $\tilde{A}_{ij}$ and $\tilde{C}_{kj}$ and the right-hand sides $\tilde{B}_i$ may be fuzzy.

As each real number d can be modelled as a fuzzy number $\tilde{D} = \{(x,f_D(x))\,|\,x\in R\}$ with the membership function

$$f_D(x) = \begin{cases} 1 & \text{if} \quad x = d \\ 0 & \text{else} \end{cases}$$

R. Slowinski and J. Teghem (eds.),
Stochastic versus Fuzzy Approaches to Multiobjective Mathematical Programming under Uncertainty, 279–299.

the fuzzy optimization model (1) contains the special cases

i. of some or all objective functions being crisp and/or

ii. some or all constraints having the soft form

$$a_{i1}x_1 + \cdots + a_{in}x_n \lesssim \tilde{B}_i \quad \text{with } a_{ij} \in \mathbb{R}. \tag{2}$$

In the approaches of TANAKA; ASAI [1984], RAMÍK; RIMANEK [1985], SLOWINSKI [1986], ROMMELFANGER [1987], DELGADÓ; VERDEGAY [1989], ROMMELFANGER; HANUSCHECK; WOLF [1982] a fuzzy constraint

$$\tilde{A}_{i1}x_1 + \cdots + \tilde{A}_{in}x_n \lesssim \tilde{B}_i \tag{3}$$

will be replaced by one or two crisp inequalities.
This proceeding has the disadvantage that the decision maker has no further chance to interprete the fuzzy constraints more precisely. Whereas the procedure of SLOWINSKI expresses a very optimistic behavior, the other methods reveal a more pessimistic attitude. The latter will be obvious in the special case that all coefficients of a constraint i are crisp but not the right-hand side $\tilde{B}_i$.

Then a soft constraint (2) is to be replaced by the crisp inequality

$$a_{i1}x_1 + \cdots + a_{in}x_n \leq b_i \quad . \tag{4}$$

In this paper we present a method for solving fuzzy (multicriteria) linear programs, which is based on a new interpretation of the inequality-relation "$\lesssim$". In this approach a fuzzy constraint will be replaced by a crisp inequality <u>and</u> a fuzzy objective function (utility function). The decision procedure is modelled as an interactive man-machine process, called FULPAL (Fuzzy Linear Programming based on Aspiration Levels). This general method includes e.g. the well-known procedure for solving LP-problems with soft constraints, proposed by ZIMMERMANN [1978]; ROMMELFANGER [1984]; WERNERS [1984].
An illustrative numerical example demonstrates the efficiency of the proposed algorithm, which has meanwhile become available as comfortable PC-software.

2. Modelling of the fuzzy parameters

In practise all relevant fuzzy data in (1) can be described by fuzzy numbers or fuzzy intervals of the L-R-type, see DUBOIS; PRADE [1980]. Thus, in order to quantify a fuzzy parameter, the decision maker (DM) has to specify

i. a relevant point of reference,

ii. the type of the reference functions,

iii. the support of this fuzzy set, e.g. by determining the spreads.

Yet, the DM has to discriminate between the right-hand sides $\tilde{B}_i$ and the coefficients $\tilde{A}_{ij}$ and $\tilde{C}_{kj}$.

As to <u>soft constraints</u> $\tilde{B}_i$, a suitable reference point is the greatest real number b_i, which with certainty belongs to the set of available values. Furthermore, the left spread of $\tilde{B}_i$ is to be set equal to zero because it is unessential to the satisfying of a constraint (4) whether the left side $a_{i1}x_1 + \cdots + a_{in}x_n$ equals b_i or really is smaller than b_i. In our opinion it is inadequate to model the $\tilde{B}_i$ as fuzzy intervals, as this was done by TANAKA;ASAI [1984], RAMIK; RIMANEK [1985], CARLSSON; KORHONEN [1986].

A greater problem is the specification of the graph of the membership function $f_{Bi}(y)$ for $y > b_i$. In our opinion there exist only few cases, where the DM is capable to describe the precise form of a fuzzy number $\tilde{B}_i = (b_i, 0, \bar{\beta}_i)_{RR}$ by specifying a spread $\bar{\beta}_i$ and a reference function R. In general he only knows some salient points $(y, f_{Bi}(y))$ of f_{Bi}.

As a practical way of getting suitable membership functions, we propose that the DM begin with the specification of some prominent membership levels. E.g.

$\alpha = 1$: $f_{Bi}(y) = 1$ means that the value y with certainty belongs to the set of available values,

$\alpha = \lambda_A$: $f_{Bi}(y) \geq \lambda_A$ means that the DM is willing to accept y as an available value for the time being. A value y with $f_{Bi}(y) \geq \lambda_A$ has a good chance of belonging to the set of available values. Corresponding values of y are relevant to the decision. Obviously, a value y with $f_{Bi}(y) = \lambda_A$ is a sort of aspiration level.

$\alpha = \varepsilon$: $f_{Bi}(y) < \varepsilon$ means, that y has only a very little chance of belonging to the set of available values. The DM is willing to neglect the values y with $f_{Bi}(y) < \varepsilon$.

Subsequently the DM has to fix values $\bar{b}_i^{\lambda A}$ and $\bar{b}_i^{\varepsilon}$ such that $f_{Bi}(\bar{b}^{\lambda A}) = \lambda_A$ and $f_{Bi}(\bar{b}_i^{\varepsilon}) = \varepsilon$.

Then the polygon line from $(b_i, 1)$ over $(\bar{b}^{\lambda A}, \lambda_A)$ to $(\bar{b}_i^{\varepsilon}, \varepsilon)$ is a suitable approach to f_{Bi} on the interval $[b_i, \bar{b}_i^{\varepsilon}]$.

For all $y \notin [b_i, \bar{b}_i^{\varepsilon}]$ we set $f_{Bi}(y) = 0$; see figure 1.

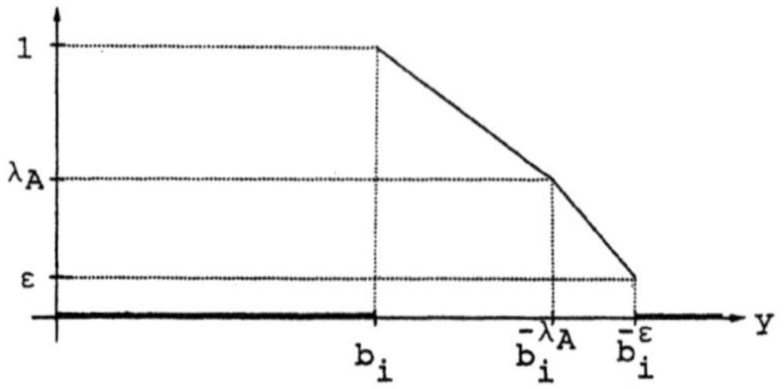

Figure 1: Membership function of B_i

Taking pattern from L-R-type fuzzy numbers we symbolize a fuzzy number with this special membership function by $\tilde{B}_i = (b_i;0,0;\bar{\beta}_i^{\lambda A},\bar{\beta}_i^{\varepsilon})^{\lambda A,\varepsilon}$ where

$\bar{\beta}_i^{\lambda A} = \bar{b}_i^{\lambda A} - b_i$ and $\bar{\beta}_i^{\varepsilon} = \bar{b}_i^{\varepsilon} - b_i$.

If required the DM may specify additional membership levels and additional points $(y,f_{B_i}(y))$ on the polygon line.

On the other hand, fuzzy coefficients $\tilde{A}_{ij}$ or $\tilde{C}_{kj}$ do not include elements that may be realized with certainty. In this case, an appropriate point of reference for a fuzzy set $\tilde{A}_{ij}$ is the subset $[\underline{a}_{ij},\bar{a}_{ij}] \subset \mathbb{R}$ consisting of the real numbers with the highest chance of realization. Furthermore, the lower the information of the DM the larger is the length of this interval. Thus, fuzzy coefficients can be modelled as fuzzy intervals

$\tilde{A}_{ij} = (\underline{a}_{ij};\bar{a}_{ij};\underline{\alpha}_{ij};\bar{\alpha}_{ij})_{LR}$. This formulation contains the special case, where $\tilde{A}_{ij}$ is a fuzzy number with $\underline{a}_{ij} = \bar{a}_{ij}$. But we think it less realistic to assume that all coefficients $\tilde{A}_{ij}$ are fuzzy sets with exactly one mean value a_{ij}, as this was done by RAMIK; RIMANEK [1985] and SLOWINSKY [1986].

If all coefficients of a fuzzy constraint

$$\tilde{A}_{i1}x_1 \oplus \tilde{A}_{i2}x_2 \oplus \cdots \oplus \tilde{A}_{in}x_n \lesssim \tilde{B}_i \qquad (5)$$

are fuzzy intervals of the same L-R-type, the left side of (5) can be summarized with the extension principle to the fuzzy interval

$$\tilde{A}_i(x) = (\underline{a}_i(x),\bar{a}_i(x);\ \underline{\alpha}_i(x);\bar{\alpha}_i(x))_{LR}$$

with $\underline{a}_i(x) = \sum_{j=1}^{n} \underline{a}_{ij}x_j$, $\quad \bar{a}_i(x) = \sum_{j=1}^{n} \bar{a}_{ij}x_j$,

$\underline{\alpha}_i(x) = \sum_{j=1}^{n} \underline{\alpha}_{ij}x_j$, $\quad \bar{\alpha}_i(x) = \sum_{j=1}^{n} \bar{\alpha}_{ij}x_j$,

which is of the same L-R-type.

In real problems however, not all the coefficients $\tilde{A}_{ij}$ of a constraint (5) will have the same reference functions. Moreover, it can not be expected that the DM knows the precise form of all coefficients $\tilde{A}_{ij}$. Therefore we propose to describe the coefficients $\tilde{A}_{ij}$ by piecewise linear membership functions based on the prominent membership values $\alpha = 1, \lambda_A, \varepsilon$; see figure 2.

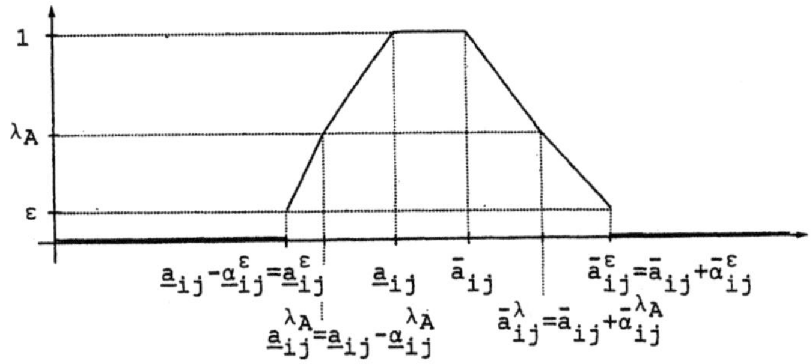

Figure 2: $\tilde{A}_{ij} = (\underline{a}_{ij},\bar{a}_{ij};\underline{\alpha}_{ij}^{\lambda_A},\underline{\alpha}_{ij}^{\varepsilon};\bar{\alpha}_{ij}^{\lambda_A},\bar{\alpha}_{ij}^{\varepsilon})^{\lambda_A,\varepsilon}$

Coefficients $\tilde{A}_{ij}$ with piecewise linear membership functions can easily be multiplyed with real numbers x_j and the terms $\tilde{A}_{ij}x_j$ can easily be added by computing the values on the selected α-levels and linking these points by straight lines, see [DUBOIS; PRADE 1980, page 42-44] or [ROMMELFANGER 1988, p.227-229] . Then we obtain the fuzzy interval

$$\tilde{A}_i(X) = \tilde{A}_{i1}x_1 \oplus\cdots\oplus \tilde{A}_{in}x_n$$
$$= (\underline{a}_i(X),\bar{a}_i(X);\underline{\alpha}_i^{\lambda_A}(X),\underline{\alpha}_i^{\varepsilon}(X);\bar{\alpha}_i^{\lambda_A}(X),\bar{\alpha}_i^{\varepsilon}(X))^{\lambda_A,\varepsilon}$$

with $\underline{a}_i(X) = \sum_{j=1}^{n} \underline{a}_{ij}x_j$, $\bar{a}_i(X) = \sum_{j=1}^{n} \bar{a}_{ij}x_j$

$$\underline{\alpha}_i^{\lambda_A}(X) = \sum_{j=1}^{n} \underline{\alpha}_{ij}^{\lambda_A}x_j \ , \quad \bar{\alpha}^{\lambda_A}(X) = \sum_{j=1}^{n} \bar{\alpha}_{ij}^{\lambda_A}x_j$$

$$\underline{\alpha}_i^{\varepsilon}(X) = \sum_{j=1}^{n} \underline{\alpha}_{ij}^{\varepsilon}x_j \ , \quad \bar{\alpha}_i^{\varepsilon}(X) = \sum_{j=1}^{n} \bar{\alpha}_{ij}^{\varepsilon}\, x_j \ .$$

If for all coefficients $\tilde{A}_{ij}$ of the constraint i the spreads $\underline{\alpha}_{ij}^{\varepsilon}$ and $\bar{\alpha}_{ij}^{\varepsilon}$ are very small in relation to the mean values $\underline{a}_{ij}$ and $\bar{a}_{ij}$ the membership value λ_A can be neglected. In this case it is sufficient to approach the membership function f_{A_i} by the polygon line from $(\underline{a}_i-\underline{\alpha}_i^{\varepsilon},\varepsilon)$ over $(\underline{a}_i,1),(\bar{a}_i,1)$ to $(\underline{a}_i+\underline{\alpha}_i^{\varepsilon},\varepsilon)$. A fuzzy interval $\tilde{A}_i(X)$ with this kind of membership function will be symbolized as

$$\tilde{A}_i(X) = (\underline{a}_i(X),\bar{a}_i(X);\underline{\alpha}_i^{\varepsilon}(X);\bar{\alpha}_i^{\varepsilon}(X))^{\varepsilon} \ .$$

3. Inequality Relations

In order to determine a solution to the fuzzy optimization problem (1) the main question to be answered lies in comparing left and right-hand sides of the constraints

$$\tilde{A}_i(X) \lesssim \tilde{B}_i \quad , \tag{6}$$

i.e. the comparison of two (flat) fuzzy numbers.

In literature many concepts are proposed for comparing fuzzy sets, see e.g. [DUBOIS; PRADE 1983], [BORTOLAN;DEGAN 1985], [ROMMELFANGER 1986]. But all these preference orderings appear to be inefficient in fuzzy mathematical programming. Apart from some simple ranking methods with weak selectivity, the most procedures need extensive calculations because the membership function of $\tilde{A}_i(X)$ depends on the decision variable $X' = (x_1,\ldots,x_n)$.

Special interpretations of the inequality relation "$\lesssim$" in fuzzy constraints of type (6) are proposed by NEGOITA; SULARIA [1976], TANAKA; ASAI [1984], RAMÍK;RIMANEK [1985], SLOWINSKI [1986], ROMMELFANGER [1987] In all of these approaches the fuzzy constraints (6) will be replaced by few crisp inequalities, mostly two crisp constraints. But this proceeding has the disadvantage of the DM not having any further opportunity to interprete the fuzzy constraints more precisely.

A detailed survey of inequality relations in fuzzy constraints is given in [ROMMELFANGER 1988,1989].

A more flexible interpretation of "$\lesssim$" was proposed by [ROMMELFANGER 1988]. He retains the "pessimistic" relation

$$\bar{a}_i(X) + \bar{\alpha}_i^\varepsilon(X) = \sum_{j=1}^{n} (\bar{a}_{ij}+\bar{\alpha}_{ij}^\varepsilon)x_j \le b_i + \bar{\beta}_i^\varepsilon \quad , \tag{7}$$

which is also used by SLOWINSKI [1986] and other authors. This means that the maximal available amount $b_i + \beta_i^\varepsilon$ should not be exceeded at any rate.

Furthermore, Rommelfanger suggests to value the needed quantity $\bar{a}_i(X) = \sum_{j=1}^{n} \bar{a}_{ij}x_j$ by $f_{D_i}(\bar{a}_i(X))$, where f_{D_i} is definied as

$$f_{D_i}(y) = \begin{cases} 1 & \text{if } y < b_i \\ f_{B_i}(y) & \text{if } y \ge b_i \end{cases} \tag{8}$$

and f_{B_i} is the membership function of the right-hand side $\tilde{B}_i$.

Obviously the values $f_{D_i}(\bar{a}_i(X))$ may be interpreted as the subjective valuation of $\bar{a}_i(X)$ by the DM with regard to the right-hand side $\tilde{B}_i$. The DM's goal is to maximize $f_{D_i}(\bar{a}_i(X))$. In an optimization model of the type (1), however this goal will be in contradiction to the given goals $\tilde{Z}_k(X) \longrightarrow \text{Max}$

Thus an appropriate interpretation of "$\lesssim$" is

$$\tilde{A}_i(X) \lesssim_R \tilde{B}_i \iff \begin{cases} \sum_{j=1}^{n} (\bar{a}_{ij}+\bar{\alpha}^{\varepsilon}_{ij})x_j \le b_i + \bar{\beta}^{\varepsilon}_i \\ f_{D_i}(\sum_{j=1}^{n} \bar{a}_{ij}x_j) \rightarrow \text{Max} \end{cases} \tag{9}$$

The inequality relation "$\lesssim_R$" has the advantage, that a possible surplus $\bar{a}_i(X)-b_i$ directly influences the decision process. Moreover "$\lesssim_R$" coincides with the usual interpretation of the inequality relation in soft constraints (2), see e.g. [ZIMMERMANN 1978], [ROMMELFANGER 1984],[WERNERS 1984]; and in the special case of deterministic inequalities it corresponds to the classical "$\le$"-relation. Thus "$\lesssim_R$" is a general definition for inequality relations in optimization models. When using the inequality relation "$\lesssim_R$", it is sufficient for all $\tilde{A}_{ij}$ to specify only the values $\bar{a}_{ij}$ and $\bar{a}^{\varepsilon}_{ij} = \bar{a}_{ij} + \alpha^{\varepsilon}_{ij}$.

4. Maximizing the objective functions

At first, we consider the special case of the DM having only one goal, i.e. K = 1.

If all the coefficients of the objective function

$$\tilde{Z}_1(X) = \tilde{C}_{11}x_1 \oplus \tilde{C}_{12}x_2 \oplus \cdots \oplus \tilde{C}_{1n}x_n \rightarrow \widetilde{\text{Max}} \tag{10}$$

may be described by fuzzy intervals of the type

$$\tilde{C}_{1j} = (\underline{c}_{1j},\bar{c}_{1j};\underline{\gamma}^{\lambda_A}_{1j},\underline{\gamma}^{\varepsilon}_{1j};\bar{\gamma}^{\lambda_A}_{1j},\bar{\gamma}^{\varepsilon}_{1j})^{\lambda_A,\varepsilon} \quad ,$$

$\tilde{Z}_1(X)$ can be written as the fuzzy interval

$$\tilde{Z}_1(X) = (\underline{c}_1(X),\bar{c}_1(X);\underline{\gamma}^{\lambda_A}_1(X),\underline{\gamma}^{\varepsilon}_1(X);\bar{\gamma}^{\lambda_A}_1(X),\bar{\gamma}^{\varepsilon}_1(X))^{\lambda_A,\varepsilon} \tag{11}$$

with

$$\underline{c}_1(X) = \sum_{j=1}^{n} \underline{c}_{1j}x_j, \qquad \bar{c}_1(X) = \sum_{j=1}^{n} \bar{c}_{1j}x_j$$

$$\underline{\gamma}^{\lambda_A}_1(X) = \sum_{j=1}^{n} \underline{\gamma}^{\lambda_A}_{1j}x_j, \qquad \bar{\gamma}^{\lambda_A}_1(X) = \sum_{j=1}^{n} \bar{\gamma}^{\lambda_A}_{1j}x_j$$

$$\underline{\gamma}^{\varepsilon}_1(X) = \sum_{j=1}^{n} \underline{\gamma}^{\varepsilon}_{1j}x_j, \qquad \bar{\gamma}^{\varepsilon}_1(X) = \sum_{j=1}^{n} \bar{\gamma}^{\varepsilon}_{1j}x_j \quad .$$

The fuzzy goal (10) obviously implies that the six goals

$$\underline{c}_1(X) \rightarrow \text{Max} \qquad \bar{c}(X) \rightarrow \text{Max}$$

$$\underline{c}_1 - \underline{\gamma}^{\lambda_A}_1(X) \rightarrow \text{Max} \qquad \bar{c}_1 + \bar{\gamma}^{\lambda_A}_1(X) \rightarrow \text{Max}$$

$$\underline{c}_1 - \gamma^{\varepsilon}_1(X) \rightarrow \text{Max} \qquad \bar{c}_1 + \bar{\gamma}^{\varepsilon}_1(X) \rightarrow \text{Max}$$

should be satisfied simultaneously.

Since an ideal solution of this multicriteria problem does not exist in general, we think it better to look for a satisfactory solution.

Let $\tilde{N}_1$ be a fuzzy aspiration level of the goal (10). In analogy to the description of the right sides $\tilde{B}_i$, the aspiration level $\tilde{N}_1$ can be described by a fuzzy number of the type

$$\tilde{N}_1 = (n_1; \underline{\nu}_1^{\lambda_A}, \underline{\nu}_1^{\varepsilon}; 0,0)^{\lambda_A,\varepsilon} \quad , \tag{12}$$

where the right spreads are equal zero.

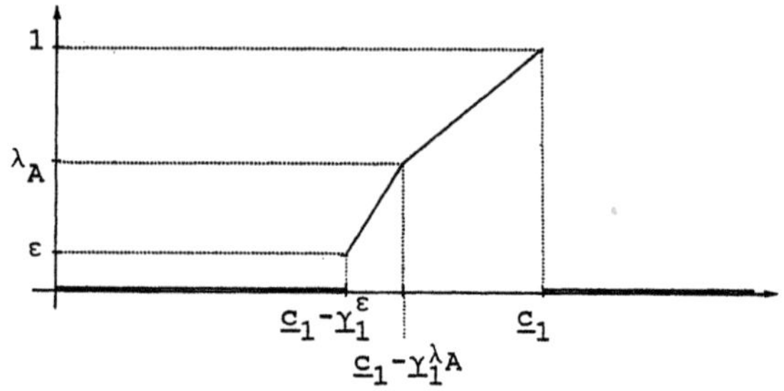

Figures 3: Membership function of f_{N_1}

When using the inequality relation "$\lesssim_R$", the condition

$$\tilde{N}_1 \lesssim_R \tilde{Z}_1(X) \tag{13}$$

is equivalent to

$$n_1 - \underline{\nu}_1^{\varepsilon} \le \underline{c}_1(X) - \underline{\gamma}_1^{\varepsilon}(X) = \sum_{j=1}^{n} (\underline{c}_{1j} - \underline{\gamma}_{1j}^{\varepsilon}) x_j \tag{14}$$

$$f_{Z_1}(\underline{c}_1(X)) = f_{Z_1}(\sum_{j=1}^{n} \underline{c}_{1j} x_j) \longrightarrow \text{Max} \quad , \tag{15}$$

where f_{Z_1} is the piecewise linear membership function

$$f_{Z_1}(y) = \begin{cases} 0 & \text{if } \; y < n_1 - \underline{\nu}_1^{\varepsilon} \\ f_{N_1}(y) & \text{if } \; n_1 - \underline{\nu}_1^{\varepsilon} \le y \le n_1 \\ 1 & \text{if } \; n_1 < y \end{cases} \quad . \tag{16}$$

The problem of specifying the parameter n_1 and $\underline{\nu}_1^{\varepsilon}$ of the fuzzy aspiration level $\tilde{N}_1$ will be discussed in the section "Fuzzy optimization", because these data depend on the other restrictions of the optimization problem, too.

In a multicriteria problem with several objective functions $\tilde{Z}_k(X)$, k = 1,...,K, we propose to proceed in the same way as above for all objective functions. But now, the aspiration levels depend on each other additionally.

5. Fuzzy optimization

The results in the preceding section can be used directly to transform the fuzzy optimization problem (1) into a deterministic multicriteria mathematical program.
Employing the inequality relation "$\lesssim_R$", instead of (1) we obtain the crisp optimization system

$$(f_{Z_1}(\underline{c}_1'x),\ldots,f_{Z_K}(\underline{c}_K'x),f_{D_1}(\bar{a}_1'x),\ldots,f_{D_m}(\bar{a}_m'x)) \longrightarrow \text{Max}$$

subject to

$$(\underline{c}_k-\underline{\gamma}_k^{\varepsilon})'\cdot x = \sum_{j=1}^{n}(\underline{c}_{kj}-\underline{\gamma}_{kj}^{\varepsilon})x_j \geq \underline{n}_k - \underline{\nu}_k^{\varepsilon} \quad , \quad k = 1,\ldots,K \qquad (17)$$

$$(\bar{a}_i+\bar{\alpha}_i^{\varepsilon})'\cdot x = \sum_{j=1}^{n}(\bar{a}_j+\alpha_{ij}^{\varepsilon})x_j \leq b_i + \bar{\beta}_i^{\varepsilon} \quad , \quad i = 1,\ldots,m$$

$$x_j \geq 0 \quad , \quad i = 1,\ldots,m$$

where $\underline{c}_k' = (\underline{c}_{k1},\ldots,\underline{c}_{kn})$, $k = 1,\ldots,K$,

$\underline{\gamma}_k^{\varepsilon\,\prime} = (\underline{\gamma}_{k1}^{\varepsilon},\ldots,\underline{\gamma}_{k1}^{\varepsilon})$, $k = 1,\ldots,K$,

$\bar{a}_i' = (\bar{a}_{i1},\ldots,\bar{a}_{in})$, $i = 1,\ldots,m$,

$\bar{\alpha}_i^{\varepsilon\,\prime} = (\bar{\alpha}_{i1}^{\varepsilon},\ldots,\bar{\alpha}_{i1}^{\varepsilon})$, $i = 1,\ldots,m$, and

$x' = (x_1,\ldots,x_n)$.

For abbreviation we define

$$X_R = \{x\in\mathbb{R}_o^n \,|\, (\bar{a}_i+\bar{\alpha}_i^{\varepsilon})'\cdot x \leq b_i + \bar{\beta}_i^{\varepsilon} \text{ for any } i = 1,\ldots,m\} \quad .$$

Corresponding to the definition of the data $\bar{\alpha}_i^{\varepsilon}$ and $\bar{\beta}_i^{\varepsilon}$ at best decisions $x\in X_R$ are feasible solutions of (1).

With regard to the aspiration levels $\tilde{N}_k$, the set of feasible solutions must be restricted further to the set

$$X_C = \{x\in X_r \,|\, (\underline{c}_k-\underline{\gamma}_k^{\varepsilon})'\,x \geq \underline{n}_k - \underline{\nu}_k^{\varepsilon} \text{ for any } k = 1,\ldots,K\} \ .$$

As to practical problems a DM is not interested in getting the complete solution of the system (17), i.e. the set of all pareto optimal solutions, but he needs a procedure which generates a so-called compromise solution.

To determine a compromise solution, in accordance with many other methods on fuzzy optimization we introduce ZADEH's min-operator as an appropriate preference function, see e.g. [ZIMMERMANN 1978],[SAKAWA;YANO 1987]. Obviously the compromise objective function

$$\lambda(x) = \text{Min}(f_{Z_1}(\underline{c}_1'x),\ldots,f_{Z_k}(\underline{c}_K'x),f_{D_1}(\bar{a}_1'x),\ldots,f_{D_m}(\bar{a}_n'x))$$

expresses the total satisfaction of the DM on x .

In the case of X_C not being empty, NEGOITA and SULARIA [1976] have shown that the decision problem

$$\underset{X \in X_C}{\text{Max}} \text{ Min } (f_{Z_1}(\underline{c}_1' X), \ldots, f_{Z_K}(\underline{c}_K' X), f_{D_1}(\bar{a}_1' X), \ldots, f_{D_m}(\bar{a}_m' X)) \tag{18}$$

is equivalent to the optimization problem

$\lambda \longrightarrow \text{Max}$

subject to

$$\begin{aligned} &\lambda \le f_{Z_k}(\underline{c}_k' \cdot X) && k = 1, \ldots, K \\ &\lambda \le f_{D_i}(\bar{a}_i' \cdot X) && i = 1, \ldots, m \\ &X \in X_C \; ; && \varepsilon \le \lambda \le 1 \quad . \end{aligned} \tag{19}$$

Moreover, if all membership functions f_{Z_k} and f_{D_i} are piecewise linear functions, which are concave on $[\underline{n}_k - \underline{\nu}_k^\varepsilon, +\infty[$ or $[-\infty, b_i + \bar{\beta}_i^\varepsilon]$ it has been shown in [ROMMELFANGER 1983] that the system (19) is equivalent to a crisp LP-model. In this special case, the solution of (19) is also a pareto optimal solution of the multicriteria optimization system (17), see [RAMÍK 1987] or [ROMMELFANGER 1988] .

If the membership functions f_{Z_k} and f_{D_i} are constructed as piecewise linear functions with the interpolation nodes

$\{(n_k - \underline{\nu}_k^\varepsilon, \varepsilon), (z_k^A = n_k - \underline{\nu}_k^{\lambda_A}, \lambda_A), (n_k, 1)\}$ or

$\{(b_i, 1), (g_i^A = b_i + \bar{\beta}_i^{\lambda_A}, \lambda_A), (b_i + \bar{\beta}_i^\varepsilon, \varepsilon)\}$

they can be written as

$$f_{Z_k}(\underline{c}_k' X) = \begin{cases} 0 & \text{if} \quad \underline{c}_k' X < n_k - \underline{\nu}_k^\varepsilon \\ \varepsilon + \dfrac{\underline{c}_k' X - n_k + \nu_k^\varepsilon}{z_k^A - n_k + \underline{\nu}_k^\varepsilon}(\lambda_A - \varepsilon) & \text{if} \quad n_k - \underline{\nu}_k^\varepsilon \le \underline{c}_k' X \le z_k^A \\ \lambda_A + \dfrac{\underline{c}_k' X - z_k^A}{n_k - z_k^A}(1 - \lambda_A) & \text{if} \quad z_k^A < \underline{c}_k' X \le n_k \\ 1 & \text{if} \quad n_k < \underline{c}_k' X \end{cases}$$

and

$$f_{D_i}(\bar{a}_i' X) = \begin{cases} 1 & \text{if} \quad a_i' X \le b_i \\ 1 - \dfrac{\bar{a}_i' X - b_i}{g_i^A - b_i}(1 - \lambda_A) & \text{if} \quad b_i < \bar{a}_i' X \le g_i^A \\ \lambda_A - \dfrac{\bar{a}_i X - g_i^A}{b_i + \beta_i^\varepsilon - g_i^A}(\lambda_A - \varepsilon) & \text{if} \quad g_i^A < \bar{a}_i' X \le b_i + \bar{\beta}_i^\varepsilon \\ 0 & \text{if} \quad b_i + \bar{\beta}_i^\varepsilon < \bar{a}_i' X \; . \end{cases}$$

The new parameters z_k^A and g_i^A are not just used for abbreviation, but we will control the interactive process for determining a compromise solution of (1) by modifying these parameters step by step. Using an interactive procedure, we shall attempt to compensate the limited information of the DM and the disadvantages of the min-operator.

If one of the membership functions f_{Z_k} or f_{D_i}, defined by (20) or (21), is not concave, we can always derive a concave one by a suitable reduction of the ε-level set. This correction, which is demonstrated in figure 4, does not influence the solution of the optimization problem because the DM only accepts solutions with $\lambda \geq \lambda_A$.

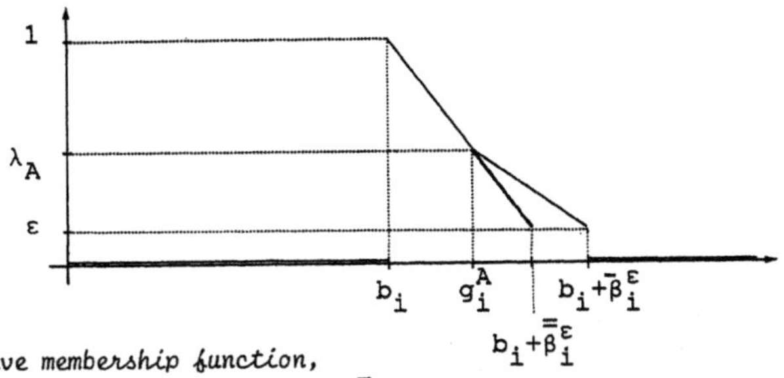

Figure 4: To get a concave membership function, the ε-cut of $\tilde{B}_i$ is reduced to $[b_i, b_i+\bar{\bar{\beta}}_i^{\varepsilon}]$

With concave, piecewise linear membership functions the mathematical program (19) can be written as the LP-model

$\lambda \longrightarrow$ Max

subject to (LP.*)

$\lambda \leq 1\,, \qquad X \geq 0$

$$\frac{z_k^A - n_k + \underline{\nu}_k^{\varepsilon}}{\lambda_A - \varepsilon}\,\lambda - \underline{c}_k' X \leq -n_k + \underline{\nu}_k^{\varepsilon} + \varepsilon\cdot\frac{z_k^A - n_k + \underline{\nu}_k^{\varepsilon}}{\lambda_A - \varepsilon} \qquad (22)$$

$$\frac{n_k - z_k^A}{1-\lambda_A}\,\lambda - \underline{c}_k'\cdot X \leq -z_k^A + \lambda_A\cdot\frac{n_k - z_k^A}{1-\lambda_A} \qquad k = 1,\dots,K \qquad (23)$$

$$(\underline{c}_k - \underline{\gamma}_k)'X \leq -n_k + \underline{\nu}_k^{\varepsilon} \qquad (24)$$

$$\frac{g_i^A - b_i}{1-\lambda_A}\,\lambda + \bar{a}_i'\cdot X \leq b_i + \frac{g_i^A - b_i}{1-\lambda_A} \qquad (25)$$

$$\frac{b_i + \bar{\beta}_i^{\varepsilon} - g_i^A}{\lambda_A - \varepsilon}\,\lambda + \bar{a}'\cdot X \leq g_i^A + \lambda_A\cdot\frac{b_i + \bar{\beta}_i^{\varepsilon} - g_i^A}{\lambda_A - \varepsilon} \qquad i = 1,\dots,n \qquad (26)$$

$$(\bar{a}_i + \bar{\alpha}_i^{\varepsilon})'\cdot X \leq b_i + \bar{\beta}_i^{\varepsilon} \qquad (27)$$

Obviously, for crisp coefficient-vectors $\mathbf{c}_k$ or $\mathbf{a}_i$ the constraints (24) or (27) are satisfied by all x, which fulfill the constraints (22),(23) or (25),(26). Moreover, in the special case of not just the coefficient vector $\mathbf{a}_i$ but the right-hand side b_i also being crisp, the three restrictions (25),(26) and (27) coincide.

An open problem is the determination of the fuzzy aspiration level $\tilde{N}_k$ with the characteristic values n_k, z_k^A and $n_k - \underline{\nu}_k^\varepsilon$.
Evidently an upper limit for n_k is

$$\bar{z}_k = \underline{\mathbf{c}}_k' x_k^{**} = \underset{x \in X_R}{\operatorname{Max}} \ \underline{\mathbf{c}}_k' x \ . \tag{28}$$

On the other hand, $n_k - \nu_k^\varepsilon$ should not be smaller then

$$\underline{z}_k = \operatorname{Min}\Big(\underset{\substack{l=1\\ l\neq k}}{\overset{K}{\operatorname{Min}}}(\underline{\mathbf{c}}_k - \underline{\gamma}_k^\varepsilon)' x_l^*, \ \underset{\substack{l=1\\ l\neq k}}{\overset{K}{\operatorname{Min}}}(\underline{\mathbf{c}}_k - \underline{\gamma}_k^\varepsilon)' x_l^{**}\Big) \tag{29}$$

where x_k^* is the solution of the LP-system

$$(\underline{\mathbf{c}}_k - \underline{\gamma}_k^\varepsilon)' x \longrightarrow \operatorname{Max} \tag{30}$$

subject to

$$\bar{\mathbf{a}}_i' x_i \leq b_i$$

$$(\bar{\mathbf{a}}_i + \bar{\alpha}_i^\varepsilon)' x \leq b_i + \bar{\beta}_i^\varepsilon \qquad i = 1,\ldots,m$$

$$x \geq \mathbf{o}$$

The complex form in (29) is necessary because it has been been shown in [WERNERS 1984, p.84] and in [ROMMELFANGER 1988, p.251] that it is not always the first term in (29) that leads to the smallest values.

Moreover, if a solution of (28) or (30) is not determined unequivocally, we propose to use the solution of $\{x_l^*, x_l^{**}\}$ which leads to the greatest value of $\underline{z}_k$ in (29).

In the special case with only one objective function the equation (29) can be reduced to

$$\underline{z}_1 = (\underline{\mathbf{c}}_1 - \underline{\gamma}_1^\varepsilon)' x_1^* \quad . \tag{31}$$

In the interaction procedure FULPAL (**FU**zzy **L**inear **P**rogramming based on **A**spiration **L**evels), which is described in the next section by a flowchart we set $n_k = \bar{z}_k$ and $n_k - \nu_k^\varepsilon = \underline{z}_k$. In doing so we have enough play for controlling the process by changing the aspiration levels z_k^A, $k = 1,\ldots,K$ and g_i^A, $i = 1,\ldots,m$.

6. FULPAL

In the flowchart of the procedure FULPAL additional restrictions are introduced for ensuring the convergence of the solution process. Auxilary means of this kind are the thresholds W_{z_k}, W_i and the lower and upper bounds $\check{z}, \hat{z}, \check{g}, \hat{g}$ for the levels of aspiration.
In real problems these remedies can be neglected. But we think, it is better to preserve these bounds and make moderate improvements at each iteration step.

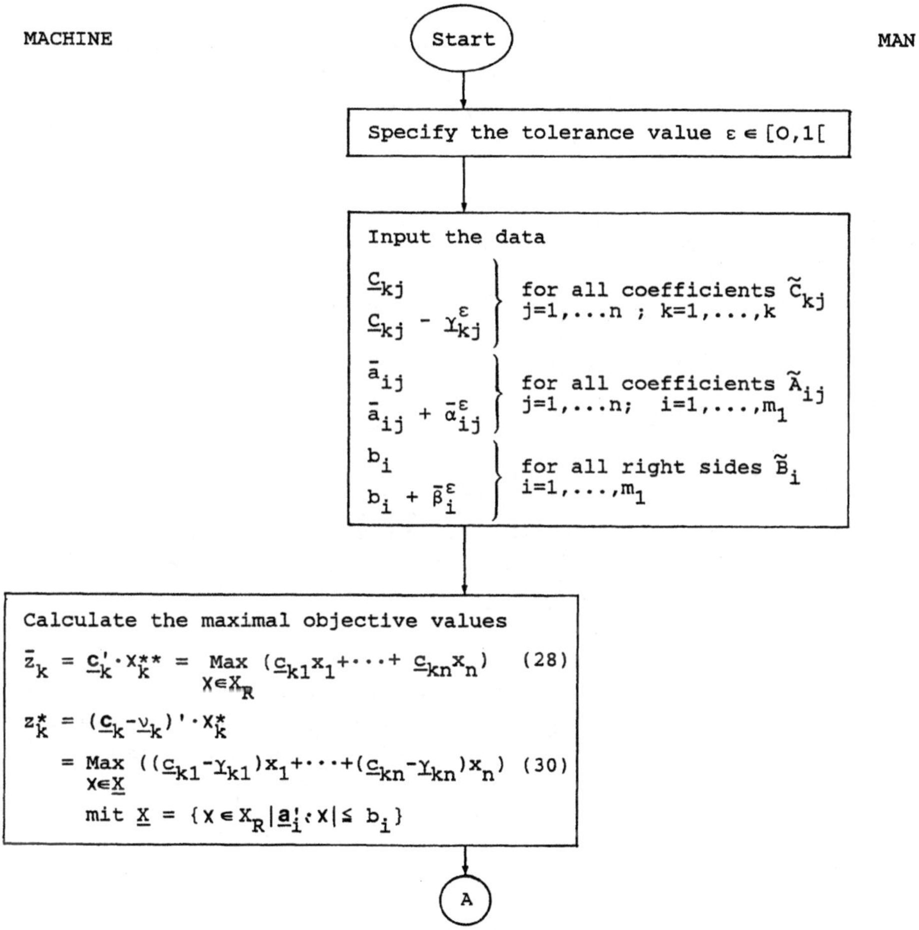

MACHINE | (A) | MAN

Calculate the low bounds

$$\underline{z}_k = \mathrm{Min}\Big(\mathop{\mathrm{Min}}_{\substack{l=1\\ l\neq k}}^{K}(\underline{c}_k-\underline{\nu}_k)'\cdot x_l^*),\ \mathop{\mathrm{Min}}_{\substack{l=1\\ l\neq k}}^{K}(\underline{c}_k-\underline{\nu}_k)'\cdot\ x_l^{**})\Big) \qquad (29)$$

Ist K = 1, so ist $\underline{z}_1 = (\underline{c}_1-\underline{\ }_1)'x_1^*$

Specify the critical membership value $\lambda_A \in]\varepsilon,1[$

Input the tolerance thresholds

$w_{Z1},\dots,w_{ZK}$ for the objectives

$w_1,\dots,w_{m_1}$ for the constraints

Fix the lower and upperbounds for the aspiration levels

$\check{\mathbf{z}} = (\check{z}_1,\dots,\check{z}_K) := (\underline{z}_1,\dots,\underline{z}_K)$

$\hat{\mathbf{z}} = (\hat{z}_1,\dots,\hat{z}_K) := (\bar{z}_1,\dots,\bar{z}_K)$

$\check{\mathbf{g}} = (\check{g}_1,\dots,\check{g}_{m_1}) := (b_1,\dots,b_{m_1})$

$\hat{\mathbf{g}} = (\hat{g}_1,\dots,\hat{g}_{m_1}) := (b_1+d_1,\dots,b_{m_1}+d_{m_1})$

Specify the levels of aspiration

$z_k^A[1] \in]\check{z}_k,\hat{z}_k[$ for all objectives k=1,..,K

$g_i^A[1] \in]\check{g}_i,\hat{g}_i[$ for all constraints $i=1,..,m_1$

If need be, specify additional points $(z_k,f_{Zk}(z_k))$ and/or $(g_i,f_i(g_i))$ which have to be considered by the construction of the membership functions f_{Zk}, k=1,...,K, or f_i, $i=1,\dots,m_1$

Setze r = 0

(B)

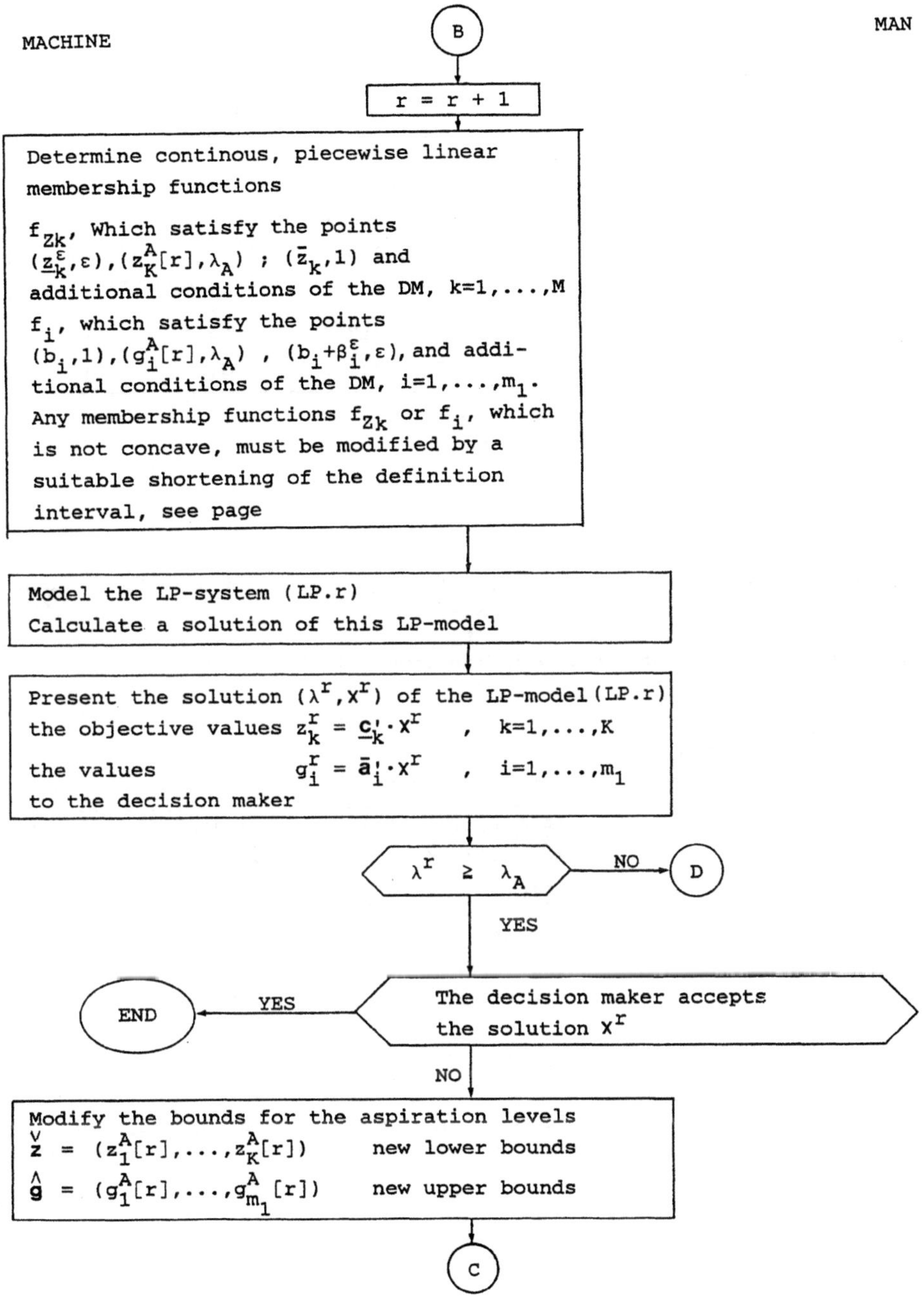
MACHINE
MAN
B
$r = r + 1$
Determine continous, piecewise linear membership functions
f_{Zk}, Which satisfy the points
$(\underline{z}_k^{\varepsilon}, \varepsilon), (z_K^A[r], \lambda_A)$; $(\bar{z}_k, 1)$ and
additional conditions of the DM, k=1,...,M
f_i, which satisfy the points
$(b_i, 1), (g_i^A[r], \lambda_A)$, $(b_i+\beta_i^{\varepsilon}, \varepsilon)$, and additional conditions of the DM, $i=1,\ldots,m_1$.
Any membership functions f_{Zk} or f_i, which is not concave, must be modified by a suitable shortening of the definition interval, see page
Model the LP-system (LP.r)
Calculate a solution of this LP-model
Present the solution (λ^r, x^r) of the LP-model (LP.r)
the objective values $z_k^r = \underline{c}_k' \cdot x^r$, $k=1,\ldots,K$
the values $g_i^r = \bar{a}_i' \cdot x^r$, $i=1,\ldots,m_1$
to the decision maker
$\lambda^r \geq \lambda_A$
NO
D
YES
END
YES
The decision maker accepts the solution x^r
NO
Modify the bounds for the aspiration levels
$\check{z} = (z_1^A[r],\ldots,z_K^A[r])$ new lower bounds
$\hat{g} = (g_1^A[r],\ldots,g_{m_1}^A[r])$ new upper bounds
C

MACHINE | (C) | MAN

Specify the levels of aspiration

$z_k^A[r+1] \in [\check{z}_k, \hat{z}_k]$, $k=1,\ldots,K$

$g_i^A[r+1] \in [\check{g}_i, \hat{g}_i]$, $i=1,\ldots,m_1$.

In doing so, the DM has to take care that $z_k^A[r+1] > z_k^r$ or $g_i^A[r+1] < g_i^r$ for at least an objective k or a constraint i.

If $\hat{z}_k - \check{z}_k < W_{ZK}$, then $\check{z}_k^A[r+1] = \check{z}_k$

If $\hat{g}_i - \check{g}_i < W_i$, then $g_i^A[r+1] = \hat{g}_i$

(B)

MACHINE | (D) | MAN

Modify the bounds for the aspiration levels

$\hat{z} = (z_1^A[r],\ldots,z_K[r])$

$\check{g} = (g_1^A\ r\ ,\ldots,g_{m_1}^A[r])$

Specify the levels of aspiration

$z_k^A[r+1] \in [\check{z}_k, \hat{z}_k]$, $k=1,\ldots,K$

$g_i^A[r+1] \in [\check{g}_i, \hat{g}_i]$, $i=1,\ldots,m_1$.

In doing so, the DM has to take care that $z_k^A[r+1] < z_k^r$ or $g_i^A[r+1] > g_i^r$ for at least an objective k or a constraint i.

If $\hat{z}_k - \check{z}_k < W_{ZK}$, then $z_k^A[r+1] = \check{z}_k$

If $\hat{g}_i - \check{g}_i < W_i$, then $g_i^A[r+1] = \hat{g}_i$

(B)

7. Numerical example

In order to illustrate the application of the concepts introduced above, a simple numerical example is presented.

Let us consider the following fuzzy LP-problem

$$\left.\begin{array}{l}\tilde{Z}_1(X) = \tilde{C}_{11}x_2 + \tilde{C}_{12}cx_2 \qquad\qquad + 2x_4 \\ \tilde{Z}_2(X) = \quad x_1 + \qquad\qquad + 2x_3 + 3x_4\end{array}\right] \rightarrow \widetilde{\text{MAX}}$$

subject to (32)

$$\begin{array}{rl} 3x_1 + \quad 3x_2 + \quad 4x_3 + \quad 3x_4 & \tilde{\leq} \ 18 \\ 9x_1 + \quad 4x_2 + \quad 4x_3 + \quad 6x_4 & \tilde{\leq} \ (35;0;4)^{\varepsilon} \\ \tilde{A}_{31}x_1 + \tilde{A}_{32}x_2 + \tilde{A}_{33}x_3 + \tilde{A}_{34}x_4 & \tilde{\leq} \ (60;0;5)^{\varepsilon} \\ 8x_1 + \tilde{A}_{42}x_2 + \tilde{A}_{43}x_3 + \quad 5x_4 & \tilde{\leq} \ (50;0;6)^{\varepsilon} \\ X = (x_1,x_2,x_3,x_4)' & \geq 0 \end{array}$$

where

$$\begin{array}{llll} \tilde{C}_{11} = (4,4.5;0.5;0.5)^{\varepsilon}, & \tilde{C}_{12} = (3, \ 3;0,2;0,3)^{\varepsilon} \\ \tilde{A}_{31} = (6,6.5;0.3;0,3)^{\varepsilon}, & \tilde{A}_{32} = (7.5, \ 8;0.5;0.6)^{\varepsilon} \\ \tilde{A}_{33} = (15, \ 16; \ 1; \ 1)^{\varepsilon}, & \tilde{A}_{14} = (10, \ 11;0.5;0.5)^{\varepsilon} \\ \tilde{A}_{42} = (15, \ 16; \ 1; \ 1)^{\varepsilon}, & \tilde{A}_{43} = (9.3,10;0.5;0.3)^{\varepsilon} \end{array}$$

with $\varepsilon = 0.1$.

Using the interactive procedure for calculating a compromise solution of this multiobjective fuzzy optimization problem not all of the data in (12) are necessary. In the PC-expert-system FULP the needed values will only be required, see figure 5.

```
                    Input of the data of the problem

  Please specify the coefficients of the objective function.
---------------------------------------------------------------------
  1. Left (minimal) value of c[2, 4 ] on level alpha = 1    : 3.000000

  2. Left (minimal) value of c[2, 4 ] on level Epsilon      : 3.000000

  Please specify the coefficients of the constraints.
---------------------------------------------------------------------
  1. Right (maximal) value of a[4,4] on level alpha = 1     : 5.000000

  2. Right (maximal) value of a[4,4] on level Epsilon       : 5.000000

  Please specify the right sides of the constraints
---------------------------------------------------------------------
  1. Certain right side b[.4 ] on level alpha = 1           : 50.00000

  2. Maximal right side b[ 4 ] on level alpha = Epsilon     : 56.00000________
```

Figure 5: Data input

Computing the LP-models (28) and (30) and using (29) we get

$$\bar{z}_1 = 19.6364 \qquad \underline{z}_1 = 10.9091$$
$$\bar{z}_2 = 17.2566 \qquad \underline{z}_2 = 3.3571 \quad .$$

The DM specifies the critical level λ_A as $\lambda_A = 0.5$ and sets the thresholds for all objectives and all constraints equal 0.01.
For the first iteration of the interactive decision process, the DM intends, to get more information about the solution space. Therefore he begins with the low aspiration levels

$$z_1^A = 12 \quad , \quad z_2^A = 5$$
$$g_2^A = 39 \ , \quad g_3^A = 64 \ , \quad g_4^A = 54 \quad .$$

Using the PC-expert-system "FULP", we get the solution

The LP-problem has the following solution :

The critical membership value Lamda-1 = 0.761330

with the following x-values :
x1 = 1.344922 x2 = 1.301371 x3 = 0.000000 x4 = 3.353707

the objective values,	improvement compand with the level of aspiration	and the supply,	exceeding of the certain right sides
z[1] = 15.991215	52.27 %	g[1] = 18.000000	Fixed variable.
z[2] = 11.406043	52.27 %	g[2] = 37.432023	47.73 %
		g[3] = 56.043738	Not exceeded !
		g[4] = 48.349849	Not exceeded !

Figure 6: First compromise solution

As $\lambda^1 = 0.761330 > \lambda^A$, the levels of aspiration can be increased. The DM decides to change all levels and he specifies

$$z_1^A = 16 \ , \quad z_2^A = 10$$
$$g_2^A = 38.5 \ , \quad g_3^A = 61 \ , \quad g_4^A = 52 \quad .$$

In doing so the DM gets helpful assistance by the PC-expert-system FULP, see figure 7.

```
Please specify the new levels of aspiration:

concerning the objectives  z(k)
concerning the constraints g(i)

Concerning the objective z(2)  the level of aspiration must belong to the

interval  [5.0000,17.2566]  .

Aspiration level of objective  z(2)        =    10.00000__________

      z-Min[k]           zA[k]            z[k]           z-Max[k]
      |------------------|----------------|-----------------|
     3.3571              |                |              17.2566
                      5.0000              |
                                       11.4060
```

Figure 7: Specification of the new aspiration level z_2^A

Calculating the new LP-model (LP.2), we get the result:

```
The LP-problem has the following solution :

The critical membership value Lamda-2 = 0.560660

with the following x-values :
x1 = 1.582718    x2 = 1.275726    x3 = 0.000000    x4 = 3.141556

the objective values,  improvement        and the supply,    exceeding of the
                     compand with the                       certain right
                    level of aspiration                     sides
z[1] = 16.441162      12.13 %            g[1] = 18.000000    Fixed variable.
z[2] = 11.007386      13.88 %            g[2] = 38.196701     87.87 %
                                         g[3] = 55.050591    Not exceeded !
                                         g[4] = 48.781142    Not exceeded !
```

Figure 8: Second compromise solution

As $\lambda^A = 0.560660$ indicates that there is only a little margin for further improvements, the DM keeps constant the levels z_2^A , g_2^A and g_3^A . He merely changes the two remaining levels to

$$z_1^A = 17 \quad \text{and} \quad g_4^A = 51 \quad .$$

The new LP-model (LP.3) has the solution:

The LP-problem has the following solution :

The critical membership value Lamda-3 = 0.502732

with the following x-values :
x1 = 1.815236 x2 = 1.422234 x3 = 0.000000 x4 = 2.743379

the objective values,	improvement compand with the level of aspiration	and the supply,	exceeding of the certain right sides
z[1] = 17.014406	0.55 %	g[1] = 17.942550	Fixed variable.
z[2] = 10.045374	0.63 %	g[2] = 38.486339	99.45 %
		g[3] = 53.354083	Not exceeded !
		g[4] = 50.994535	99.45 %

Figure 9: Third compromise solution

Now the DM accepts this solution.

8. References

Bortolan, G. and Degani, R. (1985) 'A review of some methods for ranking fuzzy subsets', FSS 15, 1-19.

Carlson, C. and Korhonen, P. (1986) 'A parametric approach to fuzzy linear programming', FSS 20, 17-30.

Delgado, M., Verdegay, J.L., and Vila, M.A. (1989) 'A general model for fuzzy linear programming', FSS 29, 21-30.

Dubois, Didier and Prade, Henri (1980) 'Fuzzy sets and systems: Theory and applications', Academic Press, London.

Dubois, D. and Prade, H. (1983) 'Ranking of fuzzy numbers in the setting of possibility theory', Information Sciences 30, 183-224.

Luhandjula, M.K. (1986) 'On possibilistic linear programming', FSS 18, 1-16.

Luhandjula, M.K. (1989) 'Fuzzy optimization: An appraisal', FSS 30, 257-282.

Negoita, C.V., Minoiu, S., and Stan, E. (1976) 'On considering imprecision in dynamic linear programming', Economic Computation and Economic Cybernetics Studies and Research 3, 83-95.

Negoita, C.V., and Sularia, M. (1976) 'On fuzzy mathematical programming and tolerances in planning', Economic Computation and Economic Cybernetics Studies and Research 3, 3-15.

Orlovski, S.A. (1985) 'Mathematical programming problems with fuzzy parameters', in J. Kacprzyk and R.R. Yager (eds.), Management decision support systems using fuzzy sets and possibility theory, TÜV Rheinland, Köln, pp. 136-145.

Ramik, J. and Rimanek, J. (1985) 'Inequality between fuzzy numbers and its use in fuzzy optimization', FSS 16, 123-138.

Rommelfanger, H. (1983) 'Lineare Ersatzmodelle für lineare Fuzzy-Optimierungsmodelle mit konkaven Zugehörigkeitsfunktionen', Diskussionspapier des Instituts für Statistik und Mathematik, Universität Frankfurt.

Rommelfanger, H. (1984) 'Concave membership functions and their applications in fuzzy mathematical programming', Proceedings of the Workshop on the Membership Function, ed. by the European Institute for Advanced Studies in Management (EIASM), Brussels, pp. 88-101.

Rommelfanger, H. (1986) 'Rangordnungsverfahren für unscharfe Mengen', OR-Spektrum 8, 219-228.

Rommelfanger, H. (1988) 'Entscheiden bei Unschärfe. Fuzzy Decision Support-Systeme', Springer Verlag, Berlin Heidelberg.

Rommelfanger, H. (1989) 'Interactive decision making in fuzzy linear optimization problems', EJOR 41, 210-217.

Rommelfanger, H., Hanuscheck, R., and Wolf, J. (1989) 'Linear programming with fuzzy objectives', FSS 29, 31-48.

Sakawa, M. and Yano, H. (1989) 'Interactive fuzzy decision making for generalized multiobjective linear programming problems with fuzzy parameters', FSS 29, 315-326.

Slowinski, R. (1986) 'A multicriteria fuzzy linear programming method for water supply system development planning', FSS 19, 217-237.

Tanaka, H. and Asai, K. (1984) 'Fuzzy linear programming with fuzzy numbers', FSS 13, 1-10.

Werners, Brigitte (1984) 'Interaktive Entscheidungsunterstützung durch ein flexibles mathematisches Programmierungssystem', Minerva Publikation, München.

Yazenin, A.V. (1987) 'Fuzzy and stochastic programming', FSS 22, 171-180.

Zadeh, M. (1982) 'Fuzzy sets', Information and Control 8, 338-353.

Zeleny, M. (1982) 'Multi criteria decision-making',McGraw Hill, New York.

Zimmermann, H.J. (1978) 'Fuzzy programming and linear programming with several objective functions', FSS 1, 45-55.

Zimmermann. H.J. (1985) 'Fuzzy set theory - and its applications', Kluwer-Nijhoff Publishing, Boston Dordrecht Lancaster.

Part III
Chapter 6

MULTIPLE OBJECTIVE LINEAR PROGRAMMING PROBLEMS IN THE PRESENCE OF FUZZY COEFFICIENTS

M. K. LUHANDJULA
Institut d'Informatique, Université de Tizi-Ouzou
B.P. 531, 15000, Tizi-Ouzou, Algeria

MASATOSHI SAKAWA
Department of Computer Science, Faculty of Engineering,
Iwate University, Ueda, Morioka 020 Japan

ABSTRACT. This paper grapples with the problem of incorporating fuzzy coefficients in the objective functions as well as in the constraints of a multiobjective linear program. Concepts of satisfying solutions based on the bounded rationality principle are introduced, and necessary and sufficient conditions for a potential action to be a satisfying one are established. Efficient methods for singling out satisfying alternatives are also described together with illustrative numerical examples. We end up with a number of concluding remarks and indicate lines for further developments.

1. INTRODUCTION

Many organizations within both the private and public sectors are confronted with problems which can be casted into a multiple objective mathematical programming framework. Examples are : Scheduling of nursing resources, evaluation of urban policy, macro-economic policy, corporate financial management, forest management, water resources management, regional planning, etc. The literature is rich with articles describing analytical results as well as algorithms for finding efficient solutions of these problems. Many excellent surveys or books can be found including Cohon and Marks (1975), Roy (1971) , Chankong and Haimes (1983), Seo and Sakawa (1988) and so forth.

However one often neglected (yet critical) aspect is the incorporation of intrinsic or informational imprecision in these models. This paper grapples with this problem in the case where the above mentioned imprecision is modelled by fuzzy

R. Slowinski and J. Teghem (eds.),
Stochastic versus Fuzzy Approaches to Multiobjective Mathematical Programming under Uncertainty, 301–320.

numbers. Concepts of satisfying solutions based on the bounded rationality principle are introduced and necessary and sufficient conditions for a potential action to be a satisfying one are established. Linkups between imprecision and infinity in an optimization context emerge throughout.

Potential opened up by semi-infinite optimization and Interval mathematical programming are gainfully exploited to push forward approaches for yielding satisfying solutions. For the paper to be somewhat self-contained, some preliminaries including the statement of the problem are given in the next section. In section 3, satisfying solutions for the problem at hand are introduced and rigourosly characterized. Section 4 centers around the algorithmic point of view. Efficient methods for singling out some satisfying alternatives are described. In section 5, illustrative numerical examples are demonstrated. Finally we draw some conclusions and indicate lines for further developments.

2. PRELIMINARIES

2.1 Fuzzy sets : Elementary notions

A fuzzy set $\tilde{A}$ of a universal set X is characterized by its membership function

$$\Pi_{\tilde{A}} : X \rightarrow [0,1],$$

where $\Pi_{\tilde{A}}(x)$ may represent, according to the context, the degree of belongness to a set without well-defined boundaries, the degree of similarity with a prototype, the level of compatibility with vague concept and so on.

Several basic definitions concerning fuzzy sets are as follows:

(1) The support of a fuzzy set $\tilde{A}$ is defined by:

$$supp \quad \tilde{A} = \{x \in X | \Pi_{\tilde{A}}(x) > 0\}.$$

(2) The α-cut of $\tilde{A}$ is defined as:

$$\tilde{A}^{\alpha} = \{x \in \tilde{X} | \Pi_{\tilde{A}}(x) \geq \alpha\}.$$

(3) A fuzzy set $\tilde{A}$ is normalized if there is $x \in X$ such that $\Pi_{\tilde{A}}(x) = 1$.

(4) A fuzzy set $\tilde{A}$ is convex if

$$\Pi_{\tilde{A}}(\lambda x^1 + (1-\lambda)x^2) \geq \min(\Pi_{\tilde{A}}(x^1), \Pi_{\tilde{A}}(x^2))$$

for all $x^1, x^2 \in X$ and $\lambda \in [0,1]$.

(5) A fuzzy number is a convex normalized fuzzy set A of $\Re$ such that

$$\exists x^0 \quad \text{satisfying} \quad \Pi_{\tilde{A}}(x^0) = 1.$$

A fuzzy number suits well for modelizing an ill-defined quantity.

2.2 Statement of the problem

In order to couple their perceptions with hard statistical data, experts involved in the formulation of a given multiobjective program, may feel more comfortable in specifying these data in a fuzzy rather than a precise numerical form. The resulting problem can be described as :

$$\begin{array}{lll} \max & (\tilde{c}^1 x, \cdots, \tilde{c}^k x) & (P) \\ \text{subject to} & \tilde{A}x \leq \tilde{b} & \\ & x \geq 0 & \end{array}$$

where $\tilde{c}^j$ $(j = 1, \cdots, k)$ are n-vectors, $\tilde{b}$ is an m vector and $\tilde{A}$ is an $m \times n$ matrix whose components are fuzzy numbers, and the membership functions of which are denoted by $\Pi_{\tilde{c}^1_j}$, $\Pi_{\tilde{b}_i}$ and $\Pi_{\tilde{a}_{ij}}$ respectively. The reminder of this paper is devoted to the analysis of this problem as well as to the development of efficient numerical techniques for finding related satisfying solutions.

3. SATISFYING SOLUTIONS FOR PROBLEM (P)

3.1 $(\alpha, \beta)_1$-satisfying solution for (P)

Definition 1 (α-possible feasibility)
x^0 is said to be α-possibly feasible for (P), if

$$Poss[\tilde{A}x^0 \leq \tilde{b}] \geq \alpha,$$

where Poss denotes Possibility.

By virtue of the extension principle ,

$$Poss[\tilde{A}x^0 \leq \tilde{b}] = \sup_{\substack{A,b \\ Ax^0 \leq b}} \min(\Pi_{\tilde{A}}(A), \Pi_{\tilde{b}}(b)),$$

where $\Pi_{\tilde{A}}$ and $\Pi_{\tilde{B}}$ are $m \times n$-ary and m-ary possibility distribution respectively (Zadeh 1978).

It is an easy matter to see that this concept extend quite naturally that of feasibility. As a matter of fact, a feasible (infeasible) alternative for a deterministic program is nothing but a 1-possibly (0-possibly) feasible action in this sense. The necessity measure can also be used for extending the concept of feasibility. For details on these matters, the readers are referred to Sakawa and Yano(1988).

Definition 2 (α-possible domination)

Consider $x^1, x^2 \in \Re^n$, with $x^1 \geq 0, x^2 \geq 0$. x^1 is said to be α-possibly dominates x^2, in symbol $x^1 \geq_\alpha x^2$, if

$$Poss[\tilde{c}^1 x^1 \geq \alpha, \cdots, \tilde{c}^k x^1 \geq \alpha] \geq Poss[\tilde{c}^1 x^2 \geq \alpha, \cdots, \tilde{c}^k x^2 \geq \alpha].$$

Definition 3 (α-possible non-domination)
Consider X as a some deterministic version of the constraints of (P). $x \in X$ is α-possibly non-dominated , if there is no $\bar{x} \in X$ such that $\bar{x} \geq_\alpha x$.

Definition 4 ($(\alpha, \beta)_1$-satisfying solution)
x is said to be an $(\alpha, \beta)_1$-satisfying solution for (P), if x is α-possibly feasible for (P) and x is β-possibly non-dominated.

It is clear that an $(\alpha, \beta)_1$-satisfying solution for (P) with α and β close to 1 are interested from the standpoints of feasibility and optimality for (P), because such an action is possibly feasible and possibly non-dominated to a great extent. Let us now turn to the characterization of such an action.

To this end we introduce the following parameters sets :

Let $T^1, T^2, S^1, \cdots, S^k$ be parameters sets which are in univocal correspondence with $Supp\ \tilde{A}, Supp\ \tilde{b}, Supp\ \tilde{c}^1, \cdots, Supp\ \tilde{c}^k$ respectively, and let $A(t_1)$,$b(t_2)$,$c^1(s_1)$,$\cdots$,$c^k(s_k)$ be the corresponding bijections.

In order to alleviate notations we will write merely :

$$A(t), b(t), c^1(s), \cdots, c^k(s)$$

instead of

$$A(t_1), b(t_2), c^1(s_1), \cdots, c^k(s_k)$$

respectively.

Consider now the subdivisions :

$$(T^{\alpha_i})_{i=1,\cdots,p+1} \quad \text{of} \quad T = T^1 \times T^2 \quad \text{and} \quad S^{\beta^j_n}_{n=1,\cdots,v+1} \quad \text{of} \quad S^j.$$

defined as follows :

$$\begin{aligned} T^{\alpha_i} &= \{t | \alpha_{i-1} \leq \mu(t) \leq \alpha_i\}, \\ \mu(t) &= \min\left(\Pi_{\tilde{A}}(A(t)), \Pi_{\tilde{b}}(b(t))\right), \\ S^{\beta^j_n} &= \{s | \beta^j_{n-1} \leq \Pi_{\tilde{c}^j}(c^j(s)) \leq \beta^j_n\}, \end{aligned}$$

where α_i, β^j_n are real numbers such that :

$$\begin{aligned} 0 = \alpha_0 < \alpha_1 < \cdots < \alpha_p < \alpha_{p+1} = 1, \\ 0 = \beta^j_0 < \beta^j_1 < \cdots < \beta^j_v < \beta^j_{v+1} = 1. \end{aligned}$$

Let now real numbers $(\delta_i)_{i=1,\cdots,p+1}, (\gamma_n^j)_{n=1,\cdots,v+1}$ such that :

$$0 = \delta_{p+1} \leq \cdots \leq \delta_1 ,$$
$$0 = \gamma_{v+1}^j \leq \cdots \leq \gamma_1^j ,$$

chosen to penalize $t \in T, s \in S^j$ such that $A(t), b(t), c^j(s)$ are not compatible with restrictions defined by $\tilde{A}$ and $\tilde{b}, \tilde{c}^j$ respectively.

Consider now the following mathematical program :

$$\max \quad \theta \qquad (P')$$

subject to

$$\begin{array}{rlll} \theta - c^j(s)x - \gamma_{v+1}^j & \leq 0 & ; \; s \in S^{\beta_{v+1}^j} & j = 1, \cdots, k \\ \theta - c^j(s)x - \gamma_v^j & \leq 0 & ; \; s \in S^{\beta_v^j} & j = 1, \cdots, k \\ & \vdots & & \\ \theta - c^j(s)x - \gamma_1^j & \leq 0 & ; \; s \in S^{\beta_1^j} & j = 1, \cdots, k \\ A(t)x - b(t) & \leq \delta_{p+1}^m & ; \; t \in T^{\alpha_{p+1}} & \\ A(t)x - b(t) & \leq \delta_p^m & ; \; t \in T^{\alpha_p} & \\ & \vdots & & \\ A(t)x - b(t) & \leq \delta_1^m & ; \; t \in T^{\alpha_1} & \\ x & \geq 0 & & \end{array}$$

where (δ_r^m) is an m-vector with both components equal to δ_j.

(P') is a semi-infinite mathematical program intimately related to (P).

As a matter of fact , an optimal solution of (P') is such that the more elements in the support of fuzzy quantities in (P) have high membership degrees , the less the violation of corresponding constraints and the achievement of low level for corresponding objective functions are tolerated.

It is also worth noticing that in passing from (P) to (P') , all elements in the support of involved fuzzy quantities are taken into consideration with their respective membership degrees. In this sense, the deterministic program (P') outline more properly the original fuzzy program (P) than approaches consisting in considering only elements with higher membership degrees , or both elements in the supports with equal degree of compatibility. Furthermore , we can state and prove the following result.

Theorem 1
If (x^0, θ^0) is an optimal solution of (P'), then x^0 is an $(\alpha_p, \theta^0)_1$-satisfying solution for (P).

Proof

For any $t \in T^{\alpha_{p+1}}$, we have

$$u(t) = \min\Big(\Pi_{\tilde{A}}(A(t)) \ ; \ \Pi_{\tilde{b}}(b(t))\Big) \geq \alpha_p$$

and

$$A(t)x^0 \leq b(t).$$

Hence :

$$Poss[\tilde{A}x^0 \leq \tilde{b}] = \sup_{\substack{t \\ A(t)x^0 \leq b(t)}} \min[\Pi_{\tilde{A}}(A(t)), \Pi_{\tilde{b}}(b(t))] \geq \alpha_p,$$

i.e., x^0 is α_p-possibly feasible for (P').

Let us now show that x^0 is β-possibly non-dominated action for (P').

$$\begin{aligned} Poss[\tilde{c}^1 x^0 \geq \alpha^0, \cdots, \tilde{c}^k x^0 \geq \alpha^0] \\ = \sup_{\substack{s^1,\cdots,s^k \\ s^1 x^0 \geq \theta^0 \\ \vdots \\ s^k x^0 \geq \theta^0}} \min[\Pi_{\tilde{c}^1}(c^1(s^1)), \cdots, \Pi_{\tilde{c}^k}(c^k(s^k))] \end{aligned}$$

Consider $\bar{s}^j$ such that :

$$\Pi\tilde{c}^j(c^j(\bar{s}^j)) = 1 \quad j = 1, \cdots, k,$$

then

$$\Pi_{\tilde{c}^j}(c^j(s^j)) \leq \Pi_{\tilde{c}^j}(c^j(\bar{s}^j)) \quad \forall s^j \in S^j, j = 1, \cdots, k,$$

and

$$\min_j\Big(\Pi_{\tilde{c}^j}(c^j(s^j))\Big) \leq \min_j\Big(\Pi_{\tilde{c}^j}(c^j(\bar{s}^j))\Big) = \Pi_{\tilde{c}}(c(\bar{s})).$$

Hence

$$\sup_{s^1,\cdots,s^k} \min\left[\Pi_{\tilde{c}^1}(c^1(s^1)), \cdots, \Pi_{\tilde{c}^k}(c^k(s^k))\right] \leq \Pi_{\tilde{c}}(c(\bar{s})). \tag{1}$$

We have also :

$$\bar{s}^j \in S^{\beta^j_{v+1}}$$

and

$$c^j(\bar{s}^j)x^0 \geq \theta^0 \quad j = 1, \cdots, k. \tag{2}$$

By (1), we have :

$$\sup_{\substack{s^1,\cdots,s^k \\ c^1(s^1)x^0 \geq \theta^0 \\ \vdots \\ c^k(s^k)x^0 \geq \theta^0}} \min_j \Pi_{\tilde{c}^j}(c^j(s^j)) \leq \sup_{s^1,\cdots,s^k} \min_j \Pi_{\tilde{c}^j}(c^j(s^j)) \leq \Pi_{\tilde{c}}(c(\bar{s})). \tag{3}$$

Combining (1) and (2) we get

$$\Pi_{\tilde{c}}(c(\bar{s})) = \sup_{\substack{s^1,\cdots,s^k \\ c^1(s^1)x^0 \geq \theta^0 \\ \vdots \\ c^k(s^k)x^0 \geq \theta^0}} \min_j \Pi_{\tilde{c}^j}(c^j(s^j)) = Poss[\tilde{c}^1 x^0 \geq \theta^0, \cdots, \tilde{c}^k x^0 \geq \theta^0]. \quad (4)$$

Let now x be a feasible action for (P') , then from (3) and (4)

$$\begin{aligned} Poss[\tilde{c}^1 x \geq \theta^0, \cdots, \tilde{c}^k x \geq \theta^0] &= \sup_{\substack{s^1,\cdots,s^k \\ c^1(s^1)x \geq \theta^0 \\ \vdots \\ c^k(s^k)x \geq \theta^0}} \min_j \Pi_{\tilde{c}^j}(c^j(s^j)) \leq \Pi^j_{\tilde{c}}(c(\bar{s})) \\ &= Poss[\tilde{c}^1 x^0 \geq \theta^0, \cdots, c^k x^0 \geq \theta^0] \end{aligned}$$

as desired.

3.2 $(\alpha, \beta)_2$-Satisfying Solution for (P)

Definition 5 (β-possible efficiency)
x^0 is said to be β-possibly efficient for the program :

$$\max_{x \in X \subseteq \Re^n} (\tilde{c}^1 x, \cdots, \tilde{c}^k x)$$

if there is no $x \in X$ and $l \in \{1, \cdots, k\}$ such that

$$\begin{aligned} Poss[\tilde{c}^1 x \geq \tilde{c}^1 x^0, \cdots, &\tilde{c}^{l-1} x \geq \tilde{c}^{l-1} x^0, \tilde{c}^l x > \tilde{c}^l x^0, \\ &\tilde{c}^{l+1} x \geq \tilde{c}^{l+1} x^0, \cdots, \tilde{c}^k x \geq \tilde{c}^k x^0] \geq \beta. \end{aligned}$$

The above definition absorb that of efficiency (Pareto Optimality) as a special case. As a matter of fact , an efficient solution of a deterministic multiple objectives linear program is nothing but a 1-possibly efficient solution in the sense of Definition 5. An extension of the concept of Pareto optimality via the measure of necessity is discussed in Sakawa and Yano(1988).

Definition 6 ($(\alpha, \beta)_2$-satisfying solution)
x^0 is siad to be an $(\alpha, \beta)_2$-satisfying solution for (P), if x^0 is β-possibly efficient for the program :

$$\begin{array}{lll} \max & (\tilde{c}^1 x, \cdots, \tilde{c}^k x) & (P1) \\ \text{subject to} & x \in X^\alpha, x \geq 0 & \end{array}$$

where X^α is the set of α-possibly feasible actions for (P').

Let us now move to the characterization of $(\alpha, \beta)_2$ satisfying alternatives for (P').

Consider the following mathematical program :

$$\begin{array}{lll} \max & \left((\tilde{c}^1)^\beta x, \cdots, (\tilde{c}^k)^\beta x\right) & (P'') \\ \text{subject to} & x \in K = \{x \in \Re^n | \overline{\tilde{A}}^\alpha x \leq \underline{\tilde{b}}^\alpha\} & \\ & x \geq 0, & \end{array}$$

where $(\tilde{c}^i)^\beta = ((\tilde{c}^i_1)^\beta, \cdots, (\tilde{c}^i_n)^\beta)$.

$\tilde{m}^\alpha$ denotes as previously (see section 2) the α-cut of $\tilde{m}$, and

$$\overline{\tilde{m}}^\alpha = \sup_{t \in \tilde{m}^\alpha} t \; ; \; \underline{\tilde{m}}^\alpha = \inf_{t \in \tilde{m}^\alpha} t. \tag{4'}$$

As $(\tilde{c}^i_j)$ are fuzzy numbers, they are convex fuzzy sets of $\Re$, and $(\tilde{c}^i_j)^\beta; j = 1, \cdots, n; i = 1, \cdots, k$ are real closed intervals that will be denoted by:

$$[\; \bar{c}^{iL}_j \;,\; \bar{c}^{iM}_j \;].$$

Let now Φ_β be the set of $k \times n$ matrices $C = (c_{ij})$ with $c_{ij} \in [\bar{c}^{iL}_j, \bar{c}^{iM}_j]$.

Definition 7 (efficient)
x^0 is said to be efficient for (P''), if and only if there is no $C \in \Phi_\beta$ and $x \in X^\alpha$ such that $Cx \geq Cx^0$ with at least one strict inequality.

The following notations will be used in the sequel.

(1) $EF(C)$ is the set of efficient solution of the multiobjective program

$$\max_{x \in K} Cx. \tag{P2}$$

(2) EF is the set of efficient solutions for (P''), i.e.,

$$EF = \bigcap_{c \in \Phi_\beta} EF(C).$$

(3) $EF_{ex}(C), EF_{ex}$ are the set of efficient extreme points for $(P2)$ and (P'') respectively.

Theorem 2

x^0 is an $(\alpha,\beta)_2$-satisfying solution for (P), if and only if there is $\varepsilon > 0$ such that x^0 is β-possibly efficient for $(P'')_\varepsilon$, where $(P'')_\varepsilon$ is the program (P'') with β is replaced by $\beta - \varepsilon$.

Proof

It is well established in Sakawa and Yano (1988) that $X^\alpha = K$.

So we have only to show the equivalence between the β-possible efficiency for (P) and the efficiency for $(P'')_\varepsilon$.

Suppose x^0 is β-possibly efficient for (P) and not efficient for $(P'')_\varepsilon$. Then for any $\varepsilon > 0$, there is

$$x^1 \in K$$

and

$$(q^1, \ldots q^k) \in \Re^{nk}$$

with

$$q^i \in (\tilde{c}^i)^{\beta-\varepsilon} \qquad \forall \varepsilon > 0$$

such that

$$q^i x^1 \geq q^i x^0 \qquad \forall i \in \{1, \ldots, k\}$$

and

$$q^s x^1 > q^s x^0 \quad \text{for some} \quad s \in \{1, \ldots, k\}.$$

As

$$q^i \in (\tilde{c}^i)^{\beta-\varepsilon} \qquad \forall \varepsilon > 0, \quad i = 1, \ldots, k$$

we have

$$\min\left(u_{\tilde{c}^1}(q^1), \ldots, u_{\tilde{c}^k}(q^k)\right) \geq \beta - \varepsilon \quad \forall \varepsilon > 0.$$

Hence

$$\sup_{(t^1,\ldots,t^k)\in T^s} \min\left(u_{\tilde{c}^1}(c^1(t^1), \ldots, u_{\tilde{c}^k}(c^k(t^k))\right) \geq \beta - \varepsilon \quad \forall \varepsilon > 0 \tag{5}$$

where

$$\begin{aligned} T^s - \{(t^1, \ldots, t^k) \in \Re^{nk} | & c^1(t^1)x \geq c^1(t^1)x^0, \ldots, \\ & c^{s-1}(t^{s-1})x \geq c^{s-1}(t^{s-1})x^0, c^s(t^s)x > c^s(t^s)x^0, \\ & c^{s+1}(t^{s+1})x \geq c^{s+1}(t^{s+1})x^0, \ldots, c^k(t^k)x \geq c^k(t^k)x^0\}. \end{aligned}$$

By (5) we have

$$\begin{aligned} Poss(\tilde{c}^1 x \geq \tilde{c}^1 x^0, \ldots, & \tilde{c}^{s-1} x \geq \tilde{c}^{s-1} x^0, \tilde{c}^s x > \tilde{c}^s x^0, \\ & \tilde{c}^{s+1} x \geq \tilde{c}^{s+1} x^0, \ldots, \tilde{c}^k x \geq \tilde{c}^k x^0) \geq \beta. \end{aligned} \tag{6}$$

The relation (6) contradicts the β-possible efficiency of x^0 for (P) and we have established the if part of the theorem.

To show the only if part , assume x^0 is efficient for $(P'')_\varepsilon$ for some $\varepsilon > 0$ and that x^0 is not β-possibly efficient for (P). For this ε, we can not find $x^1 \in K$ and

$$q^i \in (\tilde{c}^i)^{\beta-\varepsilon} \tag{7}$$

such that

$$q^i x^1 \geq q^i x^0 \qquad \forall i \in \{1, \ldots, k\} \tag{8}$$

and

$$q^s x^1 > q^s x^0 \quad \text{for some} \quad s \in \{1, \ldots, k\}. \tag{9}$$

By (7) we have

$$\min\big(u_{\tilde{c}^1}(q^1), \ldots, u_{\tilde{c}^k}(q^k)\big) > \beta - \varepsilon \tag{10}$$

(8) , (9) and (10) entail

$$\sup_{(t^1,\ldots,t^k)\in T^s} \min\big(u_{\tilde{c}^1}(t^1), \ldots, u_{\tilde{c}^s}(t^s), \ldots, u_{\tilde{c}^k}(t^k)\big) \geq \beta - \varepsilon. \tag{11}$$

As x^0 is not β-possibly efficient for (P) , there is $x^1 \in K$ such that

$$\begin{aligned} Poss(\tilde{c}^1 x^1 \geq \tilde{c}^1 x^0, \ldots, \tilde{c}^{s-1} x^1 \geq \tilde{c}^{s-1} x^0, \tilde{c}^s x^1 > \tilde{c}^s x^0, \\ \tilde{c}^{s+1} x^1 \geq \tilde{c}^{s+1} x^0, \ldots, \tilde{c}^k x^1 > \tilde{c}^k x^0) \geq \beta \end{aligned} \tag{12}$$

i.e.,

$$\sup_{(t^1,\ldots,t^k)\in T^s} \min\big(u_{\tilde{c}^1}(t^1), \ldots, u_{\tilde{c}^s}(t^s), \ldots, u_{\tilde{c}^k}(t^k)\big) \geq \beta > \beta - \varepsilon. \tag{13}$$

This is a contradiction with the fact that there is no $x^1 \in K$ such that (11) holds.

Corollary
If x^0 is efficient for (P''), then x^0 is $(\alpha, \beta)_2$-satisfying for (P).
This corollary follows trivially from the fact that

$$(\tilde{c}^i)^\beta \subseteq (\tilde{c}^i)^{\beta-\varepsilon} \qquad \forall i.$$

4. FINDING SATISFYING SOLUTIONS FOR PROBLEM (P)

4.1 $(\alpha, \beta)_1$ - satisfying solutions

By virtue of Theorem 1 , a solutions of (P') is $(\alpha, \beta)_1$-satisfying for (P). The following result warrants a computational bonus in solving (P').

Theorem 3

(P') is equivalent to the linear program :

$$
\begin{array}{lll}
\max & \theta & \\
\text{subject to} & \theta - (\tilde{c}^j)^{\beta^j_{v+1}} x - \gamma^j_{v+1} \leq 0 & j = 1, \cdots, k \\
& \theta - (\tilde{c}^j)^{\beta^j_v} x - \gamma^j_v \leq 0 & j = 1, \cdots, k \\
& \vdots & \\
& \theta - (\underline{\tilde{c}}^j)^{\beta^j_1} x - \gamma^j_1 \leq 0 & j = 1, \cdots, k \\
& (\overline{\tilde{A}})^{\alpha_{p+1}} x - \underline{\tilde{b}}^{\alpha_{p+1}} - \delta^m_{p+1} \leq 0 & \\
& \vdots & \\
& (\overline{\tilde{A}})^{\alpha_1} x - \underline{\tilde{b}}^{\alpha_1} - \delta^m_1 \leq 0, x \geq 0 &
\end{array}
\qquad (P''')
$$

where $(\tilde{c}^j)^\alpha$, $\tilde{b}^\alpha$, $\tilde{A}^\alpha$ are respectively matrices where components are α-cuts of fuzzy numbers $\tilde{c}^j_r$, $\tilde{b}_i$ and $\tilde{a}_{ir}$ $(r = 1, \ldots, n \; ; \; i = 1, \ldots, m)$ and

$$\overline{\tilde{m}}^\alpha = \sup_{t \in \tilde{m}^\alpha} t \; ; \; \underline{\tilde{m}}^\alpha = \inf_{t \in \tilde{m}^\alpha} t.$$

Proof

As the objective function is the same for (P') and (P'''), we have to show only that the sets of feasible solutions of the two programs coincide.

Assume that (θ^0, x^0) is feasible for (P') then : $x^0 \geq 0$

$$\theta^0 - c^j(s)x^0 - \gamma^j_u \leq 0 \quad \forall \; s \in S^{\beta^j_u} \quad j = 1, \ldots, k \quad u = 1, \ldots, v+1 \qquad (14)$$

and

$$A(t)x^0 \leq b(t) + \delta^m_r \qquad \forall \; t \in T^{\alpha_r} \quad r = 1, \ldots, p+1 \qquad (15)$$

For all j , u and r , consider $s^{u^*}_j$ and $t^{\alpha^*_r} = (t^{\alpha^*_r}_1, t^{\alpha^*_r}_2)$ such that :

$$c^j(s^{u^*}_j) = (\underline{\tilde{c}}^j)^{\beta^j_u},$$
$$A(t^{\alpha^*_r}_1) = \overline{\tilde{A}}^{\alpha_r},$$
$$b(t^{\alpha^*_r}_2) = \underline{\tilde{b}}^{\alpha_r}.$$

Obviously , for any u , j and r , we have :

$$s^{u^*}_j \in S^{\beta^j_u} \quad \text{and} \quad t^{\alpha^*_r} \in T^{\alpha_r}.$$

Hence by (14) and (15) we have :

$$\theta^0 - c^j(s_j^{u^*})x^0 - \gamma_u^j \leq 0 \qquad j = 1,\ldots,k$$
$$u = 1,\ldots,v+1$$

and

$$A(t^{\alpha_r^*})x^0 \leq b(t^{\alpha_r^*}) + \delta_r^m \qquad r = 1,\ldots,p+1.$$

i.e.,

$$\theta^0 - (\underset{\sim}{\tilde{c}}^j)^{\beta_u^j} - \gamma_u^j \leq 0 \qquad j = 1,\ldots,k,$$
$$u = 1,\ldots,v+1,$$
$$(\bar{\tilde{A}})^{\alpha_r}x^0 \leq \underset{\sim}{\tilde{b}}^{\alpha_r} + \delta_r^m \qquad r = 1,\ldots,p+1,$$

hence (θ^0, x^0) is feasible for (P''').

Reciprocally if (θ^0, x^0) is feasible for (P''') then :

$$\theta^0 - (\underset{\sim}{\tilde{c}}^j)^{\beta_u^j}x^0 - \gamma_u^j \leq 0 \qquad j = 1,\ldots,k$$
$$u = 1,\ldots,v+1$$
$$(\bar{\tilde{A}})^{\alpha_r}x^0 \leq \underset{\sim}{\tilde{b}}^{\alpha_r} + \delta_r^m \qquad r = 1,\ldots,p+1$$
$$x^0 \geq 0$$

By the definition of $(\underset{\sim}{\tilde{c}}^j)^{\beta_u^j}$, $\bar{\tilde{A}}^{\alpha_r}$. $\bar{\tilde{b}}^{\alpha_r}$, for any j, u and r we have :

$$\theta^0 - \gamma_u^j \leq (\underset{\sim}{\tilde{c}}^j)^{\beta_u^j} \leq c^j(s) \quad \forall s \in S^{\beta_u^j}$$

$$A(t)x^0 \leq (\bar{\tilde{A}})^{\alpha_r}x^0 \leq \underset{\sim}{\tilde{b}}^{\alpha_r} + \delta_j^m \leq b(t) + \delta_r^m \quad \forall t \in T^{\alpha_r}.$$

i.e., (θ^0, x^0) is feasible for (P').

4.2 $(\alpha, \beta)_2$ - satisfying solutions

By virtue of Theorem 2 and its corollary, to find an $(\alpha, \beta)_2$-satisfying solution for (P) we have to solve either (P'') or $(P'')_\varepsilon$ for some $\varepsilon > 0$.

The following result due to Birtran(1980) provides the possibility of finding an $(\alpha, \beta)_2$-satisfying solution for (P) under some assumptions on the supports of involved fuzzy numbers.

Therorem 4

Assume $c_j^{iL} > 0\ \forall i, j$ and let $EF(I_n)$ be the set of efficient solutions for the program:

$$\max_{x \in K} \quad I_n x$$

where I_n is the $(n \times n)$ unity matrix, then

$$EF \subset EF(I_n).$$

Furthermore, if $x^0 \in EF(I_n)$ and $x^0 \notin EF$ then there is $y^0 \in EF(I_n)$, such that

$$Cy^0 \geq Cx^0 \qquad \text{for some} \qquad C \in \Phi_\beta.$$

This theorem tells us that EF is contained in $EF(I_n)$ and for x^0 in $EF(I_n)$ and not in EF , there is an element $y^0 \in EF(I_n)$ and a matrix C in Φ_β that can be used to eliminate x^0 from $EF(I_n)$, i.e, such that :

$$Cy^0 \geq Cx^0.$$

We are now prepared to give the following algorithm for obtaining efficient extreme points of (P'').

(1) Using any algorithm for finding efficient solutions of a linear multiple objective program, determine $EF_{ex}(I_n)$.

(2) For each $x^0 \in EF_{ex}(I_n)$ and each $x \in EF_{ex}(I_n)$, $x \neq x^0$, define the matrix $C^* = (c^*_{ij})$ as follows:

$$\begin{aligned} c^*_{0j} &= \bar{c}_0^{iM} \qquad \text{if} \quad x_j - x_j^0 \geq 0, \\ c^*_{0j} &= \bar{c}_0^{iL} \qquad \text{if} \quad x_j - x_j^0 < 0, \end{aligned}$$

where c^*_{0j} denotes the jth column of the matrix C^*. If $C^*(x - x^0) \geq 0$, then $x^0 \notin EF_{ex}$ and x^0 can be discarded from $EF_{ex}(I_n)$. Otherwise choose another elememt from $EF_{ex}(I_n)$ and repeat the procedure.

If x^0 is not discarded after being compared with the elements of $EF_{ex}(I_n)$, it follows from the previous theorem that it is an element of EF.

For the case where the above positivity assumption on c_j^{iL} does not hold, the following result (Luhandjula 1987) gives a way for solving (P'').

Theorem 5

If x^0 is a solution of the linear program :

$$\max_{x \in K, x \geq 0} \quad q^0 x,$$

where q^0 is a solution of the system :

$$\begin{aligned} V^i C^i - q = 0 \qquad \forall i \qquad \text{such that} \qquad C^i \in M_\beta \\ V^i \in \Re^k \quad , \quad V^i > 0 \quad , \quad q \in \Re^n \end{aligned}$$

where M_β is the subset of Φ_β whose elements are matrices having all elements of each column either at the upper bound or at the lower bound of intervals $(\tilde{c}_j^i)^\beta$, then x^0 is $(\alpha,\beta)_2$ - satisfying solution for (P).

5. Numerical examples.

Example 1 ($(\alpha,\beta)_1$-Satisfying solution)

To illustrate an $(\alpha,\beta)_1$-satisfying solution, consider the following two objective linear programming problem with fuzzy coefficients .

$$\begin{array}{ll} \max & \tilde{c}_1^1 x_1 + \tilde{c}_2^1 x_2 \\ \max & \tilde{c}_1^2 x_1 + \tilde{c}_2^2 x_2 \\ \\ \text{subject to} & \tilde{a}_1^1 x_1 + \tilde{a}_2^1 x_2 \leq \tilde{b}^1 \\ & \tilde{a}_1^2 x_1 + \tilde{a}_2^2 x_2 \leq \tilde{b}^2 \\ & x_1 \geq 0 \quad , \quad x_2 \geq 0. \end{array} \qquad (P\text{-}E1)$$

The membership functions of the fuzzy numbers $\tilde{a}_1^1, \cdots, \tilde{a}_2^2, \tilde{b}^1, \cdots, \tilde{b}^2, \tilde{c}_1^1, \cdots, \tilde{c}_2^2$ are defined by the triangular functions :

$$\Pi_{\tilde{m}} = \max\left\{0, 1 - \frac{|m^m - m\,|}{m^s}\right\}$$

and explained in Table 1 , where m^m and m^s denote the mean and spread parameters respectively.

Table 1. Fuzzy numbers for example 1

fuzzy coefficient	*mean*	*spread*
$\tilde{a}_1^1$	3	2
$\tilde{a}_2^1$	4	2
$\tilde{a}_1^2$	2	1
$\tilde{a}_2^2$	1.5	0.5
$\tilde{b}^1$	13	3
$\tilde{b}^2$	7.5	5
$\tilde{c}_1^1$	5	2
$\tilde{c}_2^1$	2	1
$\tilde{c}_1^2$	3	1
$\tilde{c}_2^2$	5	2

An $(\alpha,\beta)_1$-satisfying solution in case of $v = p = 2$, $\alpha_0 = \beta_0 = 0$, $\alpha_1 = \beta_1 = 0.5$, $\alpha_2 = \beta_2 = 1$, $\gamma_1^1 = 2.0$, $\gamma_2^1 = 1.5$, $\gamma_3^1 = 1.0$, $\gamma_1^2 = 1.5$, $\gamma_2^2 = 1.0$, $\gamma_3^2 = 0.5$, $\delta_1^1 = 3.0$, $\delta_2^1 = 2.5$, $\delta_3^1 = 2.0$, $\delta_1^2 = 2.5$, $\delta_2^2 = 2.0$, and $\delta_3^2 = 1.5$, of (P-$E1$) can be obtained by solving the following linear programming problem as was shown in Theorem 3.

$$
\begin{array}{ll}
\max & \theta \\
\text{subject to} & \theta - 5x_1 - 2x_2 \leq 1 \\
& \theta - 4x_1 - 1.5x_2 \leq 1.5 \\
& \theta - 3x_1 - 1x_2 \leq 2 \\
& \theta - 3x_1 - 5x_2 \leq 0.5 \\
& \theta - 2.5x_1 - 4x_2 \leq 1 \\
& \theta - 2x_1 - 3x_2 \leq 1.5 \\
& 3x_1 + 4x_2 \leq 16 \\
& 4x_1 + 5x_2 \leq 15 \\
& 5x_1 + 6x_2 \leq 14 \\
& 2x_1 + 1.5x_2 \leq 10 \\
& 2.5x_1 + 1.75x_2 \leq 8 \\
& 3x_1 + 2x_2 \leq 6 \\
& x_1 \geq 0 \quad , \quad x_2 \geq 0.
\end{array}
$$

The optimal solution (x^0, θ^0) of this problem becomes (1.375, 0.9375, 7.0625). Hence, from Theorem 1, we can understand that $x^0 = (\ 1.375\ ,\ 0.9375\ )$ is a $(\ 0.5,\ 7.0625\ )_1$-satisfying solution.

Example 2 ($(\alpha,\beta)_2$-satisfying solution)
Consider the following two objective linear programming problem with fuzzy coefficients.

$$
\begin{array}{ll}
\max & \tilde{c}_1^1 x_1 + \tilde{c}_2^1 x_2 \\
\max & \tilde{c}_1^2 x_1 + \tilde{c}_2^2 x_2 \\
\text{subject to} & \tilde{a}_1^1 x_1 + a_2^1 x_2 \leq \tilde{b}^1 \\
& \tilde{a}_1^2 x_1 + \tilde{a}_2^2 x_2 \leq \tilde{b}^2 \\
& \tilde{a}_1^3 x_1 + \tilde{a}_2^3 x_2 \leq \tilde{b}^3 \\
& \tilde{a}_1^4 x_1 + \tilde{a}_2^4 x_2 \leq \tilde{b}^4 \\
& x_1 \geq 0 \quad x_2 \geq 0.
\end{array}
\qquad (P\text{-}E2)
$$

The membership functions of the fuzzy numbers $\tilde{a}_1^1, \cdots, \tilde{a}_2^2, \tilde{b}^1, \ldots, \tilde{b}^4, \tilde{c}_1^1, \cdots, \tilde{c}_2^2$ are defined by the triangular functions :

$$\Pi_{\tilde{m}} = \max\left\{0, 1 - \frac{|\, m^m - m\,|}{m^s}\right\}$$

and explained in Table 1, where m^m and m^s denote the mean and spread parameters respectively.

Table 2. Fuzzy numbers for example 2

fuzzy coefficient	*mean*	*spread*
$\tilde{a}_1^1$	-1.5	1
$\tilde{a}_2^1$	1.5	1
$\tilde{a}_1^2$	1.25	0.5
$\tilde{a}_2^2$	0.5	1
$\tilde{a}_1^3$	3.5	3
$\tilde{a}_2^3$	0.5	1
$\tilde{a}_1^4$	9.0	2.0
$\tilde{a}_2^4$	0.75	0.5
$\tilde{b}^1$	32	4
$\tilde{b}^2$	44	8
$\tilde{b}^3$	32	4
$\tilde{b}^4$	120	20
$\tilde{c}_1^1$	1.5	1.25
$\tilde{c}_2^1$	3.5	1.25
$\tilde{c}_1^2$	3.5	1.25
$\tilde{c}_2^2$	1.5	1.25

From Theorem 2, an $(\alpha, \beta)_2$-Satisfying solution for this example can be obtained as EF of the problem :

$$\max \quad Cx \ , \ C \in \Phi_{0.6} \qquad (P''\text{-}E2)$$

$$\text{subject to} \qquad \overline{\tilde{A}}_\alpha x \leq \underline{\tilde{b}}_\alpha$$

$$x \geq 0.$$

For example, consider an $(\alpha, \beta)_2$-satisfying solution for $\alpha = 0.5, \beta = 0.6$. Then $\Phi_{0.6}$ is a 2×2 matrix with each component c_j^i are :

$$c_1^1 \in \tilde{c}_1^1{}_{0.6} = [1, 2], \qquad c_2^1 \in \tilde{c}_2^1{}_{0.6} = [3, 4],$$
$$c_1^2 \in \tilde{c}_1^2{}_{0.6} = [3, 4], \qquad c_2^2 \in \tilde{c}_2^2{}_{0.6} = [1, 2],$$

and a 4×2 matrix $\overline{\tilde{A}}_{0.5}$ and right side vector $\underline{\tilde{b}}_{0.5}$ become as :

$$\overline{\tilde{A}}_{0.5} = \begin{bmatrix} -1 & 2 \\ 1.5 & 2 \\ 5 & 1 \\ 10 & 1 \end{bmatrix}, \quad \underline{\tilde{b}}_{0.5} = \begin{bmatrix} 30 \\ 40 \\ 30 \\ 110 \end{bmatrix}.$$

With these values , (P''-$E2$) reduces to the following multiobjective linear programming problem :

$$\begin{array}{lll} \max & c_1^1 x_1 + c_2^1 x_2 & c_1^1 \in [1,2], c_2^1 \in [3,4] \\ \max & c_1^2 x_1 + c_2^2 x_2 & c_1^2 \in [3,4], c_2^2 \in [1,2] \end{array}$$

$$\text{subject to} \quad \begin{aligned} -x_1 + 2x_2 &\le 30 \\ 1.5x_1 + 2x_2 &\le 40 \\ 2x_1 + x_2 &\le 30 \\ 10x_1 + x_2 &\le 110 \\ x_1 \ge 0 \ , \ x_2 &\ge 0. \end{aligned}$$

Now let the coefficient matrix of the objective function be C, and denote the corresponding Pareto optimal solution set by $EF(C)$.

As can be easily understood from Fig.1, EF becomes the line segment $[x^1, x^2]$, where $x^1 = (4,17)$ and $x^2 = (8,14)$.

Using the proposed algorithm , we can derive a $(0.5, 0.6)_2$-satisfying solution as follows:

(1) In this example , from Fig.1, it is evident $EF_{ex}(I_n)$ consists of the following five efficient extreme points:

$$x^1 = (4,17), x^2 = (8,14), x^3 = (10,10), x^4 = (11,0).$$

(2) For example, selecting $x^1 = (4,17)^T$ as x^0 and let us check whether x^1 is a $(0.5, 0.6)_2$-satisfying solution or not. First select x^2 as x, then $x^2 - x^1 = (4,-3)^T$.

Thus a 2×2 matrix C^* becomes :

$$C^* = \begin{bmatrix} 2 & 3 \\ 4 & 1 \end{bmatrix}.$$

Then

$$C^*(x^2 - x^1) = \begin{bmatrix} 2 & 3 \\ 4 & 1 \end{bmatrix} \begin{bmatrix} 4 \\ -3 \end{bmatrix} = (-1, 13)^T.$$

Next, select x^3 as x, then

$$C^*(x^3 - x^1) = (-9, 17)^T.$$

Finally, select x^4 as x, then

$$C^*(x^4 - x^1) = (-37, 9)^T.$$

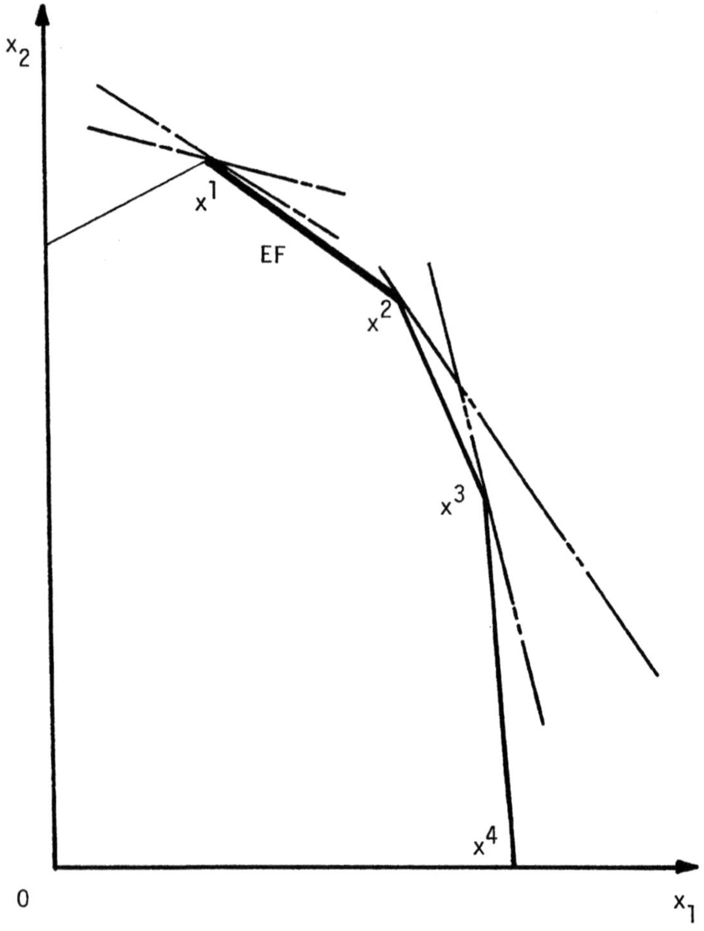

Fig. 1 $(0.5,0.6)_2$ - satisfying solutions for example 2

As can be understood from these results, for $x^1 = (4,17)^T$, the condition $C^*(x - x^1) \leq 0$ dose not hold for all the elements of $EF_{ex}(I_n)$. Therefore,$x^1 = (4,17)^T$ is a $(0.5, 0.6)_2$-satisfying solution.

To show the case where x^0 in $EF_{ex}(I_n)$ but not in EF_{ex}, we select $x^3 = (10,10)^T$ as x^0 and $x^2 = (8,14)^T$ as x. Then we have

$$C^* = \begin{bmatrix} 1 & 4 \\ 3 & 2 \end{bmatrix}$$

and

$$C^*(x^2 - x^3) = \begin{bmatrix} 1 & 4 \\ 4 & 2 \end{bmatrix} \begin{bmatrix} -2 \\ 4 \end{bmatrix} = (14,2)^T > 0.$$

Hence ,$x^3 \notin EF_{ex}$. Therefore , $x^3 = (8,14)^T$ is not a $(0.5, 0.6)_2$-satisfying solution.

6. CONCLUDING REMARKS

We have considered, in this paper, multiple objective linear programming problems with the additional complication brought about by the incorporation of fuzzy numbers. It is worth mentioning that Bellman and Zadeh (1970) introduced a general framework within which the analysis of a large spectrum of problems including deterministic multiple objective linear program was possible (Zimmermann 1978). Two solution concepts reflecting an optimistic attitude because based on possibility measures were introduced and characterized. Linkups between imprecision and infinity in an optimization context emerge from these characterizations.

As a matter of fact , the resulting deterministic programs are infinite structured problems.

Potential extensions and topic for further research are :

(1) The determination of optimal values (in the Pareto sense) for α and β as well as the development of a method for generating the whole set of (α, β) - satisfying alternatives.

(2) The consideration of both possibility and necessity measures for specifying optimistic-pessimistic satisfying solutions and the determination of easy to check necessary and sufficient conditions for the existence of such solutions.

(3) The development of algorithms to solve integer or boolean multiple objective linear programming problems with fuzzy coefficients.

Acknowledgement

The first author is indebted to the Matsumae International Foundation who has Supported him during the realization of this work.

REFERENCES

[1] Bellman, R. and Zadeh, L.A., (1970) 'Decision making in a fuzzy environment', Management Science 17, 141-164.

[2] Bitran, G.R., (1980) 'Linear Multiple Objective problems with interval coefficients', Management Science 26, 694-706.

[3] Chankong, V. and Haimes, Y.Y.(1983) Multiobjective Decision Making : Theory and Methodology, North-Holland.

[4] Cohon, J. L. and Marks, D. M. (1975) 'Multiobjective programming techniques', a review and evaluation, Multiobjective Programming Techniques 11, 208-220.

[5] Luhandjula, M.K. (1987) 'Multiple objective programming problems with possibilistic coefficients', Fuzzy Sets and Systems 21, 135-145.

[6] Luhandjula, M.K. , Ichihashi, H. and Inuiguchi, M. (1989) 'Fuzzy and semi-infinite mathematical programming,' Information Science (to appear).

[7] Roy, B. (1971) 'Problems and methods with multiple objective functions', Mathematical Programming 1, 239-266.

[8] Sakawa, M. and Yano, H. (1986) 'Interactive decision making for multiobjective linear programming problems with fuzzy parameters', Large Scale Modelling and Interactive Decision Analysis, (Eds. G. Fandel, M. Grauer, A. Kurzhanski and A.P.Wierzbicki) Springer Verlag, pp. 88-96.

[9] Sakawa, M. and Yano, H. (1988) 'A unified approach for multiobjective linear programming problems with fuzzy coefficients', Presented at the 13th International Symposium on Mathematical Programming, Tokyo Japan.

[10] Sakawa, M. and Yano, H. (1988) 'New solution concept for multiobjective linear programming problems with fuzzy parameters and its application', Proceedings of the International Workshop on Fuzzy Systems Applications, Fukuoka, Japan, pp. 163-164.

[11] Seo, F. and Sakawa, M. (1988) Multiple Criteria Decision Analysis in Regional Planning : Concepts, Methods and Applications, D. Reidel Publishing Company.

[12] Zadeh, L. A. (1978) 'Fuzzy sets as a basis for a Theory of Possibility' , Fuzzy Sets and Systems 1, 3-28.

[13] Zimmermann, H.-J., (1978) Fuzzy programming and linear programming with several objective functions, Fuzzy Sets and Systems 1, 46-55.

Part III
Chapter 7

INEQUALITY CONSTRAINTS BETWEEN FUZZY NUMBERS AND THEIR USE IN MATHEMATICAL PROGRAMMING

MARC ROUBENS
Faculte Polytechnique de Mons and *Institute of Mathematics,*
Rue de Houdain, 9 *University of Liege*
7000 Mons *Avenue des Tilleuls, 15*
4000 Liege

ABSTRACT. A definition of the degree to which a normalized convex fuzzy number is greater or equal to another fuzzy number is proposed which satisfies the Ferrer's condition and consequently the max-min transitivity.

It is based on the compensation of areas determined by the membership functions.

Using the fuzzy preference relation, one can handle inequality constraints, the related conditions being free from any parametrization.

The particular case of $L-R$ fuzzy numbers is considered and its is proved that in this case, comparison of areas is reduced to the comparison of upper and lower bounds of α-cuts.

Two examples related to comparison of fuzzy numbers and fuzzy optimization are also presented.

1. Introduction

We consider a set of normalized convex fuzzy numbers $A : \{a\}$ defined by the membership functions $\{\mu_a(x), x \in R\}$ with

$$\begin{cases} \bigvee_x \mu_a(x) = 1 \\ \mu_a(y) \geq \mu_a(x) \wedge \mu_a(z), \ \ \forall x \geq y \geq z \in R, \end{cases}$$

where $\vee$ and $\wedge$ represent MAX and MIN operators.

The α-level sets $I(a,\alpha) = \{x \in R : \mu_a(x) \geq \alpha\}$ are convex subsets of R and there is m in R such that $\mu_a(m) = 1$.

The lower and upper limits of any α-level set $I(a,\alpha)$ are represented by $\inf_{x\in R} I(a,\alpha)$ and $\sup_{x\in R} I(a,\alpha)$. We suppose that both limits are finite.

Dubois and Prade [2] and Ramik and Rimanek [4] proposed to define

R. Slowinski and J. Teghem (eds.),
Stochastic versus Fuzzy Approaches to Multiobjective Mathematical Programming under Uncertainty, 321–330.

$$a \underset{L}{\geq} b \text{ iff } a \underset{L_\alpha}{\geq} b, \forall \alpha \in [0,1],$$

$$a \underset{R}{\geq} b \text{ iff } a \underset{R_\alpha}{\geq} b, \forall \alpha \in [0,1],$$

where

$$a \underset{L_\alpha}{\geq} b \text{ iff } \inf I(a,\alpha) \geq \inf I(b,\alpha),$$

$$a \underset{R_\alpha}{\geq} b \text{ iff } \sup I(a,\alpha) \geq \sup I(b,\alpha).$$

These definitions give transitive relations $\geq$ for L_α, R_α, and are pessimistic compared with the definition obtained using the possibility index :

$$\tilde{a} \geq \tilde{b} \text{ iff Poss.}(a \geq b) = \bigvee_{x>y} [\mu_a(x) \wedge \mu_b(y)] > 0$$

which is however not necessarily transitive (the degree of possibility is shown to be negatively transitive : see Roubens and Vincke [6]).

We define a transitive compensatory $\geq$ in the sense that values of α for which $\inf I(a,\alpha) + \sup I(a,\alpha) \geq \inf I(b,\alpha) + \sup I(b,\alpha)$ are compensating those values for which the inequality is not satisfied.

2. Definition

Let a, b be normalized convex fuzzy numbers and

$$S_L(a \geq b) = \int_{U_1(a,b)} [\inf I(a,\alpha) - \inf I(b,\alpha)] d\alpha \tag{2.1}$$

$U_1(a,b)$ being the subset of $[0,1]$: $\{\alpha \mid \inf I(a,\alpha) \geq \inf I(b,\alpha)\}$

$$S_R(a \geq b) = \int_{V_1(a,b)} [\sup I(a,\alpha) - \sup I(b,\alpha)] d\alpha$$

$V_1(a,b)$ being the subset of $[0,1]$: $\{\alpha \mid \sup I(a,\alpha) \geq \sup I(b,\alpha)\}$.

The degree to which $\tilde{a} \geq \tilde{b}$ is naturally considered as

$$\begin{aligned} C(a \geq b) &= C(a,b) = S_L(a \geq b) + S_R(a \geq b) - S_L(b \geq a) - S_R(b \geq a), \text{if positive} \\ &= 0 \text{ otherwise} \end{aligned} \tag{2.2}$$

The idea of compensation by integration can also be found in Baldwin and Guild [1] or Nakamura [3].

We will proof that the preference relation C is a fuzzy interval order.

3. C is a fuzzy interval order

Let A be a finite set. A fuzzy binary relation on A is a mapping from $A \cdot A$ to the interval $[0, \infty[$. This relation is defined by its membership function C, $C : A \cdot A \to [0, \infty[$.

We shall say that a fuzzy binary relation C is

irreflexive if $C(a,a) = 0$ for every a in A
antisymmetric if $C(a,b) \wedge C(b,a) = 0$ for all a and b in A
transitive if $C(a,c) \geq C(a,b) \wedge C(b,c)$ for all a, b and c in A
Ferrers if $C(a,b) \vee C(c,d) \geq C(a,d) \wedge C(c,b)$ for all a, b, c and d in A.

By definition, a fuzzy preference relation on the set A is a fuzzy binary relation on this set. For any given a and b in A, $C(a,b)$ is regarded as the degree to which a is greater or equal to b.

A fuzzy interval order is an antisymmetric Ferrers fuzzy binary relation.

From definition (2.2), $C(a,b)$ is an antisymmetric relation.

Proposition 3.1 *C is a Ferrers relation.*

Proof. From (2.2), we have :

$$\begin{aligned} 0 &< C(a,b) \\ &= \int_{U_1(a,b)} [\inf I(a,\alpha) - \inf I(b,\alpha)]d\alpha + \int_{V_1(a,b)} [\sup I(a,\alpha) - \sup I(b,\alpha)]d\alpha \\ &\quad - \int_{[0,1]\setminus U_1(a,b)} [\inf I(a,\alpha) - \inf I(b,\alpha)]d\alpha - \int_{[0,1]\setminus V_1(a,b)} [\sup I(a,\alpha) - \sup I(b,\alpha)]d\alpha \\ &= \int_0^1 [\inf I(a,\alpha) - \inf I(b,\alpha)]d\alpha + \int_0^1 [\sup I(a,\alpha) - \sup I(b,\alpha)]d\alpha. \end{aligned}$$

If we call $\inf I(a,\alpha) + \sup I(a,\alpha) = \mathrm{sum}I(a,\alpha)$,

$$\begin{aligned} C(a,b) &= \int_0^1 [\mathrm{sum}I(a,\alpha) - \mathrm{sum}I(b,\alpha)]d\alpha, \text{ if positive} \\ &= 0, \text{ otherwise} \end{aligned} \tag{3.1}$$

We will prove that $C(a,b) \vee C(c,d) \geq C(a,d) \wedge C(c,b)$ for any given a, b, c and d in A.

If $C(a,d) \vee C(c,b) = 0$, the Ferrers condition is satisfied. We now consider that $C(a,d) > 0$ and $C(c,b) > 0$ and we suppose without any restriction that

$$C(a,b) \geq C(c,d). \tag{3.2}$$

Case 1 : $C(c,d) > 0$

Let

$$0 < C(a,b) < C(a,d) \wedge C(c,b). \tag{3.3}$$

One obtains $C(a,b) < C(a,d)$ and $C(a,b) < C(c,b)$ or

$$\int_0^1 [\mathrm{sum}I(a,\alpha) - \mathrm{sum}I(b,\alpha)]d\alpha < \int_0^1 [\mathrm{sum}I(a,\alpha) - \mathrm{sum}I(d,\alpha)]d\alpha$$

and

$$\int_0^1 [\mathrm{sum}I(a,\alpha) - \mathrm{sum}I(b,\alpha)]d\alpha < \int_0^1 [\mathrm{sum}I(c,\alpha) - \mathrm{sum}I(b,\alpha)]d\alpha$$

or

$$\int_0^1 [\mathrm{sum}I(b,\alpha) - \mathrm{sum}I(d,\alpha)]d\alpha > 0 \tag{3.4}$$

$$\int_0^1 [\mathrm{sum}I(c,\alpha) - \mathrm{sum}I(a,\alpha)]d\alpha > 0. \tag{3.5}$$

Considering (3.4) and (3.5), one can see that

$$\int_0^1 [\mathrm{sum}I(c,\alpha) - \mathrm{sum}I(d,\alpha)]d\alpha > \int_0^1 [\mathrm{sum}I(a,\alpha) - \mathrm{sum}I(b,\alpha)]d\alpha > 0,$$

and $C(c,d) > C(a,b) > 0$ which refutes (3.2).

Case 2 : $C(c,d) = 0$

$C(c,d) = 0$ implies $C(a,b) = 0$ – see (3.2) – and $C(b,a) \geq 0$ and $C(d,c) \geq 0$ (antisymmetry), or

$$\int_0^1 [\mathrm{sum}I(b,\alpha) - \mathrm{sum}I(a,\alpha)]d\alpha \geq 0,$$

$$\int_0^1 [\mathrm{sum}I(d,\alpha) - \mathrm{sum}I(c,\alpha)]d\alpha \geq 0.$$

From these two inequalities, one obtains

$$\int_0^1 [\mathrm{sum}I(a,\alpha) + \mathrm{sum}I(c,\alpha) - \mathrm{sum}I(b,\alpha) - \mathrm{sum}I(d,\alpha)]d\alpha \leq 0. \tag{3.6}$$

Let

$$C(a,b) = 0 < C(a,d) \wedge C(c,b). \tag{3.7}$$

Obviously, $C(a,d) + C(c,b) > 0$, which refutes (3.6). ∎

By substituting $a = b$ into antisymmetry's definition and $c = d$ into Ferrers' definition, one obtains immediately that a fuzzy interval order is irreflexive and transitive.

If we consider that

$$a \geq b \text{ iff } C(a,b) \geq 0, \tag{3.8}$$

we obtain a parameter-free definition of the inequality between two fuzzy numbers which is transitive :

$$a \geq b \text{ and } b \geq c \Rightarrow a \geq c, \quad \forall a, b \text{ and } c \in A.$$

4. The case of $L-R$ fuzzy numbers

A convenient representation for a normalized convex fuzzy number a is a quadruple $(a^-, a^+, \sigma_a^-, \sigma_a^+)$ of parameters of its membership function

$$\mu_a(x) = \begin{cases} L(\frac{a^- - x}{\sigma_a^-}) & \text{if } x < a^- \\ 1 & \text{if } x \in [a^-, a^+] \\ R(\frac{x - a^+}{\sigma_a^+}) & \text{if } x > a^+ \end{cases}$$

where L and R are reference functions $R^+ \to [0,1]$, non-increasing in $[0, \infty[$, $L(0) = R(0) = 1$ and $L(a) = R(a) = 0$ if $u > 1$.

If $L(u) \vee R(u) < 1$ for all $u > 0$, $[a^-, a^+]$ is the *kernel* of a. The *support* of a is $[a^- - \sigma_a^-, a^+ + \sigma_a^+]$ if $L(u) \wedge R(u) > 0$, for all $u < 1$; it is supposed to be finite.

Proposition 4.1 *Let a and b be $L-R$ fuzzy numbers with parameters $(a^-, a^+, \sigma_a^-, \sigma_a^+)$, $(b^-, b^+, \sigma_b^-, \sigma_b^+)$ and reference functions $(L_a, R_a), (L_b, R_b)$, $a \geq b$ iff*

$$\sup_{x \in R} I(a, \alpha_{a,R}) + \inf_{x \in R} I(a, \alpha_{a,L}) \geq \sup_{x \in R} I(b, \alpha_{b,R}) + \inf_{x \in R} I(b, \alpha_{b,L}) \tag{4.1}$$

where, if $k = a, b$,

$$\begin{aligned} \alpha_{k,R} &= R_k(\int_0^1 R_k^{-1}(\alpha) d\alpha), \\ \alpha_{k,L} &= L_k(\int_0^1 L_k^{-1}(\alpha) d\alpha). \end{aligned}$$

Proof. From (3.8), we know that $a \geq b$ iff $C(a,b) \geq 0$ or

$$\begin{aligned} \int_0^1 [\text{sum} I(a,\alpha) - \text{sum} I(b,\alpha)] d\alpha &\geq 0, \\ \int_0^1 [\inf I(a,\alpha) + \sup I(a,\alpha)] d\alpha &\geq \int_0^1 [\inf I(b,\alpha) + \sup I(b,\alpha)] d\alpha. \end{aligned} \tag{4.2}$$

(4.2) can be rewritten as :

$$\int_0^1 [a^- - L_a^{-1}(\alpha)\sigma_a^- + a^+ + R_a^{-1}(\alpha)\sigma_a^+] d\alpha \geq \int_0^1 [b^- - L_b^{-1}(\alpha)\sigma_b^- + b^+ + L_b^{-1}(\alpha)\sigma_b^+] d\alpha,$$

or

$$\begin{aligned} a^- - \sigma_a^- \int_0^1 L_a^{-1}(\alpha) d\alpha + a^+ + \sigma_a^+ \int_0^1 R_a^{-1}(\alpha) d\alpha &\geq \\ b^- - \sigma_b^- \int_0^1 L_a^{-1}(\alpha) d\alpha + b^+ + \sigma_b^+ \int_0^1 R_b^{-1}(\alpha) d\alpha \end{aligned} \tag{4.3}$$

From the definition of L_k, $k = a, b$,

$$k^- - \sigma_k^- L_k^{-1}(\alpha) = \inf I(k, \alpha)$$

and

$$L_k^{-1}(\alpha) < 1, \;\; \forall \alpha.$$

$k^- - \sigma_k^- \int_0^1 L_k^{-1}(\alpha)d\alpha$ lies inside $I(k,0)$ and $I(k,1)$, since

$$0 < \int_0^1 L_k^{-1}(\alpha)d\alpha < \int_0^1 d\alpha = 1 \quad .$$

$$k^- - \sigma_k^- \int_0^1 L_k^{-1}(\alpha)d\alpha = k^- - \sigma_k^- L_k^{-1}(\alpha_{k,L}) = \inf I(k, \alpha_{k,L})$$

and

$$\alpha_{k,L} = L_k(\int_0^1 L_k^{-1}(\alpha)d\alpha).$$

The same arguments are true for R_k, $k = a, b$. ∎

We immediately see from (4.1) that the comparison of areas is reduced to the comparison of limits of α-level sets.

If we consider R (or L) as a

- linear function ($R(\alpha) = 1 - \alpha, \alpha\epsilon[0,1]$), $\int_0^1 R^{-1}(\alpha)d\alpha = \frac{1}{2}$,
- quadratic function ($R(\alpha) = 1 - \alpha^2, \alpha\epsilon[0,1]$), $\int_0^1 R^{-1}(\alpha)d\alpha = \frac{2}{3}$,
- exponential function ($R(\alpha) = c^{-\alpha}, \alpha\epsilon[0,1]$), $\int_0^1 R^{-1}(\alpha)d\alpha = 1 - c^{-1}$,
- square root function ($R(\alpha) = \sqrt{\alpha}, \alpha\epsilon[0,1]$), $\int_0^1 R^{-1}(\alpha)d\alpha = \frac{1}{3}$.

In the case of trapezoidal fuzzy numbers,

$$a \geq b \quad \text{iff} \;\; a^- + a^+ + \tfrac{1}{2}(\sigma_a^+ - \sigma_a^-) \geq b^- + b^+ + \tfrac{1}{2}(\sigma_b^+ - \sigma_b^-) \tag{4.4}$$

$$\text{iff} \;\; \inf I(a, \cdot 5) + \sup I(a, \cdot 5) > \inf I(b, \cdot 5) + \sup I(b, \cdot 5) \tag{4.5}$$

5. Example

We consider the example taken from Roubens [5] and Roubens and Vincke [6], see figure 1 :

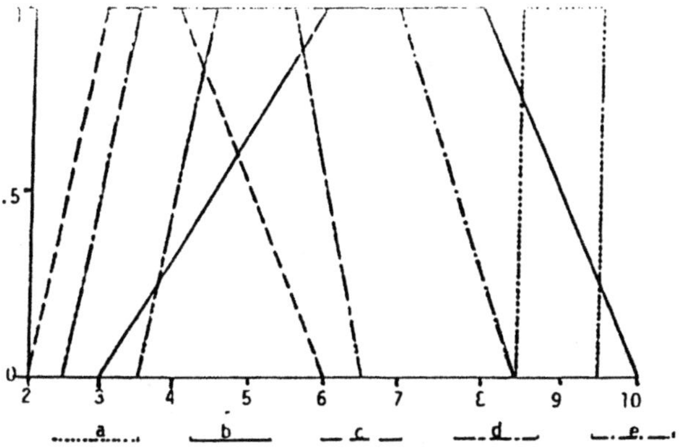

Fig. 1

$$\mu_a(x) = \begin{cases} 0 & x < 8.5 \\ 1 & 8.5 \leq x \leq 9.5 \\ 0 & x > 9.5 \end{cases}$$

$$\mu_b(x) = \begin{cases} 0 & x < 3 \\ \text{linear} & 3 \leq x \leq 6.5 \\ 1 & 6 < x < 8 \\ \text{linear} & 8.5 \leq x \leq 10 \\ 0 & x > 10 \end{cases} \qquad \mu_c(x) = \begin{cases} 0 & x < 2 \\ \text{linear} & 2 \leq x \leq 3 \\ 1 & 3 < x < 4 \\ \text{linear} & 4 \leq x \leq 6 \\ 0 & x > 6 \end{cases}$$

$$\mu_d(x) = \begin{cases} 0 & x < 3.5 \\ \text{linear} & 3.5 \leq x \leq 4.5 \\ 1 & 4.5 < x < 5.5 \\ \text{linear} & 5.5 \leq x \leq 6.5 \\ 0 & x > 6.5 \end{cases} \qquad \mu_e(x) = \begin{cases} 0 & x < 2.5 \\ \text{linear} & 2.5 \leq x \leq 3.5 \\ 1 & 3.5 < x < 7 \\ \text{linear} & 7 \leq x \leq 8.5 \\ 0 & x > 8.5 \end{cases}$$

We obtain :

a	$a^- - \frac{1}{2}\,\sigma_a^-$	$a^+ + \frac{1}{2}\,\sigma_a^+$	$\sum$
a	8.5	9.5	18
b	4.5	8	12.5
c	2.5	5	7.5
d	4	6	10
e	3	7.75	10.75

and the following order : $a \geq b \geq e \geq d \geq c$.

6. Application to fuzzy optimization

We consider the following problem [4] :

$$[\mathrm{MAX}] f(x_1, \ldots, x_n)$$

subject to :

$$a_{i1}x_1 \oplus a_{i2}x_2 \oplus \ldots \oplus a_{in}x_n \leq b_i, \quad i = 1, \ldots, m \tag{6.1}$$

$$x_j \geq 0 \ , \ j = 1, \ldots, n$$

where

$$\begin{aligned} a_{ij} &= (a_{ij}^-, a_{ij}^+, \sigma_{a_{ij}}^-, \sigma_{a_{ij}}^+)_{L-R} \\ b_i &= (b_i^-, b_i^+, \sigma_{b_i}^-, \sigma_{b_i}^+)_{L-R}. \end{aligned}$$

We obtain that

$$a_{ii}x_1 \oplus \ldots + a_{in}x_n = (\sum_j a_{ij}^- x_j, \sum_j a_{ij}^+ x_j, \sum_j \sigma_{a_{ij}}^- x_j, \sum_j \sigma_{a_{ij}}^+ x_j)_{L-R}.$$

If the functions L and R are taken as $L(u) = R(u) = \max(0, 1-u)$, $u > 0$, the system (6.1) of fuzzy contraints is equivalent to the system of ordinary inequalities :

$$\sum_j (a_{ij}^- + a_{ij}^+ - \frac{1}{2}\, \sigma_{a_{ij}}^- + \frac{1}{2}\, \sigma_{a_{ij}}^+) x_j \leq b_i^- + b_i^+ - \frac{1}{2}\, \sigma_{b_i}^- + \frac{1}{2}\, \sigma_{b_i}^+.$$

We reconsider the example given in [4] =

$$[\mathrm{MAX}] z = 5x_1 + 4x_2$$

subject to

$$\begin{cases} (4,4,2,1)_{L-L}x_1 \oplus (5,5,3,1)_{L-L}x_2 \leq (24,24,5,8)_{L-L} \\ (4,4,1,2)_{L-L}x_1 \oplus (1,1,0.5,1)_{L-L}x_2 \leq (12,12,6,3)_{L-L} \\ x_1 x_2 \geq 0 \end{cases}$$

with $L(x) = \max(0, 1-u)$, $u \geq 0$.

This fuzzy linear program is equivalent to the L.P. :

$$[\max] z = 5x_1 + 4x_2$$

subject to

$$\begin{cases} 7.5x_1 + 9x_2 \leq 49.5 & (a) \\ 8.5x_1 + 2.25x_2 \leq 22.5 & (b) \\ x_1 x_2 \geq 0 \end{cases}$$

The solution of the linear program is :

$$x_1 = 1.53, \quad x_2 = 4.22, \quad z = 23.$$

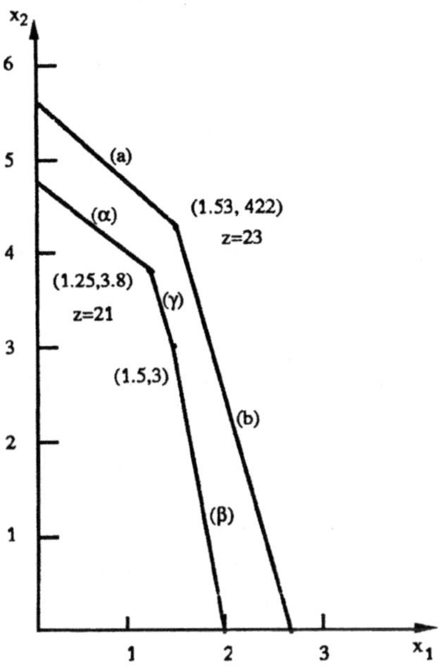

Fig. 2

If we compare our approach to the Ramik and the Rimanek's proposal, we can see that the last approach is pessimistic (re Fig. 2), because the solution of

$$[\max]z = 5x_1 + 4x_2$$

subject to

$$\begin{cases} 4x_1 + 5x_2 \leq 24 & (\alpha) \\ 4x_1 + x_2 \leq 12 & \\ 2x_1 + 2x_2 \leq 19 & \\ 3x_1 + 0,5x_2 \leq 6 & (\beta) \\ 5x_1 + 6x_2 \leq 32 & (\gamma) \\ 6x_1 + 2x_2 \leq 15 & \\ x_1, x_2 \geq 0 & \end{cases}$$

correspond to $x_1 = 1.25$, $x_2 = 3.80$, $z = 21.45$ (re Fig. 2), which constraints are contained in the polyhedron limited by ($x_1 \geq 0$, $x_2 \geq 0$, constraint (a), constraint (b)).

References

[1] Baldwin, F.F. and Guild, N.C.F. (1979) 'Comparison of fuzzy sets on the same decision space', Fuzzy Sets and Systems 2, 213-231.

[2] Dubois, D. and Prade, H. (1980) 'Systems of linear fuzzy constraints', Fuzzy Sets and Systems 3, 37-48.

[3] Nakamura, K (1986) 'Preference relations on a set of fuzzy utilities', Fuzzy Sets and Systems 20, 147-162.

[4] Ramik, J. and Rimanek, J. (1985) 'Inequality relation between fuzzy numbers and its use in fuzzy optimization', Fuzzy Sets and Systems 16, 123-138.

[5] Roubens, M. (1986) 'Comparison of flat fuzzy numbers', in N. Bandler and A. Kandel (eds.), Proceedings of NAFIPS'86, pp.462-476.

[6] Roubens, M. and Vincke, P. (1988) 'Fuzzy possibility graphs and their application to ranking fuzzy numbers', in J. Kacprzyk and M. Roubens (eds.), Non-conventional Preference Relations in Decision Making, Springer-Verlag, pp.119-128.

[7] Slowinski, R. (1986) 'A multicriteria fuzzy linear programming method for supply system development planning', Fuzzy Sets and Systems 19, 217-237.

Part III
Chapter 8

USING FUZZY LOGIC WITH LINGUISTIC QUANTIFIERS IN MULTIOBJECTIVE DECISION MAKING AND OPTIMIZATION: A STEP TOWARDS MORE HUMAN-CONSISTENT MODELS

Janusz KACPRZYK* and Ronald R. YAGER**

*Systems Research Institute
Polish Academy of Sciences
ul. Newelska 6
01 - 447 Warsaw, Poland

**Machine Intelligence Institute
Iona College
New Rochelle, NY 10801 - 1890, USA

ABSTRACT. An attempt to attain a higher human consistency of multiobjective decision making (optimization) models is considered. Basically, as opposed to virtually all conventional approaches which, explicitly or implicitly, seek an optimal option best satisfying all the objectives, which is often counterintuitive and unrealistic, we discuss some nonconventional models in which an optimal option is sought that best satisfies, say, most (or any other linguistic quantifier as almost all, much more than a half, ...) of the important objectives. Fuzzy - logic - based calculi of linguistically quantified propositions are employed. Some relations between the above new approach and the recently introduced so - called ordered weighted average (OWA) operators are discussed.

1. INTRODUCTION

The essence of virtually all formalized (mathematical) approaches to decision making, which underlies optimization

R. Slowinski and J. Teghem (eds.),
Stochastic versus Fuzzy Approaches to Multiobjective Mathematical Programming under Uncertainty, 331–350.

among many diverse fields, may be subsumed as follows. There is a set of options (alternatives, variants,...) and a value function from this set to that of outcomes (the real axis in most cases). The problem is to find an option maximizing (or minimizing) this value function. Virtually all decisions in nontrivial problems are to be made under uncertainty and imprecision as to, e.g., the set of feasible options, consequences of particular choices, form of the value function (or preferences), etc. It is easy to see that this all makes the application of classical (under certainty) models practically impossible. Many stochastic and fuzzy approaches to decision making (as well as those involving both randomness and fuzziness - see the papers in Kacprzyk and Fedrizzi, 1988b) have been therefore deviced to cope with this problem; a good source of information on imprecision related aspects in this context, which are more relevant for our purposes, are here the papers in Kacprzyk and Orlovski (1987).

However, most decisions of significance are of the multiobjective (multicriteria) type, that is an optimal option is to be found which best satisfies a number of (usually contradicting) objectives. Needless to say that this does make the decision making problem much more complicated. The essence of multiobjective decision making is that, first, outcomes of the particular options are determined in regards to individual objectives, and then these partial scores are somehow aggregated to obtain a total score (overall "value" of a particular option) which serves the purpose of an ordering function over the set of options. The option with the highest total score is chosen as the best.

This simple, maybe even trivial scheme is unfortunately pervaded with imprecision which occurs in virtually all elements of the problem formulation as, e.g., in the very meaning of objectives and their satisfaction, form of the value function (reflecting the human preferences), concept of a solution, aggregation of partial scores, etc. The necessity to account for imprecision has been recognized for a long time though first the inception of fuzzy sets theory has provided constructive formal means. Numerous fuzzy - sets - based multiobjective decision making and optimization models have been proposed, and many of them have found successful applications; for a review of earlier works, see Dubois and Prade (1980), while for an extensive survey of recent advances, see Kacprzyk and Orlovski (1987).

In this paper we are dealing with one of crucial aspects of multiobjective decision making (and its related multiobjective optimization and mathematical programming) which is concerned with the aggregation of the above mentioned partial scores.

Traditionally, this aggregation involves, explicitly or implicitly, all of them. Thus, we basically seek an optimal option which best satisfies *all* the objectives. However, it is easy to see that since, first, the objectives are usually contradictory, the satisfaction of all of them is usually impossible. Second, such a requirement may also be viewed to be not realistic since human decisionmakers rarely insist on such a rigid partial score aggregation. In most cases people are willing to accept an option whose partial scores with respect to a majority of the objectives are satisfactory even if its scores against a few of them are poor.

The above does clearly suggest that for practical reasons it might be worthwhile to replace the rigid requirement to satisfy *all* the objectives by some "milder" one as, e.g., to satisfy *most*, *almost all*, *much more than a half*, etc. of them. This was proposed, first in a slightly more particular form as to satisfy k out of p (p is the total number of objectives, and $k \leq p$) objectives in Yager (1984b), and then in a more general form as to satisfy, say, most of the important objectives by Yager (1983a); then, this approach was extended by Yager (1985a, b; 1987), and Kacprzyk and Yager (1984a, b). As a formal apparatus, fuzzy - logic - based calculi of linguistically quantified propositions were there used. Parenthetically, let us remark that a similar linguistic - quantifier - based approach proved to be useful for the "softening" in a similar spirit of: group decision making and consensus formation models (Kacprzyk, 1984, 1985a, b, c, 1986; Kacprzyk and Fedrizzi, 1986, 1988) and multistage decision making models (Kacprzyk, 1983b; Kacprzyk and Iwański, 1987).

The purpose of this paper is to further extend the above approach to multiobjective decision making in which an optimal option is sought which best satisfies, say, *most* of the *important* objectives. First, we briefly sketch two calculi of linguistically quantified propositions, one due to Zadeh (1983) and another due to Yager (1983a). Then, we present the application of these two calculi to derive the above mentioned new formulations of multiobjective decision making problem. Finally, we explore some links between this approach and a very interesting concept of the ordered weighted average (OWA) operators due to Yager (1988).

Some notational remarks. We will use a standard notation. A *fuzzy set* A in $\mathbf{X}$, $A \underset{\sim}{\subset} \mathbf{X}$, will be characterized by, and often informally equated with its *membership function* $\mu_A: \mathbf{X} \to [0, 1]$; $\mu_A(x) \in [0, 1]$ is a *degree of membership* of x in A. For a finite $\mathbf{X} = \{x_1, \ldots, x_n\}$ we will write $A = \mu_A(x_1)/x_1 + \ldots + \mu_A(x_n)/x_n$, where "$\mu_A(x_i)/x_i$" means the pair "element - its grade of membership". The (nonfuzzy) *cardinality* of A, the so - called ΣCount, is defined as $\Sigma\text{Count}(A) = \mu_A(x_1) + \ldots + \mu_A(x_n)$. For more details on the ΣCounts, fuzzy cardinalities, and some related topics, see Dubois and Prade (1985) or Zadeh (1983).

In the sequel we will also use the following binary operations:

(1) a *t - norm*, t: $[0, 1] \times [0, 1] \rightarrow [0, 1]$ such that:

(a) $a \; t \; 1 = a$

(b) $a \; t \; b = b \; t \; a$

(c) $a \; t \; b \geq c \; t \; d$ if $a \geq c$ and $b \geq d$

(d) $a \; t \; b \; t \; c = a \; t \; (b \; t \; c) = (a \; t \; b) \; t \; c$

A t - norm may be exemplified by: $a \wedge b = \min(a, b)$; ab; $1 - (1 \wedge ((1 - a)^p + (1 - b)^p)^{1/p}$, $p \geq 1$; "∧" is the most commonly used, and - if not specified otherwise - will be employed throughout the paper with the understanding that it may readily be replaced by another t - norm. We will also denote $\underset{i=1}{\overset{n}{t}} a_i = a_1 \; t \ldots t \; a_n$, and in particular $\bigwedge_{i=1}^{n} a_i = a_1 \wedge \ldots \wedge a_n$.

A t - norm generalizes the "and" operation in multivalued logic. Moreover, the *intersection* of two fuzzy sets may be in general represented by a t - norm.

(2) An *s - norm* (*co - t - norm*) s: $[0, 1] \times [0, 1] \rightarrow [0, 1]$ such that:

(a) $a \; s \; 0 = a$

(b) - (d) as for a t - norm.

An s - norm may be exemplified by: $a \vee b = \max(a, b)$; $a + b - ab$; $1 \wedge (a^p + b^p)^{1/p}$, $p \geq 1$; "∨" is the most commonly used, and - if not otherwise stated - will be employed throughout this paper with the understanding that it may readily be replaced by an s - norm.

An s - norm generalizes the "or" operation in multivalued logic. Moreover, the *union* of two fuzzy sets may be in general represented by an s - norm.

We also write $\bigvee_{i=1}^{n} a_i = \bigvee_{i \in \{1,\ldots,n\}} a_i = a_1 \vee \ldots \vee a_n$.

(3) An (generalized) *implication* (or *ply*) operation $\Rightarrow$: $[0, 1] \times [0, 1] \rightarrow [0, 1]$ such that:

(a) $a \Rightarrow 1 = 1$ $\quad 0 \Rightarrow a = 1$ $\quad 1 \Rightarrow a = a$

(b) $a \Rightarrow b \geq a \Rightarrow c$ if $b \geq c$

(c) $a \rightarrow b \geq c \rightarrow b$ if $a \leq c$

An implication operation may be exemplified by: b^a; $(1 - a)$ *s* b, in particular $(1 - a) \vee b$; and 1 *t* $(1 - a + b)$, in particular $1 \wedge (1 - a + b)$.

Belief qualification. We consider a *piece of evidence* "y is F" with which a *degree of belief*, bel $\in [0, 1]$, as to its validity is associated, written "y is F | bel"; $y \in \mathbf{Y}$, and F is a property defined as a (normalized) fuzzy set in $\mathbf{Y}$; "y is F" induces therefore a *possibility distribution* on $\mathbf{Y}$, $\pi_F(y) = \mu_F(y)$, $\forall\, y \in \mathbf{Y}$.

Now, "y is F | bel" is translated (see, e.g., Yager, 1982; or Kacprzyk and Yager, 1984a) to "y is F'" (with implied bel = 1, i.e. "y is F' | 1") whose associated possibility distribution is

$$\pi_{F'}(y) = (\pi_F(y) \; t \; \text{bel}) \; s \; (1 - \text{bel}) \qquad (1)$$

In a widely used case, when "*t*" is $\wedge$ and "*s*" is $\vee$, we have

$$\pi_{F'}(y) = (\pi_F(y) \wedge \text{bel}) \vee (1 - \text{bel}) \qquad (2)$$

Notice that bel = 1 means that we are certain that "y is F", i.e. $\pi_{F'}(y) = \pi_F(y)$, while bel = 0 means that we know nothing so that each y is possible, i.e. $\pi_{F'}(y) = 1$, $\forall\, y \in \mathbf{Y}$.

In general, the $\pi'_F(y)$ obtained should then be normalized to obtain

$$\pi_{\underline{F}}(y) = \pi'_F(y) \;/\; \max_{y \in Y} \pi'_F(y) \qquad (3)$$

2. LINGUISTICALLY QUANTIFIED PROPOSITIONS AND THEIR RELATED FUZZY - LOGIC - BASED CALCULI

A *linguistically quantified proposition* is exemplified by "most experts are convinced" and may be generally written as

Qy's are F (4)

where Q is a *linguistic quantifier* (e.g., most), $\mathbf{Y} = \{y\}$ is a *set of objects* (e.g., experts), and F is a *property* (e.g., convinced).

Importance may be added to (4) yielding

QBy's are F (5)

that is, say, "most of the important experts are convinced".

Basically, the problem is to find either truth(Qy's are F) in the case of (4) or truth(QBy's are F) in the case of

(5) knowing truth(y is F), $\forall y \in \mathbf{Y}$.

Moreover, with each "y is F" a *degree of belief*, bel $\in$ [0, 1], may be associated, written "y is F | bel"; then the linguistically quantified propositions (4) and (5) may be written as "Qy's are F | bel" and "QBy's are F | bel", and the problem becomes to find truth(Qy's are F | bel) or truth(QBy's are F | bel), respectively, knowing truth(y is F | bel), $\forall y \in \mathbf{Y}$.

Since the conventional predicate calculi do not make it possible to find the above truth values for general linguistic quantifiers as, e.g., *most*, *almost all*, *much more than a half*, etc., we will present below two fuzzy - logic - based calculi which may serve this purpose. Our presentation will be in principle limited to those elements that are needed for our next considerations. For more details, some other perspectives, etc. we will refer the reader to the literature.

2.1. The algebraic or consensory calculus

In Zadeh's method (1983), a *fuzzy linguistic quantifier* Q is assumed to be a fuzzy set in [0, 1], $Q \underset{\sim}{\subset} [0, 1]$.

For instance, Q = "most" may be given as

$$\mu_{\text{"most"}}(x) = \begin{cases} 1 & \text{for } x \geq 0.8 \\ 2x - 0.6 & \text{for } 0.3 < x < 0.8 \\ 0 & \text{for } x \leq 0.3 \end{cases} \qquad (6)$$

Particularly important in our context are the so - called *nondecreasing fuzzy quantifiers* defined as

$$\text{if } x_1 > x_2, \text{ then } \mu_Q(x_1) \geq \mu_Q(x_2), \ \forall x_1, x_2 \in \mathbb{R} \qquad (7)$$

which reflect, roughly speaking, an attitude of the type "the more the better" which is appropriate in our context; Q = "most" defined by (6) is evidently nondecreasing.

Moreover, we will mainly (except for Section 4) consider the *proportional quantifiers* exemplified by *most*, *almost all*, etc. as they are more important here than the *absolute quantifiers* exemplified by *about* 5, *much more than* 10, etc. The reasoning for the absolute quantifiers is however analogous (see, e.g., Kacprzyk, 1985a; or Yager, 1983a).

Property F is defined as a fuzzy set in $\mathbf{Y}$, $F \underset{\sim}{\subset} \mathbf{Y}$. If $\mathbf{Y} = \{y_1, \ldots, y_p\}$, then it is assumed that truth(y_i is F) = $\mu_F(y_i)$, i = 1,..., p.

The value of truth(Qy's are F) is now calculated using the (nonfuzzy) *cardinalities*, ΣCounts, of the the respective fuzzy sets in the following two steps (Zadeh, 1983):

(1) $r = \Sigma\text{Count}(F) / \Sigma\text{Count}(y) = \frac{1}{p} \sum_{i=1}^{p} \mu_F(y_i)$ (8)

(2) $\text{truth}(Qy\text{'s are } F) = \mu_Q(r)$ (9)

With *importance*, B = "important" $\underset{\sim}{\subset}$ **Y**, and $\mu_B(y_i) \in [0, 1]$ is a *degree of importance* of y_i: from 1 for definitely important to 0 for definitely unimportant, through all intermediate values.

We rewrite first "QBy's are F" as "Q(B and F)y's are B" which leads to the following counterparts of (8) and (9)

(1') $r' = \Sigma\text{Count}(B \text{ and } F) / \Sigma\text{Count}(B) =$

$$= \sum_{i=1}^{p} (\mu_B(y_i) \wedge \mu_F(y_i)) \; / \; \sum_{i=1}^{p} \mu_B(y_i) \qquad (10)$$

(2') $\text{truth}(QBy\text{'s are } F) = \mu_Q(r')$ (11)

The fuzzy cardinalities, the so - called FCounts, of the respective fuzzy sets (see, e.g., Dubois and Prade, 1985; or Zadeh, 1983) can also be in principle used in (8) and (9), and (10) and (11) but this case will not be considered here; needless to say that the use of the FCounts may lead to more complicated expressions, both conceptually and computationally, which may be prohibitive in our optimization context.

The method presented above may be viewed to provide a *consensory - like* aggregation of evidence (see, e.g., Yager, 1983a; or Kacprzyk and Yager, 1984a).

In the case of *belief qualification* in the sense of (1) - (3), the two steps (8) and (9) become

(1_{bel}) $r_{bel} = \frac{1}{p} \sum_{i=1}^{p} \mu_{\underline{F}}(y_i)$ (12)

(2_{bel}) $\text{truth}(Qy\text{'s are } F \mid bel) = \mu_Q(r_{bel})$ (13)

and (12) and (13) become

($1'_{bel}$) $r'_{bel} = \sum_{i=1}^{p} (\mu_B(y_i) \wedge \mu_{\underline{F}}(y_i)) \; / \; \sum_{i=1}^{p} \mu_B(y_i)$ (14)

($2'_{bel}$) $\text{truth}(QBy\text{'s are } F \mid bel) = \mu_Q(r'_{bel})$ (15)

where $\mu_{\underline{F}}(y_i)$ is given by (3).

<u>Example 1</u>. Let **Y** = "experts" = {X, Y, Z}, F = "convinced" = 0.1/X + 0.6/Y + 0.8/Z, Q = "most" be given by (6), B = "important = 0.2/X + 0.5/Y + 0.6/Z, and bel = 0.7 [i.e. bel(X) = 0.7, bel(Y) = 0.7 and bel(Z) = 0.7].

Then $r = 0.5$, $r' = 0.8$, $r_{bel} = 0.6$ and $r'_{bel} = 0.6$, and
truth("most experts are convinced") = 0.4
truth("most of the important experts are convinced") = 1
truth("most experts are convinced" | 0.7) =
truth("most of the important experts are convinced" | 0.7) = 0.8

2.2. The substitution or competitive calculus

This alternative calculus is due to Yager (1983a, b). As previously, $\mathbf{Y} = \{y_1, \ldots, y_p\}$ and $F \underset{\sim}{\subset} \mathbf{Y}$. A proposition "$y_i$ is F" is denoted P_i, and truth P_i = truth(y_i is F) = $\mu_F(y_i)$, $i = 1, \ldots, p$.

We introduce the set $\mathbf{V} = \{v\} = 2^{\{P_1, \ldots, P_p\}} \setminus \emptyset$, where v is "$P_{k1}, \ldots, P_{km}$" meant as the proposition "P_{k1} and ... and P_{km}"; v and "P_{k1} and ... and P_{km}" are used here interchangeably.

Each v, or its corresponding "P_{k1} and ... and P_{km}", is seen to be true to the following degree

$$\mu_T(v) = \text{truth}(P_{k1} \text{ and } \ldots \text{ and } P_{km}) = \text{truth } P_{k1} \wedge \ldots$$
$$\ldots \wedge \text{truth } P_{km} = \mu_F(y_{k1}) \wedge \ldots \wedge \mu_F(y_{km}) =$$
$$= \bigwedge_{i=1}^{p} \mu_F(y_{ki}) \tag{16}$$

i.e. we obtain a fuzzy set $T \underset{\sim}{\subset} \mathbf{V}$; evidently, any t - norm may be used instead of "$\wedge$".

The fuzzy linguistic quantifier Q is defined as a fuzzy set in $\mathbf{V}$, $Q \underset{\sim}{\subset} \mathbf{V}$. For instance, if $p = 3$, i.e. $\mathbf{V}$ = {P_1 and P_2 and P_3, P_1 and P_2, P_1 and P_3, P_2 and P_3, P_1, P_2, P_3}, then Q = "most" may be defined as

$$\mu_{\text{"most"}}(v) = 1 \quad \text{for } v \in \{P_1 \text{ and } P_2 \text{ and } P_3\}$$
$$= 0.7 \quad \text{for } v \in \{P_1 \text{ and } P_2,\ P_1 \text{ and } P_3,\ P_2 \text{ and } P_3\} \tag{17}$$
$$= 0.3 \quad \text{for } v \in \{P_1, P_2, P_3\}$$

Particularly important in our context are the so - called *monotonic* fuzzy quantifiers defined as

$$\mu_Q(v_1 \text{ and } v_2) \geq \mu_Q(v_1) \vee \mu_Q(v_2), \ \forall\, v_1, v_2 \in \mathbf{V} \qquad (18)$$

which are close in spirit to the nondecreasing ones defined by (7) though not the same. Evidently, "most" defined by (17) is monotonic since, e.g. if $v_1 \in \{P_1\}$ and $v_2 \in \{P_2$ and $P_3\}$, then "v_1 and v_2" is "P_1 and P_2 and P_3", and $1 = \mu_Q(P_1$ and P_2 and $P_3) \geq \mu_Q(P_1) \vee \mu_Q(P_2$ and $P_3) = 0.3 \vee 0.7 = 0.7$.

We have now

$$\text{truth}(Qy\text{'s are } F) = \bigvee_{v \in \mathbf{V}} (\mu_Q(v) \wedge \mu_T(v)) \qquad (19)$$

and in the case of importance

$$\text{truth}(QBy\text{'s are } F) = \bigvee_{v \in \mathbf{V}} (\mu_Q(v) \wedge \\ \wedge (\bigwedge_{i=1}^{m} (\mu_B(y_{ki}) \Rightarrow \mu_F(y_{ki})))) \qquad (20)$$

and in the most commonly used form of "$\Rightarrow$", i.e. $a \Rightarrow b = (1 - a) \vee b$, we obtain

$$\text{truth}(QBy\text{'s are } F) = \bigvee_{v \in \mathbf{V}} (\mu_Q(v) \wedge \\ \wedge (\bigwedge_{i=1}^{m} ((1 - \mu_B(y_{ki})) \vee \mu_F(y_{ki})))) \qquad (21)$$

where, as before "$\wedge$" and "$\vee$" may be replaced by any t - norm and s - norm, respectively, and another implication operator may be used.

In the case of *belief qualification*, we obtain the following counterparts of (20) and (21), respectively:

$$\text{truth}(Qy\text{'s are } F \mid \text{bel}) = \bigvee_{v \in \mathbf{V}} (\mu_Q(v) \wedge \mu_{\underline{F}}(v)) \qquad (22)$$

and

$$\text{truth}(QBy\text{'s are } F \mid \text{bel}) = \bigvee_{v \in \mathbf{V}} (\mu_Q(v) \wedge \\ \wedge (\bigwedge_{i=1}^{m} ((1 - \mu_B(y_{ki})) \vee \mu_{\underline{F}}(y_{ki})))) \qquad (23)$$

where $\mu_{\underline{F}}(y_{ki})$ is given by (3).

For more information on the substitution calculus, see Yager (1983a).

<u>Example 2</u>. For the same data as in Example 1 and Q = "most" defined by (17), we obtain

truth("most experts are convinced") = 0.6

truth("most of the important experts are convinced") = 0.8

truth("most experts are convinced" | 0.7) = 0.8

truth("most of the important experts are convinced | 0.7) = 0.9

Notice that this method may yield different results than that presented in Section 2.1.

Finally, notice that a calculus of linguistically quantified propositions is an underlying element of Zadeh's (1983, 1985, 1987) representation of *cammonsense knowledge* viewed as a collection of the so - called *dispositions*, i.e. propositions with implicit linguistic quantifier. For instance, a disposition "winter days are cold" does in fact involve an implicit linguistic quantifier as, e.g., "most" or "almost all", hence should be meant as, say, "almost all winter days are cold". In this perspective, the approach to multiobjective decision making via fuzzy linguistic quantifiers may be viewed as an attempt to introduce commonsense human perception as to the very essence of the problem into its mathematical model. This should make the multiobjective decision making models more *human - consistent*, hence should increase their *implementability*.

3. FUZZY LINGUISTIC QUANTIFIERS IN MULTIOBJECTIVE DECISION MAKING

For our purposes *multiobjective decision making under fuzziness* may be formalized as follows (see, e.g., Bellman and Zadeh, 1970; Dubois and Prade ,1980). $\mathbb{A} = \{a\} = \{a_1, \ldots, a_q\}$ is a set of possible *options* (alternatives, variants, decisions, ...) and $\mathbb{O} = \{O_1, \ldots, O_p\}$ is a set of *fuzzy objectives* (*constraints* and/or *goals*) to be satisfied. A degree to which an $a \in \mathbb{A}$ satisfies an $O_i \in \mathbb{O}$ is equal to truth(O_i is satisfied (by a)) = $\mu_{O_i}(a)$, i = 1, ..., p.

It is commonly postulated that an $a \in \mathbb{A}$ satisfy "O_1 and ... and O_p", i.e. all the fuzzy objectives; the degree of this satisfaction is given by the *fuzzy decision* (cf. Bellman and Zadeh, 1970)

$$\mu_D(a \mid \text{"all"}) = \text{truth}(O_1 \text{ and } \ldots \text{ and } O_p \text{ are satisfied (by a))} = \text{truth}(\text{"all" } O_i\text{'s are satisfied}) = $$
$$= \text{truth}(O_1 \text{ is satisfied}) \wedge \ldots \wedge \text{truth}(O_p \text{ is satisfied}) = = \mu_{O_1}(a) \wedge \ldots$$

$$\ldots \wedge \mu_{O_p}(a) = \bigwedge_{i=1}^{p} \mu_{O_i}(a) \tag{24}$$

and the problem is to find an optimal option $a^* \in \mathbf{A}$ such that

$$\mu_D(a^* \mid \text{"all"}) = \max_{a \in \mathbf{A}} \mu_D(a \mid \text{"all"}) \tag{25}$$

Notice that this formulation encompasses in its essence the conventional (nonfuzzy) formulations of multiobjective decision making.

It is quite natural that the requirement to satisfy *all* the fuzzy objectives may be viewed as too rigid, restrictive and even counterintuitive in many practical situations. The idea of replacing this rigid "all" (i.e. p in our context) by some 'milder' requirement appeared first in Yager (1984b) where the satisfaction of $k \leq p$ out of p (fuzzy) objectives was postulated. For our purposes, however, Yager's (1983a) work is more important where *all* is replaced by some (fuzzy) linguistic quantifier as, e.g., *most*, *almost all*, *much more than a half*, etc. This idea was then considerably advanced in, e.g., Kacprzyk and Yager (1984a, b) and Yager (1984a, 1985a).

In this case the fuzzy decision may be written as

$$\mu_D(a \mid Q) = \text{truth}(QO\text{'s are satisfied}) =$$

$$= (\bigwedge_{i=1}^{p} \mid Q)\ \mu_{O_i}(a) \tag{26}$$

which means that in the calculation of $\mu_D(\cdot \mid \cdot)$ only QO's are to be taken into account (but does not explicitly say which ones!).

We seek therefore an optimal option $a^* \in \mathbf{A}$ such that

$$\mu_D(a^* \mid Q) = \max_{a \in \mathbf{A}} \mu_D(a \mid Q) \tag{27}$$

i.e. such that it best satisfies Q of the objectives.

In the case of importance, $\underset{\sim}{B} \subset \mathbf{O}$, (26) becomes

$$\mu_D(a \mid Q, B) = \text{truth}(QBO\text{'s are satisfied}) =$$

$$= (\bigwedge_{i=1}^{p} \mid Q, B)\ \mu_{O_i}(a) \tag{28}$$

which means that only Q of the important O's are to be taken into account (but once again does not explicitly say which ones!). Then we seek an optimal $a^* \in \mathbf{A}$ such that

$$\mu_D(a^* \mid Q, B) = \max_{a \in \mathbf{A}} \mu_D(a \mid Q, B) \tag{29}$$

i.e. such that it best satisfies Q of the important objectives.

We will describe in the following a methodology for solving this type of problems using the two calculi of linguistically quantified propositions outlined in Section 2.

3.1. Solution by the algebraic method

We introduce first a fuzzy set S = "satisfied" $\underset{\sim}{\subset}$ $\mathbb{O}$ such that $\mu_S(O_i)$ = truth(O_i is satisfied (by a)) = $\mu_{O_i}(a)$, i = 1,..., p, and a fuzzy set B = "important" $\underset{\sim}{\subset}$ $\mathbb{O}$ such that $\mu_B(O_i) \in [0, 1]$ is a degree of importance of objective O_i, from 0 for definitely unimportant to 1 for definitely important, through all intermediate values. A fuzzy linguistic quantifier is $Q \underset{\sim}{\subset} [0, 1]$.

The fuzzy decision is therefore (28), with B = "important", and the problem is (29). To proceed to its solution, we calculate therefore first due to (10) and (11)

$$r(a) = \sum_{i=1}^{p} (\mu_B(O_i) \wedge \mu_{O_i}(a)) \Big/ \sum_{i=1}^{p} \mu_B(O_i) \qquad (30)$$

and

$$\mu_D(a \mid Q, B) = \text{truth(QBO's are satisfied)} = \mu_Q(r(a)) \qquad (31)$$

and we seek an $a^* \in \mathbb{A}$ such that

$$\mu_D(a^* \mid Q, B) = \max_{a \in \mathbb{A}} \mu_Q(r(a)) \qquad (32)$$

Notice that in the case without importance, i.e. with $\mu_B(O_i) = 1$ for each $O_i \in \mathbb{O}$, we replace (10) by

$$r(a) = \sum_{i=1}^{p} \mu_{O_i}(a) \qquad (33)$$

and we seek an $a^* \in \mathbb{A}$ such that

$$\mu_D(a^* \mid Q) = \max_{a \in \mathbb{A}} \mu_D(a \mid Q) = \max_{a \in \mathbb{A}} \mu_Q(r(a)) \qquad (34)$$

It is evidently difficult to say something about the solution of (32) [or (34)] for an arbitrary Q. However, in (7) a nondecreasing quantifier has already been defined whose essence in our context is that the more objectives are satisfied the better; quite naturally, only quantifiers of such a type are relevant for our purposes. A particularly important nondecreasing quantifier is the *linear quantifier*,

$\underset{\sim}{L} \subset [0, 1]$, defined as $\mu_L(u) = u$, $\forall\, u \in [0, 1]$.

If we use the linear quantifier, then - as shown in Yager (1983a) or Kacprzyk and Yager (1984, b) - any optimal solution of (32) for the linear quantifier is also an optimal solution of (32) for an arbitrary nondecreasing quantifier. Thus, an optimal solution of (32) for any nondecreasing Q may be found by determining an $a^* \in \mathbf{A}$ such that

$$\mu_D(a^* \mid L, B) = \max_{a \in \mathbf{A}} \mu_D(a \mid L, B) =$$

$$= \max_{a \in \mathbf{A}} \left\{ \sum_{i=1}^{p} [\mu_B(O_i) \wedge \mu_{O_i}(a)] \Big/ \sum_{i=1}^{p} \mu_B(O_i) \right\} \qquad (35)$$

<u>Example 3</u>. Let: $\mathbf{A} = \{a_1, a_2, a_3\}$, $O_1 = 1/a_1 + 0.8/a_2 + 0.3/a_3$, $O_2 = 0.6/a_1 + 1/a_2 + 0.5/a_3$, $O_3 = 0.5/a_1 + 0.8/a_2 + 1/a_3$, Q = "most" be given by (6), and B = "important" = $0.3/O_1 + 0.8/O_2 + 0.5/O_3$.

Then:

$$\mu_D(a_1 \mid \text{"most", "important"}) = 1.4/1.6 \cong 0.9$$

$$\mu_D(a_2 \mid \text{"most", "important"}) = 1.6/1.6 = 1$$

$$\mu_D(a_3 \mid \text{"most", "important"}) = 1.3/1.6 \cong 0.8$$

That is, $a^* = a_2$.

3.2. Solution by the substitution method

We introduce the propositions P_i: "O_i is satisfied (by a)", and a fuzzy set S = "satisfied" $\underset{\sim}{\subset} \mathbf{O}$ such that $\mu_S(O_i) = \mu_{O_i}(a)$ = truth P_i, i = 1, ..., p. Thus $\mathbf{V} = \{v(a)\} = \{P_{k1}$ and ... and $P_{km}\} = 2^{\{P_1,\ldots,P_p\}} \setminus \emptyset$.

If $B \underset{\sim}{\subset} \mathbf{O}$ denotes importance and $Q \underset{\sim}{\subset} \mathbf{O}$, then by (20) we obtain

$$\mu_D(a \mid Q, B) = \text{truth(QBO's are satisfied (by a))} =$$

$$= \max_{v(a) \in \mathbf{V}} \{\mu_Q(v(a)) \wedge [\bigwedge_{i=1}^{m} (\mu_B(O_{ki}) \rightarrow \mu_{O_{ki}}(a))]\} \qquad (36)$$

and we seek an $a^* \in \mathbf{A}$ such that

$$\mu_D(a^* \mid Q, B) = \max_{a \in \mathbf{A}} \mu_D(a \mid Q, B) \tag{37}$$

In general, the solution of (37) is analitycally difficult, and may be obtained via 0 - 1 programming (Yager, 1983a).

However, for a simplified, yet practically relevant form of (36) when all the objectives are of the same importance, i.e. with $\mu_B(O_i) = 1$, $\forall\, O_i \in \mathbb{O}$, the problem (37) becomes to find an $a^* \in \mathbf{A}$ such that

$$\mu_D(a^* \mid Q) = \max_{a \in \mathbf{A}} \max_{v(a) \in \mathbf{V}} [\mu_Q(v(a)) \wedge \bigwedge_{i=1}^{m} \mu_{O_{ki}}(a)] \tag{38}$$

which can be relatively easily solved under some mild assumptions.

Namely (see, e.g., Yager, 1983a; Kacprzyk and Yager, 1984 a, b; Kacprzyk, 1987b) if:

(1) the fuzzy linguistic quantifier $Q \subset \mathbf{V}$ is monotonic in the sense of (18);

(2) there is a finite number of distinct membership degrees of $\mu_Q(v(a))$, say $b_1 \le b_2 \le \ldots \le b_s$;

(3) d_i is the i - th largest element of the set of values $\{\mu_{O_1}(a), \ldots, \mu_{O_p}(a)\}$ containing for a fixed $a \in \mathbf{A}$ its membership degrees in the consecutive fuzzy objectives;

then

$$\mu_D(a \mid Q) = \max_{i \in \{1,\ldots,s\}} (d_i \wedge b_i) \tag{39}$$

and we seek an $a^* \in \mathbf{A}$ such that

$$\mu_D(a^* \mid Q) = \max_{a \in \mathbf{A}} \max_{i \in \{1,\ldots,s\}} (d_i \wedge b_i) \tag{40}$$

<u>Example 4</u>. For the same data as in Example 3, but without importance (i.e. $B = 1/O_1 + 1/O_2 + 1/O_3$), we obtain by (39):

$$\mu_D(a_1 \mid \text{"most"}) = 0.6$$
$$\mu_D(a_2 \mid \text{"most"}) = 0.7$$
$$\mu_D(a_3 \mid \text{"most"}) = 0.5$$

that is $a^* = a_2$.

On the other hand, in the case with importance assumed as in Example 3, i.e. B = "important" = $0.3/O_1 + 0.8/O_2 + 0.5/O_3$, we obtain by (36)

$\mu_D(a_1 \mid \text{"most", "important"}) = 0.6$
$\mu_D(a_2 \mid \text{"most", "important"}) = 0.8$
$\mu_D(a_3 \mid \text{"most", "important"}) = 0.7$

that is $a^* = a_2$.

Notice that the optimal solution is the same in both the cases, as well as in Example 3, although it need not be so in general.

4. MULTIOBJECTIVE DECISION MAKING AND THE ORDERED WEIGHTED AVERAGE (OWA) OPERATORS

In this section we will sketch Yager's (1988) recent approach to multiobjective decision making based on the so - called *ordered weighted average* (OWA) operators which is related to the topic of this paper.

As in Section 3, $\mathbf{A} = \{a\} = \{a_1, \ldots, a_q\}$ is the set of options, and $\mathbf{O} = \{O_1, \ldots, O_p\}$ is the set of objective. A degree to which and $a \in \mathbf{A}$ satisfies an $O_i \in \mathbf{O}$ is $\mu_{O_i}(a) \in [0, 1]$, $i = 1, \ldots, p$.

The problem is to formulate an overall decision function, called the fuzzy decision in our context [cf. (24)]

$$\mu_D(a) = F(\mu_{O_1}(a), \ldots, \mu_{O_p}(a)) \qquad (41)$$

where $\mu_D(a) \in [0, 1]$ is the degree to which an $a \in \mathbf{A}$ meets our requirements with respect to the objectives, and $F: [0, 1]^p \to [0, 1]$ is some aggregation operator.

Basically, F should first of all satisfy the following two conditions:

(1) monotonicity (or positive association), i.e. if for any $a_1, a_2 \in \mathbf{A}$, $\mu_{O_i}(a_1) \geq \mu_{O_i}(a_2)$ for all $O_i \in \mathbf{O}$, then $\mu_D(a_1) \geq \mu_D(a_2)$.

(2) symmetry (or generalized commutativity), i.e. $F(a_{i1}, \ldots, a_{ip}) = F(a_1, \ldots, a_p)$ where $(a_{i1}, \ldots, a_{ip})$ is a permutation of $(a_1, \ldots, a_p)$.

Another aspect is here the interrelationship between the objectives to be satisfied. On the one extreme, we may require *all* the objectives to be satisfied, i.e. O_1 and ... and O_p, while on the other extreme we may wish to fulfill *at least one* objective, i.e. O_1 or ...or O_p.

However, as already mentioned in Section 3, the type of aggregation required is often neither the pure "anding" nor the pure "oring" but something in between. To model this, Yager (1988) has proposed the concept of an ordered weighted average (or OWA, for short) operator which might be even more appropriately called an "orand" operator.

An OWA operator of dimension p is a mapping $F: [0, 1]^p \to [0, 1]$ if associated with F is a weighting vector $W = [w_1, \ldots, w_p]^T$ such that: $w_1 \in [0, 1]$, $w_1 + \ldots + w_p = 1$, and

$$F(x_1, \ldots, x_p) = w_1 b_1 + \ldots + w_p b_p \tag{42}$$

where b_i is the i - th largest element among $x_1, \ldots, x_p$.

B is called an ordered argument vector if each $b_i \in [0, 1]$, and $j > i$ implies $b_i \geq b_j$, $i = 1, \ldots, p$.

Then

$$F(x_1, \ldots, x_p) = W^T B \tag{43}$$

<u>Example 5</u>. Let $W^T = [0.2 \quad 0.3 \quad 0.1 \quad 0.4]$, and calculate $F(0.6, 1.0, 0.3, 0.5)$.

Thus, $B^T = [1.0 \quad 0.6 \quad 0.5 \quad 0.3]$, and $F(0.6, 1.0, 0.3, 0.5) = W^T B = 0.55$; and $F(0.0, 0.7, 0.1, 0.2) = 0.43$.

It is easy to see that the OWA operators are monotonic and symmetric; for other more relevant properties, see Yager (1988).

An important issue is how to determine the w_i's. Among some hints given in Yager (1988), the following approach is of some value for our considerations.

Suppose that we have a vector of the objectives' satisfaction $E = [e_1, \ldots e_p]$ such that $e_i = 1$ for $j \leq k$ and $e_j = 0$ for $j > k$ meant as that k (out of p) objectives are fully satisfied while the rest is fully unsatisfied. Then, if we denote $s_k = w_1 + \ldots + w_k$ (notice that $s_0 = 0$ and $s_1 = 1$), then

$$F(E) = W^T E = w_1 + \ldots + w_k = s_k \tag{44}$$

i.e. s_k is the degree of satisfaction we have if k/p portion of the objectives is satisfied. Evidently, $s_k = s_{k-1} + w_k$, i.e. w_k is the additional (or marginal) satisfaction we get when we proceed from the satisfaction of k - 1 to k objectives; obviously, $w_k = s_k - s_{k-1}$.

4.1. Quantifiers and the OWA operators

In this section we use the (linguistic) quantifier in the sense of Zadeh (cf. Section 2.1).

Suppose now that as in Section 3.1 we wish to satisfy Q (a linguistic quantifier) objectives; however, Q is an absolute type quantifier which is defined in [0, p] where p = card Φ. Moreover, for a $y \in [0, p]$, let $\mu_Q(y) \in [0, 1]$ denote the degree to which we are satisfied with y objectives being fulfilled (naturally, if we require Q of them to be satisfied). Notice that, obviously:

(1) $\mu_Q(0) = 0$, i.e. we get absolutely no satisfaction with no objective satisfied;

(2) $\mu_Q(p) = 1$, i.e. we are completely satisfied if all the objectives are satisfied;

(3) if $r_1 > r_2$, then $\mu_Q(r_1) \geq \mu_Q(r_2)$, i.e. the more objectives are satisfied the better [cf. the definition of a nondecreasing quantifier (7)].

Quantifier Q as defined above has therefore the same properties as the F aggregation function, so that it is our conjecture that the weighting vector W is a manifestation of a quantifier underlying the process of aggregating the objectives. In particular, if we wish to satisfy Q (out of p) objectives, then $w_k = \mu_Q(k) - \mu_Q(k - 1)$, $k = 1, \ldots, p$; $\mu_Q(0) = 0$.

Moreover, the OWA operator seems to be a manifestation of a nondecreasing quantifier defined by (7).

Just to give some examples of the w_i's associated with the particular quantifiers, notice that:

(1) if $w_p = 1$, and $w_i = 0$ for each $i \neq p$, then this corresponds to Q = "all";

(2) if $w_i = 1$ for $i = 1$, and $w_i = 0$ for each $i \neq 1$, then this corresponds to Q = "at least one".

The cases in between, i.e. between *all* and *at least one*, which correspond to such quantifiers as, e.g., *a half*, *most*, *more than* 75%, *a few*, *almost all*, etc. may be therefore obtained by a suitable choice of the w_i's, between those given in the above two extremes.

The OWA operators are therefore an interesting and promising class of aggregation operators which may be of use in many real life multiobjective decision making problems.

5. CONCLUDING REMARKS

The new multiobjective decision making (optimization) models discussed in this paper may be viewed as a further step to attain more human - constistent models. The main feature of the models presented is an attempt at making the aggregation of partial scores (resulting from the satisfaction of the particular objective) less rigid in that only, say, most of the important objective (and not all as in virtually all conventional models) are accounted for.

A high human consistency of the models seems to promise a higher human acceptability of those models' results, hence their higher implementability. This implementability aspect is crucial, and in many cases a serious limiting factor, in virtually all systems meant for supporting real life decision making.

BIBLIOGRAPHY

Bellman R.E. and L.A. Zadeh (1970) Decision - making in a fuzzy environment. Management Science 17, 151 - 169.

Dubois D. and Prade H. (1980) Fuzzy Sets and Systems: Theory and Applications. Academic Press, New York.

Dubois D. and H. Prade (1985) Fuzzy cardinality and the modelling of imprecise quantification. Fuzzy Sets and Systems 16, 199 - 230.

Kacprzyk J. (1983a) Multistage Decision - Making under Fuzziness. Verlag TUV Rheinland, Cologne.

Kacprzyk J. (1983b) A generalization of fuzzy multistage decision making and control via linguistic quantifiers. International Journal of Control 38, 1249 - 1270.

Kacprzyk J. (1984) Collective decision making with a fuzzy majority. Proceedings of 6th WOGSC Congress, AFCET, Paris, pp. 153 - 159.

Kacprzyk J. (1985a) Zadeh's commonsense knowledge and its use in multicriteria, multistage and multiperson decision making, in M.M. Gupta et al. (Eds.): Approximate Reasoning in Expert Systems, North - Holland, Amsterdam, pp. 105 - 121.

Kacprzyk J. (1985b) Group decision making with a fuzzy majority via linguistic quantifiers. Part I: A consensory - like pooling, Part II: A competitive - like pooling. Cybernetics and Systems: an International Journal 18, 135 - 145 (Part I), 147 - 160 (Part II).

Kacprzyk J. (1985c) Some 'commonsense' solution concepts in group decision making via fuzzy linguistic quantifiers, in J. Kacprzyk and R.R. Yager (Eds.): Management Decision Support Systems Using Fuzzy Sets and Possibility Theory,

Verlag TÜV Rheinland, Cologne, pp. 125 - 135.
Kacprzyk J. (1986) Group decision making with fuzzy linguistic majority. Fuzzy Sets and Systems 18, 105 - 118.
Kacprzyk J. (1987a) On some fuzzy cores and 'soft' consensus measures in group decision making, in J.C. Bezdek (Ed.): The Analysis of Fuzzy Information. CRC Press, Boca Raton, pp. 119 - 130.
Kacprzyk J. (1987b) Towards 'human - consistent' decision support systems through commonsense - knowledge - based decision making and control models: a fuzzy logic approach. Computers and Artificial Intelligence 6, 97 - 122.
Kacprzyk J. and Fedrizzi M. (1986) 'Soft' consensus measures for monitoring real consensus reaching processes under fuzzy preferences. Control and Cybernetics 15, 309 - 323.
Kacprzyk J. and Fedrizzi M. (1988a) A 'soft' measure of consensus in the setting of partial (fuzzy) preferences. European Journal of Operational Research 34, 315 - 325.
Kacprzyk J. and Fedrizzi M., Eds. (1988b) Combining Fuzzy Imprecision with Stochastic Uncertainty in Decision Making. Springer - Verlag, Berlin.
Kacprzyk J. and Nurmi H. (1989) Linguistic quantifiers and fuzzy majorities for more realistic and human - consistent group decision making, in G. Evans, W. Karwowski and M. Wilhelm (Eds.): Fuzzy Methodologies for Industrial and Systems Engineering, Elsevier, Amsterdam, pp. 267 - 281.
Kacprzyk J. and Orlovski S.A., Eds. (1987) Optimization Models Using Fuzzy Sets and Possibility Theory. Reidel, Dordrecht.
Kacprzyk J. and Yager R.R. (1984a) Linguistic quantifiers and belief qualification in fuzzy multicriteria and multistage decision making. Control and Cybernetics 13, 155 - 173.
Kacprzyk J. and Yager R.R. (1984b) "Softer" optimization and control models via fuzzy linguistic quantifiers. Information Sciences 34, 157 - 178.
Yager R.R. (1982) Generalized probabilities of fuzzy events from fuzzy belief structures. Information Sciences 28, 45 - 62.
Yager R.R. (1983a) Quantifiers in the formulation of multiple objective decision functions. Information Sciences 31, 107 - 139.
Yager R.R. (1983b) Quantified propositions in a linguistic logic. International Journal of Man - Machine Studies 19, 195 - 227.
Yager R.R. (1984a) General multiple objective decision making and linguistically quantified statements. International Journal of Man - Machine Studies 21, 389 - 400.
Yager R.R. (1984b) Multiple objective solutions preferred by k of p objectives. Large Scale Systems 7, 79 - 84.

Yager R.R. (1985a) Aggregating evidence using quantified statements. Information Sciences 36, 179 - 206.
Yager R.R. (1985b) Reasoning with fuzzy quantified statements: Part I. Kybernetes 14, 233 - 240.
Yager R.R. (1987) On the knowledge structure of multi - solution variables, including quantified statements. International Journal of Approximate Reasoning 1, 23 - 70.
Yager R.R. (1988) An ordered weighted averaging aggregation operators in multicriteria decision making. IEEE Transactions on Systems, Man and Cybernetics SMC - 18, 183 - 190.
Zadeh L.A. (1983a) A computational approach to fuzzy quantifiers in natural languages. Computers and Mathematics with Applications 9, 149 - 184.
Zadeh L.A. (1985) Syllogistic reasoning in fuzzy logic and its application to usuality and reasoning with dispositions. IEEE Transactions on Systems, Man and Cybernetics SMC - 15, 754 - 763.
Zadeh L.A. (1987) A computational theory of dispositions. International Journal of Intelligent Systems 2, 39 - 64.

Part IV.
STOCHASTIC VERSUS FUZZY APPROACHES AND RELATED ISSUES

Part IV
Chapter 1

STOCHASTIC VERSUS POSSIBILISTIC MULTIOBJECTIVE PROGRAMMING

J.J. Buckley
Mathematics Department
University of Alabama at Birmingham
Birmingham, AL 35294
USA

ABSTRACT. This paper discusses solutions to multiobjective linear programs where some, or all, of the parameter values may be uncertain. The uncertainty is modeled using random variables (stochastic programming) or fuzzy variables (possibilistic programming). The solution is the undominated set for a multiobjective problem associated with a game against Nature. The paper explores both the similarites, and differences, between stochastic and possibilistic programming with respect to our solution procedure. A detailed example is presented illustrating the concepts.

1. Introduction.

In this paper will discuss solutions to the multiobjective linear program

$$\begin{aligned} maxZ &= (c_1x, \ldots, c_Kx) \\ \text{subject to } &: AX * b, x \geq 0, \end{aligned} \tag{1}$$

where $*$ denotes $\leq, =$ or $\geq$ for each row in the linear constraints and the numbers in c_i, A, and b are modeled as values of random variables (stochastic programming) or as values of fuzzy variables (possibilistic programming). For simplicity, we consider a linear program but our results may be easily extended to non-linear multiobjective programming. For the dimensions to the linear problem let $c_i = (c_{i1}, \ldots, c_{in}), 1 \leq i \leq K, x^t = (x_1, \ldots, x_n)$, $A = [a_{ij}]$ a $m \times n$ matrix, and $b^t = (b_1, \ldots, b_m)$. The vector x contains the non-random (non-fuzzy) decision variables. Some, or all of the parameters c_{ij}, a_{ij}, b_i are allowed to have uncertain values and we are to find a best, or an optimal, $x \geq 0$ for the problem in equation (1). Let $z_i = c_ix, 1 \leq i \leq K$, and let $c = (c_1, c_2, \ldots, c_K)$, a $1 \times nK$ vector.

In the stochastic program let $\hat{c}_{ij}, \hat{a}_{ij}$, and $\hat{b}_j$ be independent random variables whose values are the numbers we will use for c_{ij}, a_{ij}, b_j, respectively, in equation (1). The probability density[1] functions for $\hat{c}_{ij}, \hat{a}_{ij}$, and $\hat{b}_j$ are $\mu(c_{ij}|\hat{c}_{ij}), \mu(a_{ij}|\hat{a}_{ij})$, and $\mu(b_j|\hat{b}_j)$, respectively. The notation $\mu(c_{ij}|\hat{c}_{ij})$ means that $\mu(\ |\hat{c}_{ij})$ is the probability density function for $\hat{c}_{ij}$ and $\mu(c_{ij}|\hat{c}_{ij})$ is the value of this function for some possible value c_{ij} of $\hat{c}_{ij}$. Each density function is non-negative and normalized so that the area under the graph equals one. We use the densities to calculate probablilties as follows:

$$Prob[c_{ij} \in E] = \int_E \mu(c|\hat{c}_{ij})dc. \tag{2}$$

In the possibilistic program let $\hat{c}_{ij}, \hat{a}_{ij}$, and $\hat{b}_j$ be non-interactive ([11],[12]) fuzzy variables whose values are the numbers we will use for c_{ij}, a_{ij}, b_j, respectively, in equation (1). The possibility

R. Slowinski and J. Teghem (eds.),
Stochastic versus Fuzzy Approaches to Multiobjective Mathematical Programming under Uncertainty, 353–364.

distributions for $\hat{c}_{ij}, \hat{a}_{ij}$, and $\hat{b}_j$ are given by the membership functions $\mu(c_{ij}|\hat{c}_{ij}), \mu(a_{ij}|\hat{a}_{ij})$, and $\mu(b_j|\hat{b}_j)$, respectively. Usually we have fuzzy numbers $\bar{c}_{ij}, \bar{a}_{ij}$, and $\bar{b}_j$ which we use for the possibility distribution of $\hat{c}_{ij}, \hat{a}_{ij}$, and $\hat{b}_j$, respectively. Then $\mu(c_{ij}|\hat{c}_{ij}), \mu(a_{ij}|\hat{a}_{ij})$, and $\mu(b_j|\hat{b}_j)$ will be membership functions of fuzzy numbers[2]. Each membership function is non-negative and normalized so that its values are always in the unit interval $[0,1]$. We use the membership functions to find possibilities as follows:

$$Poss[c_{ij} \in E] = sup\{\mu(c|\hat{c}_{ij})|c \in E\}, \tag{3}$$

where *Poss* is an abbreviation of "possibility."

Let C be any co-t-norm and let T be any t-norm, both extended to a finite number of arguments by associativity. Appendix A contains the definitions, and basic properties, of co-t-norms and t-norms. We now form the joint distributions as follows:

$$\Pi(c) = T(\mu(c_{ij}|\hat{c}_{ij}); 1 \leq i \leq K, 1 \leq j \leq n), \tag{4}$$

$$\Pi(A,b) = T(\mu(a_{ij}|\hat{a}_{ij}), \mu(b_j|\hat{b}_j) : 1 \leq i \leq m, 1 \leq j \leq n) \text{ and}, \tag{5}$$

$$\Pi(A,b,c) = T(\mu(a_{ij}|\hat{a}_{ij}), \mu(b_j|\hat{b}_j), \mu(c_{kj}|\hat{c}_{kj}); 1 \leq i \leq m, 1 \leq j \leq n, 1 \leq k \leq K), \tag{6}$$

where T is t-norm. If T = probabilistic AND = product, we obtain joint densities[3] and if T = min, we get joint possibility distributions. For the rest of this paper we will write $P[\ ]$ for either $Prob[\ ]$ or $Poss[\ ]$.

We will model the decision problem of choosing a best $x \geq 0$ as a game against Nature [9]. The actions we may choose are the values of $x \geq 0$ and the resulting states of Nature are the resulting multiobjective linear program given in equation (1). Equation (1) is actually many different multiobjective linear programs each having a different probability (density), or possibility, given by $\Pi(A,b,c)$. In order to completely specify the game against Nature we need to define the pay-off table. That is, we need to give the gain (or loss) incurred if we choose an $x \geq 0$ and the resulting problem has A, b and c (these are the values of the parameters to be used in equation (1)). Actually, the payoff will be a vector since we now adopt three, possibly conflicting, goals of: (1) make $x \geq 0$ feasible; (2) make $x \geq 0$ undominated; and (3) make Z as large as possible. Let $f_1(x) = P[x \text{ feasible }]$, $f_2(x) = P[x \text{ undominated }]$ and $f_3(x) = P[Z \text{ big }]$ for $x \geq 0$. We will study the $f_i(x), 1 \leq i \leq 3$, in more detail in the next section but an optimal x will be any undominated vector for the multiobjective problem

$$\max_{x \geq 0}(f_1(x), f_2(x), f_3(x)). \tag{7}$$

In practice one must usually choose one x in the undominated set but we will not in this paper discuss the selection of an undominated compromise solution.

The next section contains a more detailed discussion of the goals plus a simulation method of approximating the $f_i(x), i = 1,2,3$. An example is presented in the third section and the last section contains a brief summary and our conclusions.

2. Goals.

Let us now discuss the goals of making $x \geq 0$ feasible, having $x \geq 0$ undominated, and getting Z as large as possible in more detail.

2.1 Feasible

Define, for $x \geq 0$,

$$F(x) = \{(A,b)|Ax * b\}. \tag{8}$$

Then

$$f_1(x) = P[x \text{ feasible }] = C\{\Pi(A,b)|(A,b) \in F(x)\}, \tag{9}$$

where C is an "extended" co-t-norm. An extended co-t-norm is a co-t-norm that now has an infinite (countable or uncountable) number of arguments. For example, the extension of $C = max$ is clearly $C = sup$ for supremum. The only other extended co-t-norm we must consider in this paper is the extension of LOR, which is the Lukasiewicz OR. Let $Y = \{y_1, \dots y_n\}$, $0 \leq y_i \leq 1$, and $M = \{1, 2, \dots n\}$ If $B \subset M$ then

$$LOR(B) = min\left(\sum\{y_i | i \in B\}, 1\right). \tag{10}$$

Now let $\bar{M} = \{(A,b)|A \ m \times n \text{ matrix }, b \ m \times 1 \text{ vector }\}$ and $\bar{Y} = \{\Pi(A,b)|(A,b) \in \bar{M}\}$. A natural extension of LOR to this setting is

$$LOR(F(x)) = min(\int_{F(x)} \Pi(A,b)dAdb, 1), \tag{11}$$

which equals

$$LOR(F(x)) = \int_{F(x)} \Pi(A,b)dAdb. \tag{12}$$

Therefore, we obtain $Prob[x \text{ feasible }]$ if $C = LOR$ and $Poss[x \text{ feasible }]$ when $C = sup$. In either case we wish to maximize $f_1(x)$.

In the rest of this paper we will use these two (infinite) extensions of (finite) $C = min$ and $C = LOR$.

2.2 Undominated

Define, for $x \geq 0$

$$U(x) = \{(A,b,c)|x \text{ is feasible and undominated }\}. \tag{13}$$

Given $x \geq 0$, then $(A,b,c) \in U(x)$ if and only if $Ax * b$ and x is undominated (belongs to the Pareto optimal set, is an efficient point, ...). Then

$$f_2(x) = P[x \text{ undominated }] = C\{\Pi(A,b,c)|(A,b,c) \in U(x)\}. \tag{14}$$

Our second goal is to maximize $f_2(x)$. We obtain $Prob[x \text{ undominated }]$, or $Poss[x \text{ undominated }]$, by choosing $C = LOR$, or $C = sup$, respectively. Also, if $C = sup$, then we easily see that $f_2(x) \leq f_1(x)$ since $\Pi(A,b,c) \leq \Pi(A,b)$ for any c and $(A,b,c) \in U(x)$ implies that $(A,b) \in F(x)$.

2.3 Z Big

Let $I = (z_1^*, \dots, z_K^*)$ be some ideal point. For example, z_i^* is usually picked so that it exceeds $c_i x$, for x feasible with respect to all multiobjective linear programs for all reasonable values of A, b and c. That is, if $\Pi(A,b,c) > 0$, then $z_i^* > c_i x$, for all feasible x. Next let $R(I)$ be some subset of R^K containing I. $R(I)$ will be the target region for Z and we will say that Z is "big", or Z is sufficiently large, if $Z \in R(I)$. For example, let $r_i < z_i^*, 1 \leq i \leq K$, then we could choose $R(I) = \{z | z_i \geq r_i, 1 \leq i \leq K\}$. Define, for $x \geq 0$

$$L(x) = \{c | (c_1 x, \ldots, c_K x) \in R(I)\}. \tag{15}$$

Then

$$f_3(x) = P[\text{ Z big }] = C\{\Pi(c) | c \in L(x)\}. \tag{16}$$

Our third goal is to maximize $f_3(x)$.

Obviously, the goals are conflicting since $f_1(x)$ and $f_3(x)$ are independent in that f_1 depends on A and b while f_3 depends only on c.

2.4 Simulation

We now present a simulation (Monte Carlo) method to estimate the $f_i(x), 1 \leq i \leq 3$. This will further explain the similarities, and differences, between stochastic and possibilistic multiobjective programming. The algorithms to estimate $f_1(x), f_2(x)$ and $f_3(x)$ are essentially the same so we will present the technique only for $f_1(x)$.

We may approximate $Prob[x$ feasible $]$ as follows:

(1) $\tau = 0$;
(2) Randomly generate the a_{ij} and b_j according to their probability laws;
(3) Compute Ax;
(4) If $Ax * b$ is not true, then return to (2); otherwise let $\tau = \tau + 1$ and return to (2).

After repeating (2) through (4) N times we have

$$\frac{\tau}{N} \approx Prob[x \text{ feasible }]. \tag{17}$$

To estimate $f_1(x)$ we count all the ways to "do the job," where "do the job" means find A and b so that $Ax * b$.

We can estimate $Poss[x$ feasible $]$ as follows:

(1) $\tau = 0$;
(2) Randomly generate the a_{ij} and b_j according to a uniform distribution over the support[4] of their possibility distribution;
(3) Compute Ax;
(4) If $Ax * b$ is not true, then return to (2); otherwise compute $\Pi(A, b) = \Pi$, set $\tau = max(\tau, \Pi)$ and return to (2).

After repeating (2) through (4) N times we have

$$\tau \approx Poss[x \text{ feasible }]. \tag{18}$$

To estimate $f_1(x)$ we search for the "best" way to "do the job," where "do the job" means find A and b so that $Ax * b$ and "best" means one with largest possibility.

Let us consider another explanation of finding $Poss[x$ feasible $]$. Given values of A and b we call $[A - b]$ a chain. The strength of $[A - b]$ is the strength of its weakest link so set

$$strength[A - b] = \pi(A, b), \tag{19}$$

where $C = min$. Now $F(x)$ contains all chains $[A - b]$ that will "do the job" of making x feasible. Therefore, $Poss[x$ feasible $]$ is the strength of the strongest chain that will "do the job." However, $Prob[x$ feasible $]$ sums up the strength of all chains that will "do the job," where now $strength[A - b] = \pi(A, b)$ with $C =$ product. Since product $\leq min$ we know that the probabilistic strength of a chain is always less than or equal to the possibilistic strength of the chain.

3. Example.

Consider

$$max(z_1 = c_{11}x_1 + x_2, z_2 = x_1 - c_{22}x_2)$$
$$\text{subject to } : a_{11}x_1 + a_{12}x_2 \leq 2, x_i \geq 0. \tag{20}$$

3.1 Possibilistic Programming

Let $\bar{c}_{11}$ and $\bar{c}_{22}$ be triangular fuzzy numbers $(0/1/2)$ and $\bar{a}_{11}$ and $\bar{a}_{12}$ will be triangular fuzzy numbers $(1/2/3)$. The graph of a triangular fuzzy number $(a/b/c)$ is a triangle with base on the interval $[a, c]$ and equal to one at $x = b$. Therefore, $Poss[\hat{c}_{11} = c_{11}] = \mu(c_{11}|\bar{c}_{11}), \ldots, Poss[\hat{a}_{12} = a_{12}] = \mu(a_{12}|\bar{a}_{12})$.

For the first goal we have

$$\Pi(A, b) = min\{\mu(a_{11}|\bar{a}_{11}), \mu(a_{12}|\bar{a}_{12})\}, \tag{21}$$

and

$$F(x_1, x_2) = \{(a_{11}, a_{12})|a_{11}x_1 + a_{12}x_2 \leq 2\}, \tag{22}$$

for $x_i \geq 0$. We find that

$$f_1(x_1, x_2) = \begin{cases} 0 & , \lambda \geq 2, \\ \frac{2}{\lambda} - 1 & , 1 \leq \lambda \leq 2, \\ 1 & , \lambda \leq 1, \end{cases} \tag{23}$$

where $x_1 + x_2 = \lambda, x_i \geq 0$.

We now present an argument that equation (23) is correct but will omit the details of verifying the expressions for f_2 and f_3. Let $Box = [1, 3] \times [1, 3]$ We know that:

(1) $(2, 2) \in F(x_1, x_2)$ if and only if $\lambda \leq 1$;
(2) $F(x_1, x_2) \bigcap Box$ is empty if and only if $\lambda > 2$; and
(3) $F(x_1, x_2) \bigcap Box$ is non-empty but $(2, 2)$ does not belong to $F(x_1, x_2)$ if an only if $1 < \lambda \leq 2$.

We see that $\Pi(A, b) = 1$ at $(a_{11}, a_{12}) = (2, 2)$ so $f_1(x_1, x_2) = 1$ for $\lambda \leq 1$. Also, $f_1(x_1, x_2)$ is zero when $\lambda > 2$ since $Box \bigcap F(x_1, x_2) = \emptyset$. Now suppose $1 < \lambda \leq 2$. Hold (x_1, x_2) fixed so that $x_1 + x_2 = \lambda, x_i \geq 0$ and we see that the line $a_{11}x_1 + a_{12}x_2 = 2$ intersects the line $a_{11} = a_{12}$ (in the $a_{11}a_{12}$-plane) at $(\frac{2}{\lambda}, \frac{2}{\lambda})$. The maximum of $\Pi(A, b)$ on $F(x_1, x_2)$ occurs at $(\frac{2}{\lambda}, \frac{2}{\lambda})$ in the $a_{11}a_{12}$-plane. Then we substitute $\frac{2}{\lambda}$ into the membership function for $\bar{a}_{11}$ (or $\bar{a}_{12}$) and obtain $\frac{2}{\lambda} - 1$. This proves equation (23).

We next obtain an expression for $f_2(x_1, x_2)$. Let

$$\Pi(c) = min\{\mu(c_{11}|\bar{c}_{11}), \mu(c_{22}|\bar{c}_{22})\}, \tag{24}$$

so that

$$\Pi(A, b, c) = min\{\Pi(A, b), \Pi(c)\}. \tag{25}$$

Given values of A, b and c, the undominated set of (x_1, x_2) is

$$\{(x_1, x_2)|a_{11}x_1 + a_{12}x_2 = 2\}, \tag{26}$$

if $c_{11}a_{22} < a_{11}$, and

$$\{(x_1, x_2)|x_1 = 2/a_{11}, x_2 = 0\}, \tag{27}$$

when $c_{11}a_{22} \geq a_{11}$. In both cases we assume that $1 \leq a_{11}, a_{12} \leq 3, 0 \leq c_{11} \leq 2$, and $x_i \geq 0$. Therefore

$$U(x_1,x_2) = \{(a_{11},a_{12},c_{11},c_{22})|a_{11}x_1 + a_{12}x_2 = 2$$
$$\text{if } c_{11}a_{22} < a_{11}, \text{ or } x_1 = 2/a_{11} \text{ and } x_2 = 0 \text{ if } c_{11}a_{22} \geq a_{11}\}. \tag{28}$$

We see that

$$f_2(x_1,x_2) = \begin{cases} 0 & ,\lambda \leq \frac{2}{3} \text{ or } \lambda \geq 2, \\ \frac{2}{\lambda} - 1 & ,1 \leq \lambda \leq 2, \\ 3 - \frac{2}{\lambda} & ,\frac{2}{3} \leq \lambda \leq 1, \end{cases} \tag{29}$$

where $x_1 + x_2 = \lambda, x_i \geq 0$. Actually, we find that the c_{11} and c_{22} do not really enter into the calculations in deriving equation (29) and may be set equal to one so that their membership functions also equal one.

Lastly, we obtain $f_3(x_1,x_2)$. Let $r > 0$ and we choose $R(I) = \{(z_1,z_2)|z_i \geq r, i = 1,2\}$. Then

$$L(x_1,x_2) = \{(c_{11},c_{22})|c_{11}x_1 + x_2 \geq r, x_1 - c_{22}x_2 \geq r\}. \tag{30}$$

The results for $f_3(x_1,x_2)$ are presented in Figure 1.

In order to finish the problem we need to choose a value for $r > 0$. Let $r \in (\frac{2}{3},1)$. We now put our results for f_1, f_2 and f_3 together, as shown in Figure 2, and obtain the undominated $x \geq 0$ to

$$\max_{x\geq 0}(f_1(x), f_2(x), f_3(x)). \tag{31}$$

The undominated set is easily found because there is a region in the x_1x_2-plane where all the f_i are equal to one. The solution is

$$\{(x_1,x_2)|x_1 + x_2 = 1, \frac{(1+r)}{2} \leq x_1 \leq 1,$$
$$0 \leq x_2 \leq \frac{(1-r)}{2}\}, \tag{32}$$

the only part of the x_1x_2-plane where all the $f_i(x_1,x_2) = 1$.

3.2 Stochastic Programming

Let the probability density functions for $\hat{a}_{11} \ldots \hat{c}_{22}$ be the fuzzy numbers used in possibilistic programming since the area under each graph is equal to one. We now have

$$\Pi(c) = \mu(c_{11}|\hat{c}_{11})\mu(c_{22}|\hat{c}_{22}), \tag{33}$$

$$\Pi(A,b) = \mu(a_{11}|\hat{a}_{11})\mu(a_{12}|\hat{a}_{12}), \tag{34}$$

and

$$\Pi(A,b,c) = \Pi(A,b)\Pi(c). \tag{35}$$

To find $f_1(x_1,x_2)$ we integrate (A,b) over the region $F(x_1,x_2)$ given in equation (22). To describe f_1 let $x_2 = \theta x_1, 0 < \theta < +\infty$. Then the graph of f_1, as x_1 grows from zero to $+\infty$, is shown in Figure 3. If $x_1 = \frac{2}{(1+\theta)}$, then $x_1 + x_2 = 2$ and if $x_1 = \frac{2}{3(1+\theta)}$ then $x_1 + x_2 = \frac{2}{3}$. The special cases of: (1) $\theta = 0 (x_2 = 0)$, set $\theta = 0$ in Figure 3; and (2) $\theta = +\infty (x_1 = 0)$, set $\theta = 0$ in Figure 3 and label the horizontal axis x_2.

We next argue that $f_2(x_1,x_2) = 0$ for all $x_i \geq 0$. We obtain f_2 by integrating $\Pi(A,b,c)$ over the set $U(x_1,x_2)$ given in equation (28). However, the dimension of $U(x_1,x_2)$ is less than four, since we must have $a_{11}x_1 + a_{12}x_2 = 2$ or $(x_1,x_2) = (\frac{2}{a_{11}},0)$ so the integral always is zero.

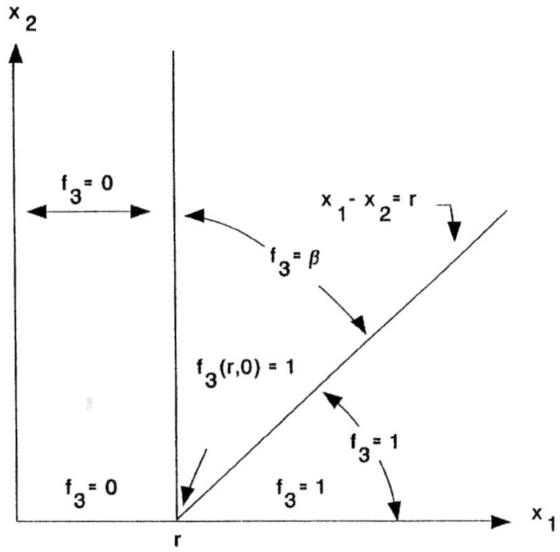

Figure 1
Values of the Third Goal Function for
the Possibilistic Programming Example
($\beta = (x_1 - r)/x_2,\ x_2 > 0$)

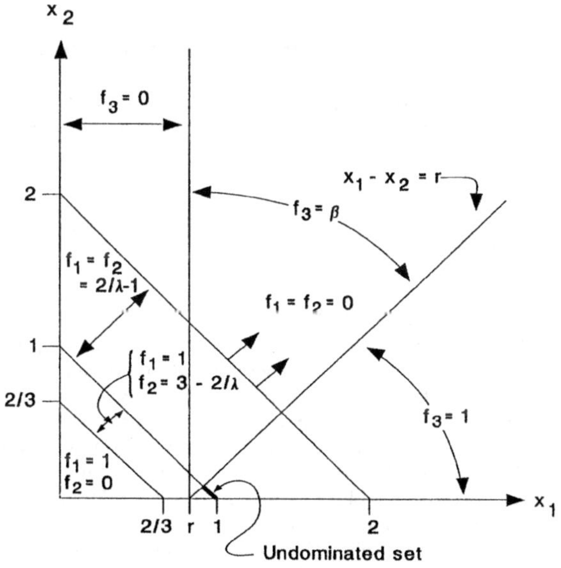

Figure 2
Solution to the Multiobjective
Possibilistic Programming Problem
($\beta = (x_1 - r)/x_2,\ x_2 > 0$ and $\lambda = x_1 + x_2$)

We employ the same region $R(I)$ as in the possibilistic programming and we calculate $f_3(x_1, x_2)$ by integrating $\Pi(c)$ over $L(x_1, x_2)$ given in equation (30). The results are displayed in Figure 4.

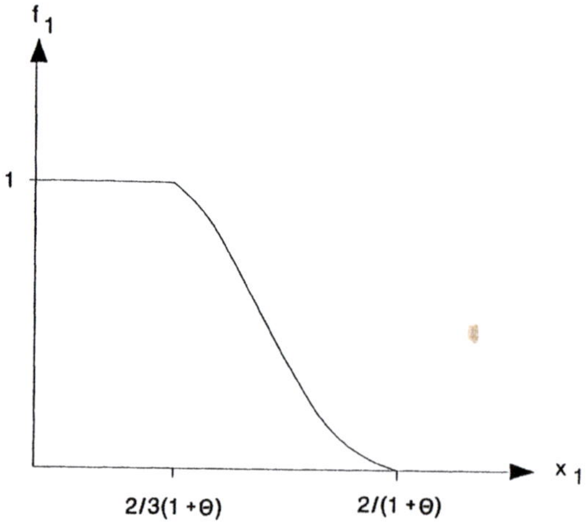

Figure 3
The Graph of the First Goal Function
in the Stochastic Programming Example.

$(x_2 = \theta\, x_1,\ 0 < \theta < +\infty)$

We now choose r in $(\frac{2}{3}, 1)$, put the results for f_1 and f_3 together in Figure 5, and find the undominated $x_i \geq 0$ to the maximization problem in equation (31). The undominated set contains the shaded regions in Figure 5 where $f_1 = 0$, $f_3 = 1$ or $f_1 = 1$, $f_3 = 0$. One reason that the undominated set is so large is that $f_2(x_1, x_2)$ is always zero and our second goal does not effect the solution.

4. Summary and Conclusions.

This paper discusses solutions to multiobjective linear programs where some, or all, of the parameter values may be uncertain. The uncertainty was modeled using either random variables (stochastic programming) or fuzzy variables (possibilistic programming). The decision problem was modeled as a game against Nature. The actions are the values of the decision vector $x \geq 0$ and the states of Nature are the actual multiobjective linear programs. The probability (density), or possibility, of any multiobjective linear program is given by a joint probability density, or a joint possibility distribution. We identified three goals: (1) maximize the probablilty (possibility) of x feasible; (2) maximize the probablilty (possibility) of x undominated; and (3) maximize the probablilty (possibility) that the value of the objective vector is sufficiently large. Our solution to the stochastic (possibilistic) multiobjective linear program was any undominated $x \geq 0$ to the multiobjective game against Nature. A detailed example was presented to illustrate these concepts.

An objective of this paper was to investigate the similarities, and differences, between stochastic and possibilistic multiobjective programming problems. This research contains the author's work

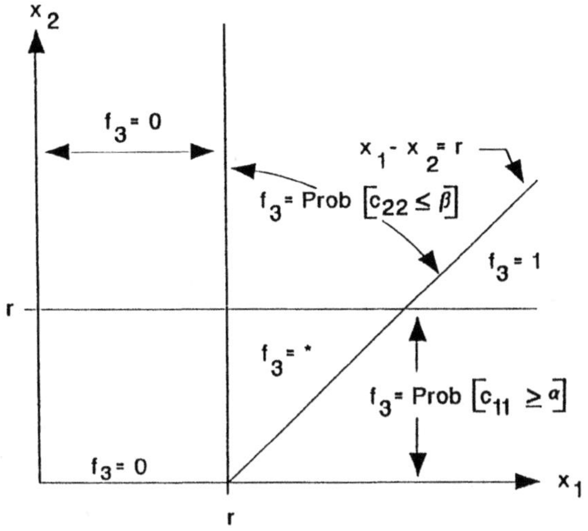

Figure 4
Values of the Third Goal Function in the Stochastic Programming Example

(* = Prob $[c_{22} \leq \beta$ and $c_{11} \geq \alpha]$,
$\alpha = (r - x_2)/x_1$, $\beta = (x_1 - r)/x_2$)

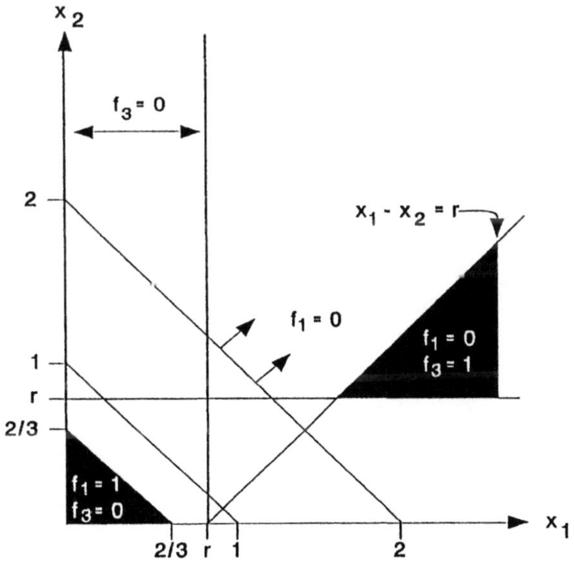

Figure 5
Solution to the Multiobfective Stochastic Programming Problem

on possibilistic programming ([1],[2],[3]) and the comparison between stochastic and possibilistic programming ([4], [5]). We see that the abstract structure of the solution method is the same for stochastic and possibilistic programming because both have as its solution the undominated set to the multiobjective problem

$$\max_{x \geq 0}(f_1(x), f_2(x), f_3(x)) \tag{36}$$

where

$$f_i(x) = P[\text{some event}], \tag{37}$$

for $i = 1, 2, 3$, for P = probability or possibility. However, the solution may be different (see the example in Section 3) since the f_i will take on different values under possibilistic and under stochastic programming.

There are only three basic differences between stochastic and possibilistic programming in this paper. They are:

(1) The μ-functions (Section 1) are normalized differently. In stochastic programming the area under the graph must be one (a probability density function) and in possibilistic programming all the values must lie in [0,1] (a membership function for a fuzzy set).

(2) In forming the joint distribution, stochastic programming employs probabilistic AND (product) for the t-norm but possibilistic programming uses *min* for the t-norm.

(3) In finding P[some event], stochastic programming uses the extended Lukasiewicz *OR* for the co-t-norm and possibilistic programming employs the extended *max* (supremum) for the co-t-norm.

Other similarities, and differences, were discussed with respect to using simulation to estimate the goal functions in Section 2.

We might consider employing other μ-functions, t-norms, and extended co-t-norms and thus produce other solutions to the multiobjective linear program. However, no interpretation, like stochastic and possibilistic programming, now exists for these other types of solutions.

Footnotes.

(1) We formulate the problem for continuous random variables with obvious changes for discrete random variables.
(2) In any case these membership functions are membership functions for fuzzy subsets of the real numbers.
(3) We need to have the μ-functions bounded between zero and one. If some μ-function exceeds one, then we use $(1/M)\mu \leq 1$, for sufficiently large $M > 0$, and then multiply Π by M.
(4) The support of a membership function, say $\mu(c_{ij}|\hat{c}_{ij})$, is the closure of the set of c_{ij}-values where it is positive. We assume that all supports are bounded sets.

Appendix A.

In this Appendix we define t-norms and co-t-norms and extend them to be functions of $n > 2$ arguments. A t-norm T [10] is a function $t : [0,1] \times [0,1] \longrightarrow [0,1]$ satisfying:

(1) $T(a,b) = T(b,a)$;
(2) $T(0,0) = 0, T(a,1) = a$;
(3) $T(a,b) \leq T(c,d)$ if $a \leq c$ and $b \leq d$; and
(4) associativity.

These axioms attempt to capture the basic properties of set intersection. T-norms are considered appropriate functions to model fuzzy set intersection ([7],[8]) and also fuzzy set Cartesian product [6] as in equations (4),(5), and (6) in the text.

The smallest t-norm is

$$T_1(a,b) = \begin{cases} a & , \text{ if } b = 1, \\ b & , \text{ if } a = 1, \\ 0 & , \text{ otherwise.} \end{cases} \tag{1}$$

The basic t-norms are $T_1(a,b)$, LAND$(a,b) = max(a+b-1,0)$, PAND$(a,b) = ab$ and $min(a,b)$ which satisfy

$$T_1(a,b) \leq LAND(a,b) \leq PAND(a,b) \leq min(a,b), \tag{2}$$

since min is the largest t-norm.

All t-norms may be extended, through associativity, to $n > 2$ arguments. The t-norm min is automatically extended and

$$PAND(z_1,\ldots,z_n) = z_1 z_2 \ldots z_n, \tag{3}$$

$$LAND(z_1,\ldots,z_n) = min\left(\sum_{i=1}^{n} z_i - n + 1, 0\right). \tag{4}$$

A co-t-norm C [10] is a function $C : [0,1] \times [0,1] \longrightarrow [0,1]$ satisfying:

(1) $C(a,b) = C(b,a)$;
(2) $C(1,1) = 1, C(a,0) = a$;
(3) $C(a,b) \leq C(c,d)$ if $a \leq c$ and $b \leq d$; and
(4) associativity.

These axioms attempt to capture the basic properties of set union. Co-t-norms are considered appropriate functions to model fuzzy set union ([7],[8]). We also usually consider using a co-t-norm to evaluate statements connected by AND. This is the major reason we employed a C in equation (9), (14), and (16), in the text.

The largest co-t-norm is

$$C_u(a,b) = \begin{cases} a & , \text{ if } b = 0, \\ b & , \text{ if } a = 0, \\ 1 & , \text{ otherwise.} \end{cases} \tag{5}$$

The basic co-t-norms are $max(a,b)$, $POR(a,b) = a + b - ab, LOR(a,b) = min(a+b,1)$ and $C_u(a,b)$ which satisfy

$$max(a,b) \leq POR(a,b) \leq LOR(a,b) \leq C_u(a,b), \tag{6}$$

since max is the smallest co-t-norm.

All co-t−norms may now be extended, through associativity, to $n > 2$ arguments. The co-t-norm max is automatically extended and

$$LOR(z_1, \ldots, z_n) = min\left(\sum_{i=1}^{n} z_i, 1\right), \tag{7}$$

$$POR(z_1, \ldots, z_n) = \sum_{i=1}^{n} z_i - \sum_{i<j} z_i z_j + \ldots (-1)^{n+1} z_1 \ldots z_n. \tag{8}$$

References

1. Buckley, J.J. (1988) 'Possibilistic Linear Programming with Triangular Fuzzy Numbers', Fuzzy Sets and Systems 26, 135-138.
2. Buckley, J.J. (1989) 'Solving Possibilistic Linear Programming Problems,' Fuzzy Sets and Systems 31,329 - 341.
3. Buckley, J.J.(1990) 'Multiobjective Possibilistic Linear Programming', Fuzzy Sets and Systems 35,23 - 28.
4. Buckley, J.J. (1990) 'Stochastic Versus Possibilistic Programming', Fuzzy Sets and Systems 34,173 - 177.
5. Buckley, J.J. 'A General Theory of Uncertainty Based on Co-t-Norms', Fuzzy Sets and Systems. Under review.
6. Dubois, D. and Prade, H. (1981) 'Additions of Interactive Fuzzy Numbers', IEEE Trans. on Automatic Control 26, 926-936.
7. Dubois, D. and Prade, H. (1984) 'Fuzzy-set-theoretic Differences and Inclusions and their Use in the Analysis of Fuzzy Equations,' Control and Cybernatics 13, 129-146.
8. Dubois, D. and Prade, H. (1985) 'A Review of Fuzzy Set Aggregation Connectives', Information Sciences 36, 85-121.
9. Luce, R.D. and Raiffa, H. (1957) 'Games and Decisions', Wiley, N.Y.
10. Schweitzer, B. and Sklar, A. (1963) 'Associative Functions and Abstract Semi-groups', Publ. Math. Debrecen 10, 69-81.
11. Zadeh, L. (1975) The Concept of a Linguistic Variable and its Application to Approximate Reasoning-I', Information Sciences 8, 199-249.
12. Zadeh, L. (1978) 'Fuzzy Sets as a Basis for a Theory of Possibility', Fuzzy Sets and Systems 1, 3-28.

Part IV
Chapter 2

A COMPARISON STUDY OF "STRANGE" AND "FLIP"

R. SLOWINSKI
Technical University of Poznan
Institute of Computer Science
60-965 Poznan
Poland

J. TEGHEM
Faculté Polytechnique de Mons
Rue de Houdain, 9
7000 Mons
Belgium

ABSRACT. The aim of this study is to compare two methods for solving multiobjective linear programming problems under uncertainty on a didactic example. The first method is STRANGE, presented in Part II, Chapter 2. It represents a stochastic approach for modeling uncertain data. The second method, FLIP, presented in Part III, Chapter 3, is based on the application of fuzzy numbers for modeling uncertainty. Both methods are interactive and at each step present to the Decision Maker (DM) a large sample of efficient solutions. A didactic example is first defined and resolved by both methods. Next, every stage of both procedures is analysed and compared on the basis of this example : handling of uncertainty, formulation of associate deterministic multiobjective programming problems, getting the first compromise, the role of the DM in the interactive process, etc. For each of these stages, possible limitations, advantages and inconveniences of both methods are emphasized. General conclusions following from this comparison are finally drawn.

1. INTRODUCTION.

Both authors of the present study become interested in multiobjective mathematical programming under uncertainty through real applications. From one side, Teghem, in collaboration with Kunsch, worked on two energy planning problems coming from a belgian engineering society. The first application (Teghem, Kunsch (1985)) concerned a comparison of two ways of producing electricity in small villages of a country of the Third World : thermodynamic electrosolar system and traditional diesel generators. The schedule of their possible installation was planned within a twenty-year horizon taking into account three objectives : the total production cost, the outside expenses, and the safety of the energy supply. The uncertainty concerned mainly future prices of fuel and solar technology appearing in the objectives, and an availability of fuel and a level of insolation appearing in the constraints. The second application (Kunsch, Teghem (1987)) concerned optimisation of the nuclear fuel cycle in Belgium, including some possibilities of

R. Slowinski and J. Teghem (eds.),
Stochastic versus Fuzzy Approaches to Multiobjective Mathematical Programming under Uncertainty, 365–393.

retreatment. The four objectives taken into account were the production cost of nuclear electricity, the importation volume of natural uranium, the commercial balance, and the employment in the nuclear industry. In the twenty-year planning horizon the price of uranium and the cost of retreatment were uncertain. In both applications, the uncertainty was handled by the definition of scenarios with associate subjective probabilities. In order to solve these problems, Teghem et al. (1986) developed the method called STRANGE (STRategy for Nuclear Generation of Electricity). A technical description of its steps has been made in Chapter II-2 of this book. In the above mentioned applications, the stochastic approach was naturally adopted because of the usual language of scenarios affected by plausibility levels, well known for Decision Makers (DM) and experts supplying data. It also allowed for taking into account a specific criterion of the safety of supply which had been required by managers responsible for the applications (cf. Chapter II-3 in this volume).

On the other side, Slowinski (1986) has developed a method called FLIP (Fuzzy LInear Programming), for solving multiobjective linear programming problems with fuzzy coefficients, in view of its application to a long-term development planning of a water supply and wastewater treatment system. The method has been described in detail in Chapter III-3 of this book. The criteria used for evaluation of development plans were the following : investment and operation costs, environmental pollution and total loss of water users whose demands were not fully satisfied. The constraints of the problem expressed a capacity of water intakes and recycling treatment plants, water demands and a capacity of main pipeline connections in the distribution network. The uncertainty of such data as cost coefficients, maximum capacities of water intakes, water pollution indices, discount factors, and users demands had an epistemological character. Statistical estimation failed because of too big subjectivity, however, the experts accepted expressing data in terms of tolerance intervals with a most-possible value or subinterval and decreasing possibility for other values within the interval. This corresponds exactly to the definition of (flat) fuzzy numbers, so, in this application, the modeling of uncertainty using fuzzy sets was quite natural.

In this way, both authors were confronted with real problems of different practical meaning but of similar mathematical character, i.e. multiobjective linear programming under uncertainty. The methods which they proposed are original and supply the DM with satisfactory solutions basing on different approaches : one stochastic and the other fuzzy. It seemed that a comparison of different aspects of these methods could be particularly worthwhile. This was precisely the goal of the first comparative study undertaken by Slowinski and Teghem (1988). Basing on a common didactic example, the following aspects of both methods have been confronted :

- modeling of uncertainty,
- formulation of an associate deterministic problem,
- getting the first compromise solution and information given to the DM,
- the role of the DM in the interactive process.

The uncertain data for the didactic example were given first in terms of the stochastic approach and then translated into the language of the fuzzy approach. In this study, we make an inverse translation of uncertain data, i.e. from the fuzzy to the stochastic approach.

In the next section, a new didactic example will be defined as a multiobjective linear programming problem with fuzzy coefficients. After translation of the fuzzy coefficients into the language of scenarios, the didactic example will be solved by STRANGE and FLIP (section 3). In section 4, every stage of STRANGE and FLIP will be analysed on the basis of the example. Final conclusions following from the general comparison will be drawn in section 5.

2. DIDACTIC EXAMPLE.

2.1. The multiobjective fuzzy linear programming program.

Using symbols instead of uncertain coefficients, the example can be formulated as follows :

$$\begin{array}{llll}
\text{(P1) maximize} & \tilde{z}_1 = \tilde{c}_{11}\, x_1 + \tilde{c}_{12}\, x_2 & & \\
 & \tilde{z}_2 = \tilde{c}_{21}\, x_1 + \tilde{c}_{22}\, x_2 & & \\
\text{s.t.} & \tilde{a}_{11}\, x_1 + \tilde{a}_{12}\, x_2 \ \tilde{\leqq}\ \tilde{b}_1 & & \text{(a)} \\
 & \tilde{a}_{21}\, x_1 + \tilde{a}_{22}\, x_2 \ \tilde{\leqq}\ \tilde{b}_2 & & \text{(b)} \\
 & x_1 \leqq 5.5 & & \text{(c)} \\
 & x_1,\ x_2 \geqq 0 & &
\end{array}$$

The uncertain coefficients of problem (P1) are defined in terms of fuzzy numbers, i.e. normal convex continuous fuzzy subsets of the real line. In other words, an uncertain coefficient is defined by an interval of possible variation and a possibility distribution within the interval; the highest degree of possibility for a value or subinterval is equal to one, and the lowest, to zero. Possibility distributions of uncertain coefficients are presented in Fig. 1.

Remark : In order to compare the two interactive methods the actions of the DM must be consistent inside each method and consistent from one to the other. To "simulate" actions of the DM going in the same direction, it is useful for our didactic example to fix a priori a target point serving as a common reference in both methods. The target point has been fixed on $x_1 = 3$, $x_2 = 3$.

2.2. Modeling of uncertainty in terms of scenarios.

The ways to define the uncertainty in FLIP and in STRANGE are really different : in the model treated by FLIP, the uncertainty is defined for each coefficient of the MOLP problem separately; in STRANGE, by the use of scenarios, the uncertainty is defined globally for all coefficients of each objective on one hand, and for all the coefficients in the constraints on the other hand. So, it is not an obvious task to translate the uncertainty from one model to the other.

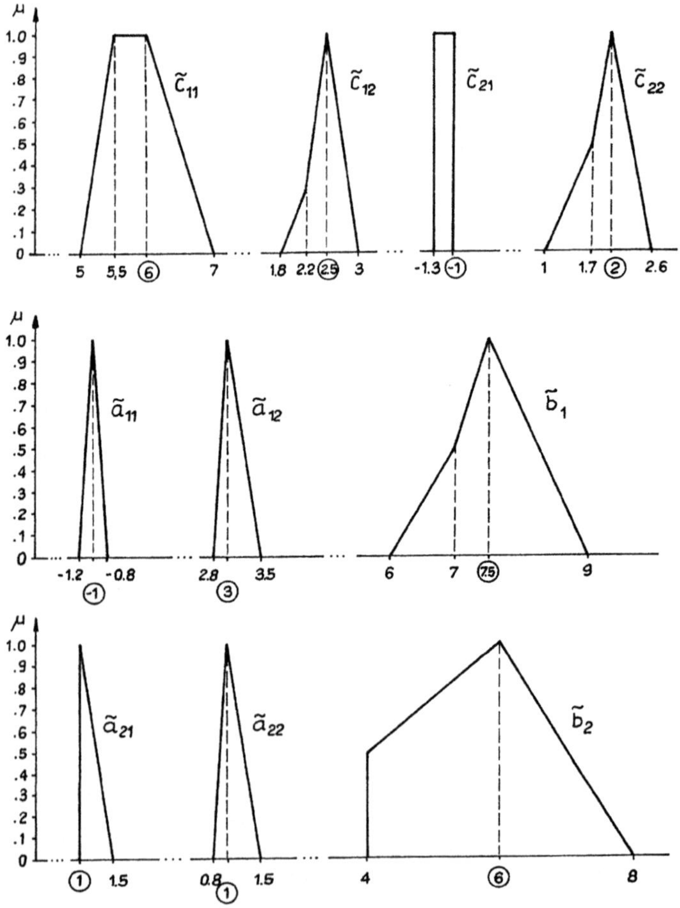

Fig. 1. Possibility distributions of uncertain coefficients.

In the first comparative study undertaken by Slowinski and Teghem (1988), the stochastic approach - with the "global uncertainty" - has been translated into the fuzzy approach - with the "separate uncertainty"; to enable this translation, the global uncertainty was destroyed, i.e. it was not taken into account that, in the definition of scenarios, some coefficients have simultaneous realizations. Of course, the main difficulty of the inverse translation, will be to recreate this global uncertainty.

In the first step, each fuzzy number is represented in terms of a discrete random variable. Each characteristic point of the possibility distribution is considered : the corresponding probability is proportionally adjusted, taking into account that when an extreme value of the fuzzy number has a degree of possibility equal to zero, a very low probability - let us say 0,1 - is associated to this value.

For instance, applying this principle, we obtain, after rounding :

. for the coefficients of the first objective :

$$P\ (c_{11} = 5) = P\ (c_{21} = 7) = 0.1$$

$$P\ (c_{11} = 5.5) = P\ (c_{11} = 6) = 0.4$$

$$P\ (c_{12} = 1.8) = P\ (c_{12} = 3) = 0.1$$

$$P\ (c_{12} = 2.2) = 0.2$$

$$P\ (c_{12} = 2.5) = 0.6$$

. for the coefficients of the first constraint :

$$P\ (a_{11} = -\ 1.2) = P\ (a_{11} = -\ 0.8) = 0.1$$

$$P\ (a_{11} = -\ 1) = 0.8$$

$$P\ (a_{12} = 2.8) = P\ (a_{12} = 3.5) = 0.1$$

$$P\ (a_{12} = 3) = 0.8$$

$$P\ (b_1 = 6) = P\ (b_1 = 9) = 0.1$$

$$P\ (b_1 = 7) = 0.25$$

$$P\ (b_1 = 7{,}5) = 0.55$$

For the definition of scenarios, it is not required and not useful, to consider all the possible combinations of the characteristic values of the coefficients (16 for the first objective; 36 for the first constraint). To apply the STRANGE method, it is more than enough to define three (possibly four or five) scenarios corresponding to two extreme situations, and to one (two or three) middle situations.

Let us consider first the case of an objective, for instance z_1. The extreme realizations are (see Fig. 2) :

$$z_{11} : 5\ x_1 + 3\ x_2$$

$$z_{13} : -7\ x_1 - 1.8\ x_2$$

To define the probabilities associated to the three scenarios z_{1k} (k = 1, 2, 3), the interval of possibles values for the ratio $\frac{c_{12}}{c_{11}}$ is considered :

$$\frac{1.8}{0.7} \approx 0.25 \leq \frac{c_{12}}{c_{11}} \leq 0.6 \quad = \frac{3}{5}$$

and is divided into three subintervals of the same length :

$$L = [0.25,\ 0.366]; \quad M = [0.366,\ 0.484]; \quad U = [0.484,\ 0.6]$$

To each of the 16 possible values of $\frac{c_{12}}{c_{11}}$ there is associated a probability

(for instance, $P\ (\frac{c_{12}}{c_{11}} = \frac{2.5}{5.5} = 0.45) = P\ (c_{11} = 5.5) \cdot P\ (c_{12} = 2.5) = 0.24$)

(an assumption of independance is thus made).

Then, the probabilities associated to z_{11}, z_{12}, z_{13} will be defined as the probabilities

$$P\ (\frac{c_{12}}{c_{11}} \in L),\ P\ (\frac{c_{12}}{c_{11}} \in M),\ P\ (\frac{c_{12}}{c_{11}} \in U), \text{ respectively.}$$

We obtain,

$$P\ (z_1 = z_{11}) = 0.26 \approx 0.25$$

$$P\ (z_1 = z_{12}) = 0.59 \approx 0.6$$

$$P\ (z_1 = z_{13}) = 0.15$$

The most probable realization of z_1,

$$5.75\ x_1 + 2.5\ x_2$$

gives a value of the ratio $\frac{c_{12}}{c_{11}}$ belonging to M; so, we choose this realization for z_{12} (see Fig. 2) (for c_{11}, we take the mean of the two modal values 5.5 and 6).

Similarly for z_2, we obtain (see Fig. 2) :

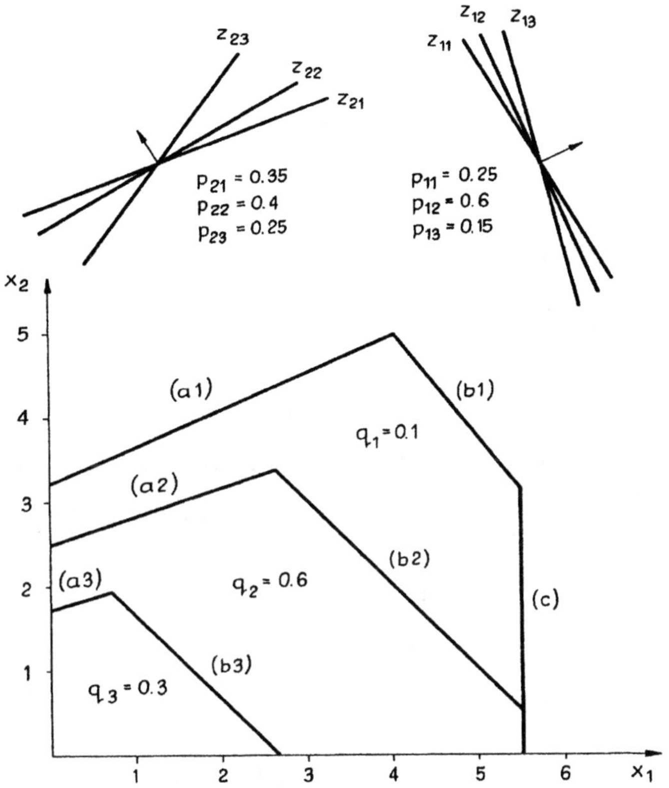

Fig. 2. Modeling of uncertainty in terms of scenarios.

z_{21} : $- x_1 + 2.6\ x_2$ $\qquad p_{21} = 0.375 \approx 0.35$

z_{22} : $- 1.15\ x_1 + 2\ x_2$ $\qquad p_{22} = 0.4$

z_{23} : $- 1.3\ x_1 + x_2$ $\qquad p_{23} = 0.225 \approx 0.25$

Let us consider the constraints.

In the first step, a similar approach is used to define scenarios associated to each constraint; this approach is illustrated for the first constraint (see Fig. 2). The less restrictive realization is obtained for the lowest values of a_{11} and a_{12} and the largest value of b_1 :

$$- 1.2\ x_1 + 2.8\ x_2 \leqq 9$$

and the most restrictive realization is obtained inversely :

$$- 0.8\ x_1 + 3.5\ x_2 \leqq 6$$

Here, the interval of values of the ratio $\dfrac{b_1}{a_{11} + a_{12}}$ is considered :

$$\frac{6}{- 0.8 + 3.5} = 2.222 \leq \frac{b_1}{a_{11} + a_{12}} \leq \frac{9}{- 1.2 + 1.8} = 5.625$$

and again, is divided into three subintervals of the same length.

L = [2.222, 3.555]; M [3.555, 4.49]; U [4.49, 5.625]

The probabilities associated to each interval are defined as before : they are respectively equal to 0.2, 0.7, 0.1 (after rounding). The most probable realization :

$$- x_1 + 3\ x_2 \leqq 7.5$$

gives a value belonging to subset M; so, this realization is chosen as a middle realization.

The first constraint is thus represented by (see Fig. 2) :

$- 1.2\ x_1 + 2.8\ x_2 \leqq 9$	0.1
$- x_1 + 3\ x_2 \leqq 7.5$	0.7
$- 0.8\ x_1 + 3.5\ x_2 \leqq 6$	0.2

Similarly, we obtain for constraint 2 (see Fig. 2) :

$x_1 + 0.8\ x_2 \leqq 8$	0.1
$x_1 + x_2 \leqq 6$	0.5
$1.5\ x_1 + 1.5\ x_2 \leqq 4$	0.4

In STRANGE, the scenarios must be globally defined for the set of constraints : thus, the probabilities q_r (r = 1, 2, 3) associated to the extreme scenarios (r = 1 the less restrictive; r = 3 the most restrictive) and to the middle scenario (r = 2), are defined as the mean value of the corresponding probabilities for each constraint.

We thus obtain (Fig. 2) :

$$r = 1,\ q_r = 0.1 \quad \begin{cases} -1.2\,x_1 + 2.8\,x_2 \leq 9 \\ x_1 + 0.8\,x_2 \leq 8 \end{cases}$$

$$r = 2,\ q_r = 0.6 \quad \begin{cases} -x_1 + 3\,x_2 \leq 7.5 \\ x_1 + x_2 \leq 6 \end{cases}$$

$$r = 3,\ q_r = 0.3 \quad \begin{cases} -0.8\,x_1 + 3.5\,x_2 \leq 6 \\ 1.5\,x_1 + 1.5\,x_2 \leq 4 \end{cases}$$

Remark : If the most probable realization does not correspond to the subinterval M, it can be a reason to introduce more than three scenarios.

3.2. FLIP.

(a) Modeling of uncertainty. Possibility distributions of uncertain coefficients are presented in Fig. 1. In order to read them as L-R type fuzzy numbers, we assume that the encircled values correspond to middle values.
(b) Associate deterministic problem. In order to define fuzzy aspiration levels, we calculate the pay-off table for problem (P1), where the fuzzy coefficients were fixed at their middle values :

$\bar{z}_1$	34.25	6.25
$\bar{z}_2$	-4.5	5.0

The fuzzy aspiration levels have been defined as :

$$\tilde{g}_1 = (34.25,\ 24.25,\ 0)_{RR}$$

$$\tilde{g}_2 = (5,\ 8,\ 0)_{RR}$$

where R is a linear function. The approximate range of variation of Θ_i at level $\eta_i = 0$ (cf. way (a) of controlling safety) is equal to :

$$\Theta_1 \in [0.3,\ 1.5]$$

$$\Theta_2 \in [-1,\ 2]$$

Since reference functions R in fuzzy objectives are linear, according to the principle explained in Chapter III-3, the objective functions of the associate deterministic problem are linear fractional. The associate biobjective linear fractional program has the form :

$$\text{(P2) maximize} \quad f_1 = 1 - \frac{34.25 - 6\, x_1 - 2.5\, x_2}{24.25 + x_1 + 0.5\, x_2}$$

$$f_2 = 1 - \frac{5 + x_1 - 2\, x_2}{8 + 0.6\, x_2}$$

$$\text{s.t.} \quad - x_1 + 3\, x_2 - 7.5 \leqq L^{-1}(\tau_1)\,(0.2\, x_1 + 0.2\, x_2 + 1.5) \qquad \text{(a1)}$$

$$x_1 + x_2 - 6 \leqq L^{-1}(\tau_2)\,(0.2\, x_2 + 2) \qquad \text{(b1)}$$

$$0.8\, x_1 - 3.5\, x_2 + 9 \geqq \Theta_1 \qquad \text{(a2)}$$

$$- 1.5\, x_1 - 1.5\, x_2 + 8 \geqq \Theta_2 \qquad \text{(b2)}$$

$$x_1 \leqq 5.5 \qquad \text{(c)}$$

$$x_1,\, x_2 \geqq 0$$

where L is a linear reference function : left for $\tilde{a}_{11}$, $\tilde{a}_{12}$, $\tilde{a}_{21}$, $\tilde{a}_{22}$, and right for $\tilde{b}_1$, $\tilde{b}_2$. The feasible region of (P2) is presented in Fig. 3 for different values of τ_1, τ_2 and Θ_1, Θ_2. Let us observe that region ABCDE is safe in the optimistic sense, i.e. $\tau_1 \geqq 1$, $\tau_2 \geqq 1$, while region ABFGH is safe in both optimistic and pessimistic sense, i.e. $\tau_1 \geqq 1$, $\tau_2 \geqq 1$ and $\Theta_1 \geqq 0$, $\Theta_2 \geqq 0$.

(c) Search for the first compromise. FLIP starts with $\tau_1 = \tau_2 = 0.8$ and $\Theta_1 = 0.9$, $\Theta_2 = 0.5$ (i.e. a half of each approximate interval of variation). The feasible region AKLM is shown in Fig. 3. Constraints (a1), (b1), (a2), (b2) take the form :

$$- 1.04\, x_1 + 2.96\, x_2 \leqq 7.8 \qquad \text{(a1)'}$$

$$x_1 + 0.96\, x_2 \leqq 6.2 \qquad \text{(b1)'}$$

$$- 0.8\, x_1 + 3.5\, x_2 \leqq 8.1 \qquad \text{(a2)'}$$

$$1.5\, x_1 + 1.5\, x_2 \leqq 7.5 \qquad \text{(b2)'}$$

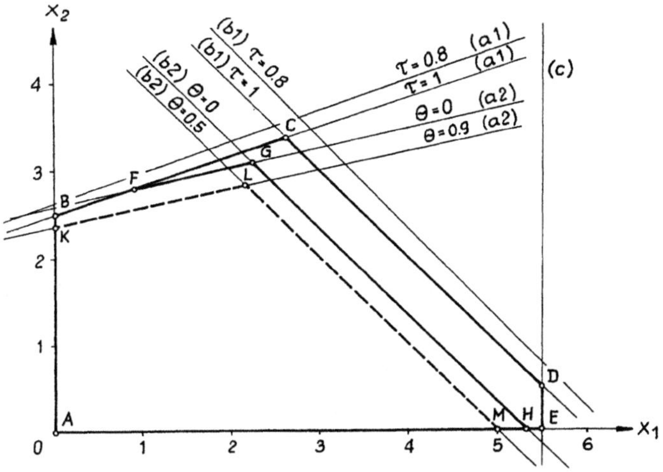

Fig. 3. FLIP-feasible region of the associate deterministic problem.

The optimal solutions of single objective optimization are the following :

for maximization of f_1 : $x_1 = 5$, $x_2 = 0$

for maximization of f_2 : $x_1 = 0$, $x_2 = 2.31$

and the corresponding pay-off table :

f_1	0.85	-0.12
f_2	-0.25	0.96

The ideal and nadir points are equal, respectively, to $\underline{u}^* = (0.85, 0.96)$ and $\underline{u}_* = (-0.12, -0.25)$. The starting point for generation of a sample of efficient points to problem (P2) with given $\tau_1, \tau_2, \Theta_1, \Theta_2$ is $\underline{y}_0 = (0.55, 0.58)$.

The first sample is composed of 8 following points :

No.	1_1	2_1	3_1	4_1	5_1	6_1	7_1	8_1
middle value of $\tilde{z}_1$	22.8	21.6	25.5	16.0	27.8	10.7	30.0	5.8
middle value of $\tilde{z}_2$	1.1	2.2	-1.1	3.8	-3.1	4.2	-5.0	4.6
x_1	2.95	2.61	3.7	1.55	4.38	0.75	5.0	0.0
x_2	2.05	2.39	1.3	2.67	0.62	2.49	0.0	2.31

The encircled point, shown graphically in Fig. 4, has been chosen as a starting point for next iteration. When choosing it, we took into account the target point at $x_1 = 3$, $x_2 = 3$ (cf. Fig. 8).

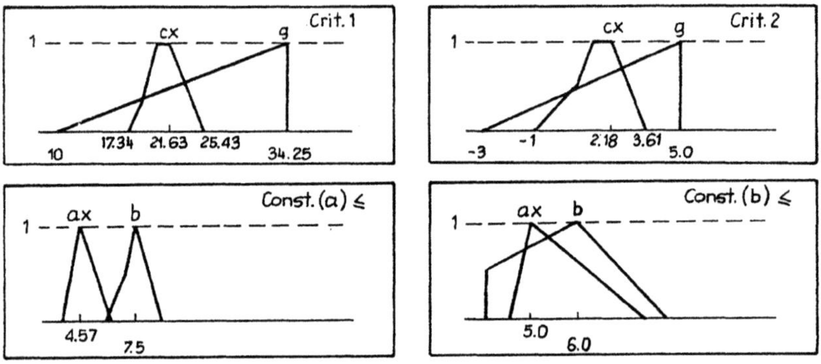

Fig. 4. FLIP-graphical representation of solution 2_1.

After solving 8 linear fractional programs, we obtain 8 new efficient points :

No.	1_2	2_2	3_2	4_2	5_2	6_2	7_2	8_2
middle value of $\tilde{z}_1$	22.1	20.1	22.6	18.6	23.1	17.2	23.6	15.8
middle value of $\tilde{z}_2$	1.7	3.4	1.3	3.6	0.9	3.7	0.5	3.8
x_1	2.75	2.18	2.9	1.95	3.04	1.73	3.17	1.52
x_2	2.25	2.81	2.1	2.76	1.96	2.71	1.83	2.66

Let point 2_2 be the first compromise. Mutual position of fuzzy goals and criteria as well as right- and left-hand-sides of fuzzy constraints in this point, is shown in Fig. 5.

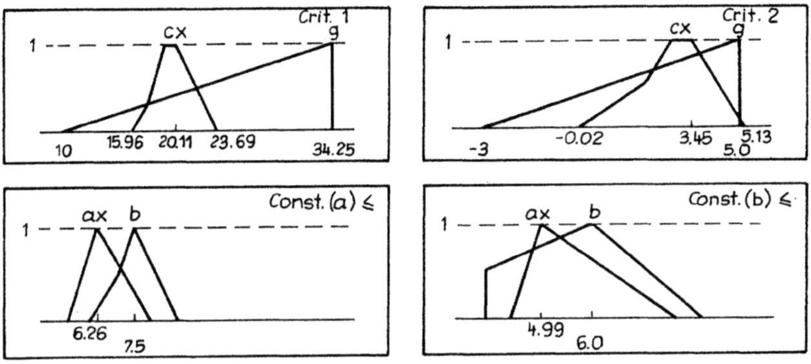

Fig. 5. FLIP-graphical representation of solution 2_2.

(d) Interactive phases. Comparing the first compromise with the target point, the DM arrives at conclusion that the compromise is not satisfactory from the viewpoint of the first criterion. Let us suppose that the DM is ready to improve the value of this criterion at the expense of augmenting risk of violation of the constraints. An analysis of graphical representation of solutions obtained in point (c), indicates that these are Θ_1 and Θ_2 which limit the possibility of improving the score on both criteria. Thus, the DM decides to relax the pessimistic safety, i.e. decrease Θ_1 and Θ_2, increasing simultaneously τ_1 and τ_2, in order to ensure at least a good optimistic safety. The four parameters are fixed on $\tau_1 = 1$, $\Theta_1 = 0$, $\tau_2 = 0.9$, $\Theta_2 = -1$.

These new values change the form of constraints (a1)', (b1)', (a2)' and (b2)' in problem (P2) :

$$- x_1 + 3 x_2 \leqq 7.5 \qquad \text{(a1)''}$$

$$x_1 + 0.98 x_2 \leqq 6.2 \qquad \text{(b1)''}$$

$$- 0.8 x_1 + 3.5 x_2 \leqq 9 \qquad \text{(a2)''}$$

$$1.5 x_1 + 1.5 x_2 \leqq 9 \qquad \text{(b2)''}$$

The corresponding feasible region is shown in Fig. 6. The single objective optimizations lead to the following pay-off table :

f_1	1.0	- 0.1
f_2	-0.14	1.0

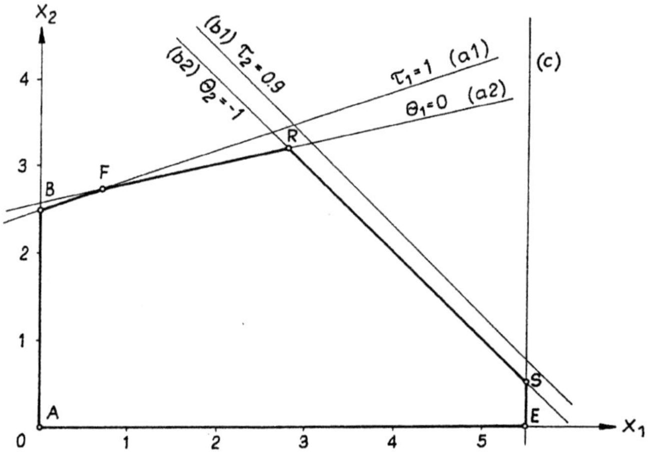

Fig. 6. FLIP-feasible region of the associate deterministic problem.

Starting from $\underset{\sim}{y}_0$ = (0.73, 0.72), the first sample of 8 efficient points has been generated :

No.	$1'_1$	$2'_1$	$3'_1$	$4'_1$	$5'_1$	$6'_1$	$7'_1$	$8'_1$
middle value of $\tilde{z}_1$	26.4	26.3	29.3	19.1	31.9	12.4	34.3	6.3
middle value of $\tilde{z}_2$	2.2	2.3	-0.2	4.1	-2.5	4.6	-4.5	5.0
x_1	3.25	3.24	4.08	1.92	4.82	0.91	5.5	0.0
x_2	2.75	2.76	1.92	3.01	1.18	2.78	0.5	2.5

The DM has selected point $4'_1$ (cf. Fig. 7), so its neighbourhood will be examined in the next iteration. The second sample is the following :

No.	$1'_2$	$2'_2$	$3'_2$	$4'_2$	$5'_2$	$6'_2$	$7'_2$	$8'_2$
middle value of $\tilde{z}_1$	31.5	24.9	29.8	25.6	28.1	26.2	26.4	26.9
middle value of $\tilde{z}_2$	-2.2	3.5	-0.7	2.9	0.8	2.4	2.2	1.8
x_1	4.72	2.84	4.22	3.03	3.73	3.21	3.25	3.39
x_2	1.28	3.16	1.78	2.97	2.27	2.79	2.75	2.61

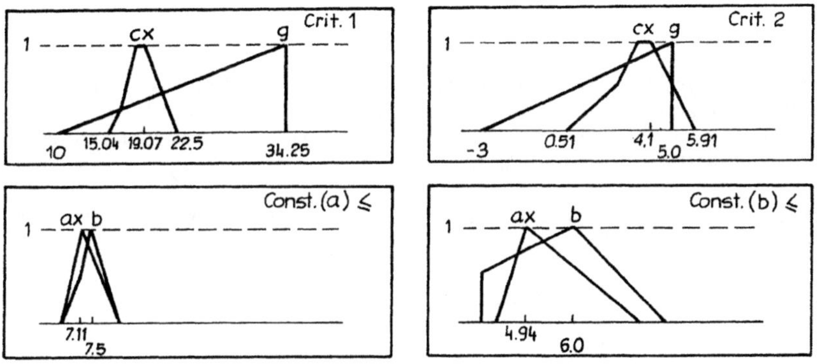

Fig. 7. FLIP-graphical representation of solution $4'_1$.

Point $4'_2$ is almost equal to the target point, so it is supposed to be the best compromise. Its graphical representation is shown in Fig. 8.

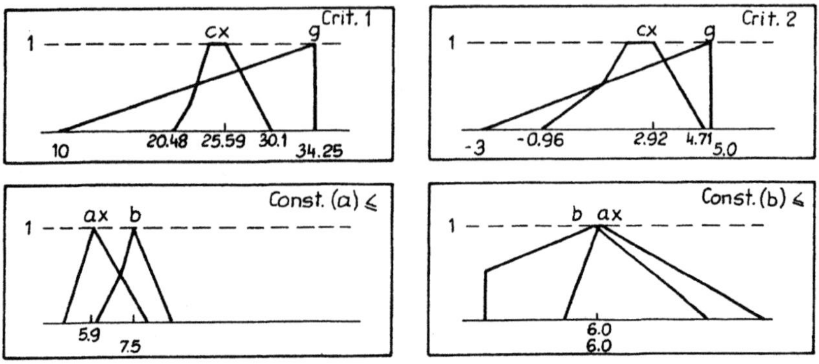

Fig. 8. FLIP-graphical representation of solution $4'_2$ - final compromise.

3.3. STRANGE.

a) Modeling of uncertainty : see 2.2.

b) Associate deterministic problem : Presently, the available software of STRANGE on a micro is designed for minimization of criteria taking non-negative values only. For these technical reasons, constants (50 and 10, respectively) are introduced in criteria z_1 and z_2.

Taking all the penalties equal to 1 in the risk criterion z_3, one obtains the following MOLP problem :

minimize

$$z_{11} = 50 - 5\ x_1 - 3\ x_2$$

$$z_{12} = 50 - 5.75\ x_1 - 2.5\ x_2$$

$$z_{13} = 50 - 7\ x_1 - 1.8\ x_2$$

$$z_{21} = 10 + x_1 - 2.6\ x_2$$

$$z_{22} = 10 + 1.15\ x_1 - 2\ x_2$$

$$z_{23} = 10 + 1.3\ x_1 - x_2$$

$$z_3 = 0.1\ (w_{1a} + w_{1b}) + 0.6\ (w_{2a} + w_{2b}) + 0.3\ (w_{3a} + w_{3b})$$

s.t.

$$- 1.2\, x_1 + 2.8\, x_2 + v_{1a} - w_{1a} = 9 \qquad (a1)$$

$$x_1 + 0.8\, x_2 + v_{1b} - w_{1b} = 8 \qquad (b1)$$

$$- x_1 + 3\, x_2 + v_{2a} - w_{2a} = 7.5 \qquad (a2)$$

$$x_1 + x_2 + v_{2b} - w_{2b} = 6 \qquad (b2)$$

$$- 0.8\, x_1 + 3.5\, x_2 + v_{3a} - w_{3a} = 6 \qquad (a3)$$

$$1.5\, x_1 + 1.5\, x_2 + v_{3b} - w_{3b} = 4 \qquad (b3)$$

$$x_1 \leqq 5.5 \qquad (c)$$

$$x_1, x_2 \geqq 0$$

$$v_{r\delta}, w_{r\delta} \geqq 0 \qquad r = 1,2,3;\ \delta = a,b$$

c) Search for the first compromise : The individual optimizations of the six objectives $z_{11}, \ldots z_{23}$ on the three polyhedra defined by

constraints (a), (b), (c) for r = 1, 2, 3, respectively, are performed. For each objective, among the three optimal solutions, the best one is determined. It is the solution :

$x_1 = 5.5;\ x_2 = 3.12$ for criteria z_{11}, z_{12}, z_{13}

$x_1 = 4.04;\ x_2 = 4.95$ for criterion z_{21}

$x_1 = 0;\ x_2 = 3.21$ for criteria z_{22}, z_{23}

The resulting pay-off table is given below (for the last column, the special technic explained in Chapter II-2 of this book, is used).

z_{11}	13.13	13.13	13.13	14.95	40.36	40.36	36.67
z_{12}	10.56	10.56	10.56	14.39	41.96	41.96	34.67
z_{13}	5.88	5.88	5.88	12.80	44.21	44.21	31.33
z_{21}	7.38	7.38	7.38	1.18	1.64	1.64	5.54
z_{22}	10.07	10.07	10.07	4.76	3.57	3.57	6.57
z_{23}	14.42	14.42	14.42	10.31	6.79	6.79	8.29
z_3	4.42	4.42	4.42	9.04	3.11	3.11	0

Using weights :

$\Pi_{11} = 0.05 \quad \Pi_{12} = 0.05 \quad \Pi_{13} = 0.05 \quad \Pi_{21} = 0.13$

$\Pi_{22} = 0.12 \quad \Pi_{23} = 0.14 \quad \Pi_{31} = 0.45$

the first compromise is obtained :

$x_1 = 3.47; \quad x_2 = 2.54$

$w_{1a} = w_{1b} = w_{2a} = 0; \quad w_{2b} = 0.1; \quad w_{3a} = 0.11; \quad w_{3b} = 5.01.$

Together with this solutions, the DM receives the following information :

z_{ks_k}	$z_{ks_k}^{(1)}$	$[M_{ks_k}, m_{ks_k}]$	$z'^{(1)}_{ks_k}$
z_{11}	25.04	[13.13, 40.36]	0.44
z_{12}	23.71	[10.56, 41.96]	0.42
z_{13}	21.15	[5.88, 44.21]	0.40
z_{21}	6.87	[1.18, 7.38]	0.92
z_{22}	8.91	[3.57, 10.07]	0.82
z_{23}	11.94	[6.79, 14.02]	0.72
z_3	1.54	[0 , 9.04]	0.17

z_k	$\bar{z}_k^{(1)}$
z_1	23.27
z_2	8.96

d) <u>Interactive phases</u> : Let us notice that the target point ($x_1=x_2=3$) corresponds to the following information :

z_{ks_k}	z_{ks_k}	z'_{ks_k}
z_{11}	26	0.47
z_{12}	25.25	0.47
z_{13}	23.6	0.46
z_{21}	5.2	0.65
z_{22}	7.45	0.60
z_{23}	10.9	0.57
z_3	2.13	0.23

z_k	$\bar{z}_k$
z_1	25.19
z_2	7.52

First interactive phase.

After comparison of the first compromise with the target point, it was decided to relax objective z_3 till the upper value m_3. The results of the parametric analysis are the following :

Absolute values of the objectives.

λ	z_{11}	z_{12}	z_{13}	z_{21}	z_{22}	z_{23}	z_3
0.40	25.04	23.71	21.15	6.87	8.91	11.97	1.54
0.50	27.93	27.00	25.12	6.38	8.33	11.27	1.38
0.72	32.90	33.22	33.34	3.71	5.77	9.05	1.89
1	40.60	42.11	44.21	2.05	3.90	6.98	2.62

Relative values of the objectives (see Fig. 9).

λ	z_{11}	z_{12}	z_{13}	z_{21}	z_{22}	z_{23}	z_3
0.40	0.44	0.42	0.40	0.92	0.82	0.72	0.17
0.50	0.54	0.52	0.50	0.84	0.73	0.62	0.15
0.72	0.73	0.72	0.72	0.41	0.34	0.31	0.21
1	1	1	1	0.14	0.05	0.03	0.29

Variables.

λ	x_1	x_2
0.40	3.47	2.54
0.50	2.91	2.51
0.72	1.6	3.03
1	0.04	3.07

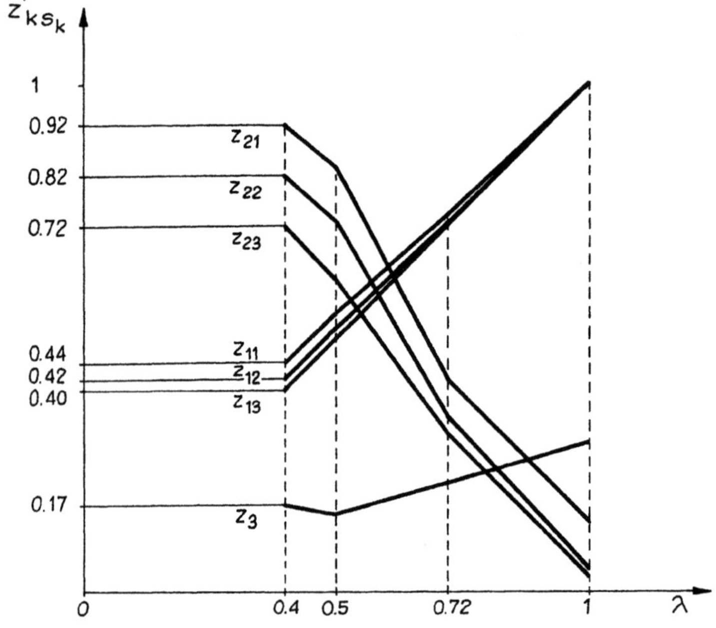

Fig. 9. Parametric analysis of the first interactive phase : Relaxation of z_{13}.

The values of λ are those corresponding to the change of basis, i.e. to bounds of stability intervals (see Fig. 9). After the comparison of different intermediate values of relaxation, it was decided to keep the value λ = 0.47 corresponding to the second compromise :

$x_1 = 3.08;\quad x_2 = 2.52$

$w_{1a} = w_{1b} = w_{2a} = w_{2b} = 0;\qquad w_{3a} = 0.35;\qquad w_{3b} = 0.40.$

z_{ks_k}	$z^{(2)}_{ks_k}$	$z'^{(2)}_{ks_k}$
z_{11}	27.03	0.51
z_{12}	25.98	0.49
z_{13}	23.89	0.47
z_{21}	6.53	0.86
z_{22}	8.51	0.76
z_{23}	11.49	0.65
z_3	1.43	0.16

z_k	$\bar{z}^{(2)}_k$
1	25.62
2	8.56

Second interactive phase.

It is decided to relax objective z_3 till the upper value 2.5. The results of this relaxation are the following :

Absolute values of the objectives.

λ	z_{11}	z_{12}	z_{13}	z_{21}	z_{22}	z_{23}	z_3
0.16	27.03	25.98	23.89	6.53	8.51	11.49	1.43
0.20	26.60	25.73	23.89	5.81	7.93	11.15	1.79
0.22	25.84	24.99	23.19	5.48	7.70	11.08	2.03
0.28	24.86	24.05	22.28	5.06	7.40	10.99	2.50

Relative values of the objectives.

λ	z_{11}	z_{12}	z_{13}	z_{21}	z_{22}	z_{23}	z_3
0.16	0.51	0.49	0.47	0.86	0.76	0.65	0.16
0.20	0.49	0.48	0.47	0.75	0.67	0.60	0.20
0.22	0.47	0.46	0.45	0.69	0.63	0.59	0.22
0.28	0.43	0.43	0.43	0.63	0.59	0.58	0.28

Variables.

λ	x_1	x_2
0.16	3.08	2.52
0.20	3.02	2.77
0.22	3.08	2.92
0.28	3.16	3.11

The choice of $\lambda = 0.24$ is made, corresponding to the third compromise :

$x_1 = 3.10;\quad x_2 = 2.98$

$w_{1a} = w_{1b} = w_{2a} = w_{2b} = 0;\qquad w_{3a} = 1.94;\qquad w_{3b} = 5.12$

z_{ks_k}	$z_{ks_k}^{(3)}$	$z_{ks_k}^{'(3)}$
z_{11}	25.55	0.46
z_{12}	24.71	0.45
z_{13}	22.92	0.44
z_{21}	5.36	0.67
z_{22}	7.61	0.62
z_{23}	11.05	0.59
z_3	2.17	0.24

z_k	$\bar{z}_k^{(3)}$
z_1	24.39
z_2	7.68

Third interactive phase.

The compromise can be improved by relaxation of z_{12}, for instance till the upper value 26. The following results are obtained :

Absolute values of the objectives.

λ	z_{11}	z_{12}	z_{13}	z_{21}	z_{22}	z_{23}	z_3
0.45	25.55	24.71	22.92	5.36	7.61	11.05	2.17
0.47	26.06	25.35	23.76	5.09	7.35	10.83	2.17
0.47	26.15	25.46	23.89	5.06	7.32	10.80	2.17
0.49	27.06	26.00	23.89	6.58	8.54	11.51	2.17

Relative values of the objectives.

λ	z_{11}	z_{12}	z_{13}	z_{21}	z_{22}	z_{23}	z_3
0.45	0.46	0.45	0.44	0.67	0.62	0.59	0.24
0.47	0.48	0.47	0.47	0.63	0.58	0.56	0.24
0.47	0.48	0.47	0.47	0.63	0.58	0.55	0.24
0.49	0.51	0.49	0.47	0.87	0.76	0.65	0.24

Variables.

λ	x_1	x_2
0.45	3.10	2.98
0.47	2.97	3.03
0.47	2.95	3.03
0.49	3.09	2.50

The first value 0.47 corresponds to a point close to the target point. It is chosen as the new compromise.

Fourth interactive phase.

This compromise can still be improved, relaxing criterion z_{21}. The level of relaxation $\lambda = 0.65$ corresponds to a point chosen as final compromise :

$x_1 = 2.99; \quad x_2 = 2.99$

$w_{1a} = w_{1b} = w_{2a} = 0; \quad w_{2b} = 0.09; \quad w_{3a} = 2.08; \quad w_{3b} = 4.97$

z_{ks_k}	$z_{ks_k}^{(5)}$	$z_{ks_k}^{'(5)}$
z_{11}	26.09	0.48
z_{12}	25.35	0.47
z_{13}	23.71	0.47
z_{21}	5.21	0.65
z_{22}	7.45	0.60
z_{23}	10.89	0.57
z_3	2.17	0.24

z_k	$\bar{z}_k^{(5)}$
z_1	25.05
z_2	7.52

4. COMPARISON OF "STRANGE" AND "FLIP".

a) Modeling of Uncertain Data.

1. The input data accepted by STRANGE use a simple and comprehensive language of scenarios well known in the fields of planning and forecasting; they are presented in a discrete way as a set of possible realizations. These scenarios are defined globally, for each objective on one hand (S_k, $\forall k$) and for the set of constraints on the other hand (R). In practice, they are supplied directly by the experts estimating some extreme scenarios and a few intermediate situations.

The data accepted by FLIP are presented in a continuous way using the language of fuzzy numbers for each uncertain coefficient separately. Fuzzy numbers have the advantage of less summary representation, however, they require a more important mathematical treatment, internal to the method.

2. Once the principle of discrete stochastic data is accepted, the only possible limitation is the number of realisations defined for imprecise coefficients. By increasing the number of scenarios one augments the number of objectives to be taken into account in the associate

deterministic problem; on the other hand, an increase of the number of outcomes of imprecise constraints implies an increase of the number of LP problems to be solved in order to set up the pay-off table. Nevertheless, it appears in practice that three- at most five- scenarios for each objective and for the constraints are sufficient.

Fuzzy numbers assume a continuous representation of uncertainty and do not imply any variation of the number of objectives; each uncertain constraint is always translated into two deterministic constraints. A limitation appears, however, at the level of fuzzy numbers admitted; it is clear that if reference functions - right of an objective and left of a goal (in the case of maximization) - are not both linear, one obtains an associate deterministic problem with nonlinear criteria. This limitation does not concern fuzzy constraints - even if all reference functions of left- and right-hand-side coefficients are nonlinear, constraints of the deterministic problem remain all linear.

3. Let us remark that the possibility distributions of uncertain coefficients shoud be unimodal, otherwise they could not be treated as L-R fuzzy numbers in the sense of the definition given in Chapter III-3. Subjective probabilities in STRANGE can have multimodal distribution.

4. Let us stress finally that independently of the approach, the modeling of uncertain data always involves some arbitrariness. In STRANGE, it is expressed by the choice of scenarios and the associate subjective probabilities, and in FLIP, by the definition of fuzzy numbers, i.e. intervals of variation and possibility distributions within the intervals. So, in the case of real applications, it is convenient to repeat calculations with slightly modified data in order to analyse the sensitivity of the obtained solution.

b) Associate Deterministic Problem.

1. In STRANGE, the DM does not intervene in the formulation of the associate deterministic problem; objective z_{ks} follows directly from criterion z_k for scenario s_k.

In FLIP, this formulation uses aspiration levels which, if possible, should be defined by the DM. This additional information, however, has no influence on the shape of the set of efficient points of the associate problem. Objectives f_k express the degree of proximity of objectives $\tilde{z}_k$ to fuzzy aspiration levels $\tilde{g}_k$ $(k = 1,\ldots, K)$.

2. The major difference between the methods is the measurement of risk of not satisfying the uncertain constraints. In STRANGE, this measurement is made globally for all uncertain constraints using the risk criterion (z_{K+1}). This criterion is easily defined using the probabilities (q_r) associated to each outcome of the constraints and the corresponding variables (W_r) expressing the violation of this outcome. The DM is then informed about the risk through the values of this criterion compared with the interval of its variation. This risk criterion is useful not only mathematically but also practically, as appeared in the course of recent applications of STRANGE (see Chapter

II-2). Nevertheless, if the DM wants to analyse the risk of violation of each uncertain constraint separately, the method STRANGE can easily be adapted by the introduction of several risk criteria, one per each uncertain constraint (e.g. $z_{K+1,a}$ with variables w_{ra} $(r = 1,2,3)$ and $z_{K+1,b}$ with variables w_{rb} $(r = 1,2,3)$). Let us remark that the subjective probabilities (p_{ks_k}) associated to the scenarios for the objectives do not appear directly in the formulation of the equivalent deterministic problem. These probabilities are just used as weights in the definition of the distance from the ideal point in the search for the compromises; thus they really play a minor role in the method.

In FLIP, the risk (or safety, in complementary terms) is considered separately for each uncertain constraint. The couple of parameters (τ_i, Θ_i) is used to transform each uncertain constraint i into two deterministic constraints $(i = 1,\ldots,m)$; the first (e.g.(a1)) is responsible for an "optimistic" and the second (e.g.(a2)) for a "pessimistic" safety of solutions. The value of τ_i may vary between 0 and 1 and it may be compared to a confidence level in stochastic analysis. The interval of variation of Θ_i is defined approximately basing on the set of solutions obtained during calculation of the pay-off table for the problem with fuzzy coefficients fixed at the middle values. A great importance for evaluation of the risk has the graphical representation of solutions - it will be discussed in point (c).

3. Finally, the character of deterministic problems is different. STRANGE leads to a classical multiobjective LP problem and FLIP to a multiobjective linear fractional programming problem (if R are linear in objectives).

c) Search for the First Compromise.

1. STRANGE automatically gives the first compromise as a solution of the minmax problem; the DM does not take part in this stage. In FLIP, however, the DM intervenes at two levels of this stage. First, when fixing τ_i and Θ_i and then in the course of the guided generation and evaluation of successive samples of efficient points of the associate deterministic problem. Of course, the DM can accept some "standard" parameters τ_i, Θ_i and come to the first compromise automatically, but in general there is no reason to delay formally an active participation of the DM in the process of searching for the best compromise. Thus, in FLIP, the first compromise is in general a result of the first interactive phase, whereas in STRANGE it is a result of a procedure which is completely independent of the DM.

2. In STRANGE, the information given to the DM with the first (and the others) compromise is composed of absolute values obtained by particular criteria for all scenarios and a value of the measure of risk, together with respective intervals of variation. The relative values within these

intervals make easier the comparison between compromises by the DM. Additional information is composed of mean values for each criterion and possibly of confidence levels corresponding to them.

FLIP informs the DM in two complementary ways. Firstly, for each generated solution the score of each fuzzy objective is displayed numerically as a middle value. This permits a quick comparison of these solutions and elimination of a large part of them from further considerations. Secondly, this information is completed graphically, showing how the values of fuzzy criteria are placed in relation to fuzzy goals, and how the left-hand-sides of the fuzzy constraints are placed in relation to the right-hand-sides for given solutions. These graphics enable an analysis of the compromise among criteria on the one hand, and an evaluation of the safety of the corresponding solution on the other hand. Thus, the total information submitted to the DM is numerical and graphical. The possibility of comparing fuzzy numbers not only on the basis of one or two numerals but taking into account the visual representation of these numbers in a common system of coordinates seems to be particularly beneficial.

3. The problem of dispersion (following from uncertainty) of the values of objectives is closely related to the point before. In STRANGE, this dispersion is perceived by comparison of different values corresponding to criterion k (mean value, values for different scenarios) and possibly, by the confidence level indicating the risk that the mean value is not attained.

In FLIP, the dispersion is analysed mainly on graphics. They show fuzzy numbers corresponding to criteria and constraints, with all dispersion produced by uncertainty.

d) Interactive Phases.

1. In both methods, the DM is asked twice to intervene in each interactive phase. In STRANGE, the DM shows direction which should be examined for possible improvement of the last compromise : he indicates an objective $(ks_k)^*$ which he is ready to relax in order to gain on other criteria. If possible, he also fixes the maximum level of this relaxation. Let us notice that the method STRANGE can easily be adapted to the improvement of a criterion indicated by the DM, as made in STRANGE MOMIX (see Chapter II-2).

In FLIP, this first intervention leads to new values of the safety coefficients τ_i and Θ_i taking into account their intervals of variation and the knowledge acquired in preceding iterations about the dependency between safety and the quality of the compromise among criteria. The second intervention takes place at the level of generating successive samples of efficient points to the associate deterministic problem; the DM then indicates an efficient point whose neighbourhood should be examined next.

2. In consequence of these indications, a calculation phase is performed. In STRANGE, a parametric analysis is carried out which gives

for each value of the relaxed criterion, between its present value and the upper acceptable value indicated by the DM, a minmax efficient solution.

In FLIP, the calculation phase consists in generating a sample of efficient points distributed in an indicated area.

3. So, in different ways, both methods propose a large sample of new possible compromise solutions from an area indicated by the DM. In STRANGE, as well as in FLIP, the choice of the DM is facilitated by an extensive use of graphics. In particular, the variation of the dispersion of the values taken by an uncertain objective can be viewed graphically. In both methods, there exists a phenomenon of zooming which consists in concentration on an efficient area of greatest interest (in STRANGE, the choice of λ's, in FLIP, the choice of a new starting point).

4. Finally, let us examine the possibility of retraction in the interactive phases of both methods. In STRANGE, two versions are proposed to the user (see Chapter II-2) :

- in the first, it is supposed that criteria which have been relaxed, will not be tightened again; so, the choice of the DM corresponds to the maximum level of relaxation for this criterion. This version presents the advantage of a kind of convergence; moreover, the software is designed such that, if the DM regrets one of his preceding choices, he always can initialise the procedure again from the beginning of the corresponding interactive phase;
- in the second, it is always possible to obtain any value of a criterion relaxed before.

In FLIP, the possibility of retraction is quite natural and does not introduce any complication. The DM may withdraw his opinion at both levels of interaction; in particular, at the second level, when he finds that a neighbourhood of an indicated point is less interesting than he expected, he may indicate for the next iteration the point refused in any preceding iteration.

5. CONCLUSIONS.

This comparison shows that the choice of one or another approach depends on the following points :

a) The Type of Information at One's Disposal.

STRANGE requires information which is sufficiently precise to define different scenarios with associate subjective probabilities. FLIP requires a less summary representation (i.e., continuous representation) of uncertainty.

b) The Way in which the DM Perceives and Defines the Imprecision of Data.

STRANGE is particularly suitable for applications in which the language of scenarios is easily accepted by the DM; FLIP is addressed to cases where the uncertainty is expressed through an imprecise definition of particular coefficients.

c) The Personality of the DM Collaborating with the Analyst.

Using a classical language of scenarios, STRANGE does not require as much explanation to the DM as FLIP does, about the concept of fuzzy numbers.

d) The Information which the DM Wishes to Obtain.

A major difference between the methods is the definition of risk (or safety); STRANGE introduces an extra criterion which is a global measure of the risk of nonfeasibility; the idea about dispersion of criteria is given by comparison of the values obtained by extreme scenarios. In FLIP, the risk is conditioned by parameters which appear in the associate deterministic problem; the graphical representation of fuzzy criteria and fuzzy constraints gives the DM a complete idea about the dispersion of criteria for each compromise and the safety of corresponding solutions. Neither method imposes the new compromise to the DM; instead, both propose a number of possibilities.

References.

. KUNCH P. and TEGHEM Jr. J. (1987), "Nuclear Fuel Cycle Optimization Using Multiobjective Stochastic Linear Programming", European Journal of Operational Research, 31, 240-249.
. SLOWINSKI R. (1986), "A Multicriteria Fuzzy Linear Programming Method for Water Supply System Development Planning", Fuzzy Sets and Systems, 19, 217-237.
. SLOWINSKI R., TEGHEM Jr. J. (1988), "Fuzzy Versus Stochastic Approaches to Multicriteria Linear Programming under Uncertainty", Naval Research Logistics, 35, 673-695.
. TEGHEM Jr. J. and KUNSCH P. (1985), "Multiobjective Decision Making under Uncertainty; an Example for Power System", in Y. Haimes and V. Changkong, Eds., Decision Making with Multiple Objectives, Springer-Verlag, Berlin, pp. 443-456.
. TEGHEM Jr. J., DUFRASNE D., THAUVOYE M. and KUNSCH P. (1986), "STRANGE - An Interactive Method for Multiobjective Linear Programming under Uncertainty", European Journal of Operational Research, 26 (1), 65-82.

Part IV
Chapter 3

MULTIOBJECTIVE MATHEMATICAL PROGRAMMING WITH INEXACT DATA

I.M.STANCU-MINASIAN
Laboratories of the Department of
Economic Cybernetics
Calea Dorobanţi 15-17
71131 Bucharest, Romania

ST.TIGAN
Territorial Computing Center
Str.Republicii 107
3400 Cluj-Napoca, Romania

ABSTRACT. The our purpose is to present two approaches for the inexact multiobjective programming with inexactness in the criteria, which correspond to the conservative and nonconservative ways for mathematical programming with set coefficients.

1. INTRODUCTION

In the classical problems of mathematical programming generally, and in the linear programming problems in particular, the coefficients of the problems are assumed to be exactly known. This assumption is not satisfied by great majority of real problems.
Usually, the coefficients of a programming problem are either subject to errors of measurement or they vary with market conditions. There are various situations in which the coefficients are known only to lie in some sets described by exact functional relations. The introduction of imprecision and uncertainty in the modelling process is an important issue of the approaching real practical problems.

A very employed approach to modelize the uncertainty is offered by the probability theory. But the probabilistic analysis, proper for phenomena which are reproducible and which already occured a sufficient number of times in the past, assumes the possibility of representation of some parameter by means of a random variable. Therefore, a probabilistic approach requires the existence of sufficient statistical data which provide information about the distribution func-

R. Slowinski and J. Teghem (eds.),
Stochastic versus Fuzzy Approaches to Multiobjective Mathematical Programming under Uncertainty, 395–418.

tions of the random variables occuring in the mathematical model.

Sensitivity analysis and stability theory for mathematical programming offer qualitative informations about the behaviour of the optimal solutions and the optimal value on the parameter perturbations.

The modelling of the data imprecision can be made also by means of the fuzzy set theory and some applications of fuzzy programming to vector maximum problem are obtained (see,/21/, /48/).

But the previously presented cases can not cover all the real situations. So, it was possible the appearance of a new chapter of mathematical programming: Inexact Mathematical Programming.

This approach assumes that the problem data are not exactly known, but only to belong to some given nonvoid sets. In such mathematical programming problems the usual inequality or equality constraints are replaced by more general constraints of inclusion type.

The first papers in the case of linear programming with set coefficients and a single objective function are due to Dantzig /9/ and Soyster /31/. Consequently, this field was developed by Soyster /32/, /33/, /34/, Falk /12/, Singh/30/, Thuente /40/, Tigan /42/, Kaur /17/, Kaul, Kaur and Lyall /16/, Bakan and Ahmetzanova /1/ among others (see references in the above papers). The multiobjective case was studied by Bitran /4/, Soyster, Lev and Di Toof /34/, Lyall /22/, Peteanu and Tigan /25/.

The our purpose is to present two approaches for the inexact multiobjective programming with inexactness in the criteria, which correspond to the conservative (Soyster /31/) and nonconservative (Dantzig /9/) ways for mathematical programming with set coefficients.

2. MAIN RESULTS ON MATHEMATICAL PROGRAMMING WITH SET COEFFICIENTS IN THE CONSTRAINTS

Dantzig and Wolfe introduced the idea of generalized linear programming (see /9/), defined by:

(GLP). Maximize

$$c_1x_1 + c_2x_2 + \dots + c_nx_n \tag{2.1}$$

subject to:

$$a_1x_1 + a_2x_2 + \dots + a_nx_n \leqslant b, \ x \geqslant 0, \tag{2.2}$$

$$a_j \in K_j \subset R^m, \ j \in J = \{1, 2, \dots, n\}, \tag{2.3}$$

where $K_j (j \in J)$ are some given sets which can be described by exact functional relations and $c = (c_1, \dots, c_n) \in R^n$, $b \in R^m$ are given vectors.

In the generalized linear programming problems one has to be determined some vectors $a_j \in K_j (j \in J)$ and a certain vector x, which maximizes the function cx on the solution set of the constraints (2.2) and (2.3). In other words, in problem (GLP) the vectors a_j $(j \in J)$ are decision variables as x_j.

It was shown /9/ that simplex method could be still used, after some modifications, to solve this problem and some situations in applications are also discussed.

The idea of generalized programming was extended by Chadha /5/, Chadha and Kaul /6/ and Wadhwa/47/ for the linear-fractional programming. Kaul and Bhatia /15/ proved that the decomposition method for linear-fractional programming can also be used for such problems. Patkar et al./24/ extended the decomposition approach further to the problems having as objective a sum of linear and linear-fractional functionals.

Now, we replace in the problem (2.1)-(2.3) the constraints (2.2) and (2.3) by a set inclusive constraint. More precisely, we shall suppose that the vectors $a_j (j \in J)$ are not constant but they belong to some nonempty convex sets K_j in R^m. Also, we assume that the vector b belongs to a nonempty convex set K in R^m. Thus, we consider the following problem:

(I) Find $\sup \{c_1 x_1 + \ldots + c_n x_n \mid x_1 K_1 + \ldots + x_n K_n \subseteq K, \; x \geqslant 0\}$

We mention that for any A, B subsets of R^m and $a \in R$, we denoted: $A + B = \{z : z = x + y, \; x \in A, \; y \in B\}$, $aA = \{z : z = ax, \; x \in A\}$.

The problem (I) is called by Soyster /31/ "linear programming with set-inclusive constraints". Analogously to usual case, we call K_j as activity sets and K as the resource set. We denote by S the set of the feasible solutions of the problem (I).

Let $b \in R^m$ and $K(b) = \{y \in R^m : y \leqslant b\}$. Then we consider the following particular case of problem (I).

I(b). Find sup cx
subject to:

$$(2.4) \qquad x_1 K_1 + \ldots + x_n K_n \subseteq K(b), \; x \geqslant 0.$$

The feasible set of problem I(b) can be also written as:

$$S(b) = \{x \in R^n : x \geqslant 0, \; x_1 a_1 + \ldots + x_n a_n \leqslant b, \text{ for all } a_j \in K_j, \; j \in J\}.$$

Hence, for problem I(b), one has to be determined those vectors x, which maximize $c\,x$ and verify the constraints (2.2), for all $a_j \in K_j$ $(j \in J)$. From this point of view, Soyster's approach represents a conservative strategy for *the inexact programming, while Dantzig's generalized linear programming a nonconservative strategy.*

Soyster /31/ has shown that the set S of the feasible solutions of the problem (I) (as well as S(b) for I(b)) is a convex set if K is a convex set. Hence the problem (I) can be considered as a convex programming problem.

The particularization of the resource set K = K(b) permits to obtain an optimal solution for problem I(b), via an auxiliary linear programming problem. Let $A'=(a'_1,\dots,a'_n)$ *be the technological matrix of this program, where the activity vectors* a'_j $(j\in J)$ *are obtained via the support functionals of the convex sets* $K_j (j\in J)$. *Hence, we get the components of the activity vectors* a'_j *by:*

$$(2.5)\qquad a'_{ij} = \sup\left\{e_i a_j : a_j \in K_j\right\},\ i\in I = \left\{1,2,\dots,m\right\},\ j\in J,$$

where e_i *is a unit row vector with 1 in the i-th component.*

The auxiliary linear program will be:

$$LP(A').\ \max\left\{cx : A'x \leqslant b,\ x \geqslant 0\right\}.$$

If there exists $i\in I$ *and* $j\in J$ *such that* $a'_{ij}=\infty$, *then (2.4) implies that* $x'_j = 0$ *and consequently, we can delete the activity set* K_j *from I(b). We shall assume that such sets* K_j *have been omitted, so that we have:* $a'_{ij}<\infty$, *for all* $i\in I$ *and* $j\in J$.

This condition is always true if the activity sets K_j *are compact.*

The following theorem shows that any optimal solution to problem LP(A') is also an optimal solution to I(b).

<u>*THEOREM 2.1*</u> *(/31/) If x' is a feasible (optimal) solution to problem LP(A'), then x' is a feasible (optimal) solution for I(b) and conversely.*

The relation between the problem LP(A') and I(b) can be more general, namely in the definition of K(b) we can replace some inequalities "$y_i \leqslant b_i$*" by "*$y_i \geqslant b_i$*" or "*$y_i = b_i$*".*

In this case, if $y_i \geqslant b_i$, *then* $a'_{ij} = \inf\left\{e_i a_j : a_j \in K_j\right\}$, *for all* $j\in J$, *and the correspondent constraint in LP(A') will be of the type "*$\geqslant$*". If in the definition of K(b) we have* $y_i = b_i$, *then* $x'_j = 0$ *and* K_j *can be omitted from I(b) if the i-th component of* K_j *is not constant.*

From Theorem 2.1 it results that the optimal solutions of problem I(b) can be obtained by finding the optimal solutions of the linear programming problem LP(A'). Some numerical results and applications of this strategy to a class of fuzzy programming problems are given by Negoiţă, Flondor,

Sularia /23/.

From a practical viewpoint the main dificulty in solving the problem I(b) via the auxiliary linear programming problem LP(A′) consists in obtaining the activity vectors a_j' $(j \in J)$ *by solving the problems (2.5), whose nature depends upon the nature of the activity sets* K_j. *Thus, if the sets* K_j $(j \in J)$ *are polyhedral sets then* a_{ij}' *can be obtained by linear programming methods. If the resource set K is non-polyhedral, no finite linear programming representation for problem (I) is generally possible. Soyster /33/ transforms problem (I) into a semi-infinite program, which permits him to give a method for determining a sequence of upper and lower bounds for the optimal value of problem (I).*

Now, we mention that Gould /13/ considered another type of linear programming problem with inexact data. He developed for a set $K = [b' - \varepsilon,\ b' + \varepsilon]$, $b', \varepsilon \in R^m$, *a simplex type algorithm for efficiently solving the following "proximate linear programming problem" : Find a nonnegative* x' *in* R^n, *such that for some* $b \in K$:

(i) $Ax' \leq b$;

(ii) $cx' \geq cx$, *for every nonnegative* x *such that* $Ax \leq b$.

Next, we present some generalizations of the problem I(b). Thus , Tigan /42/, /43/, Soyster /33/ and Stancu-Minasian, Tigan /37/ have been considered a more general case of the problem I(b), namely the case when the resource set K is a polyhedral convex set of the form:

$$K = K(D,b) = \left\{ y \in R^m \ :\ \sum_{i=1}^{m} d_{ki} y_i \leq b_k,\ k \in L \right\}$$

where $D = (d_{ki})$ *is a given real matrix and* $L = \{1, 2, \ldots, r\}$.

Let us consider the problem:

$$I(D,b).\ \sup \left\{ cx \ :\ x_1 K_1 + \ldots + x_n K_n \subseteq K(D,b)\ ,\ x \geq 0 \right\}.$$

We suppose that K_j $(j \in J)$ *and K(D,b) are nonvoid sets. Let S(D,b) be the feasible set of the problem I(D,b).*

Now, we show that I(D,b) can be stated as a problem I(b).

Indeed, if $x \in S(D,b)$ *then for all* $a_j \in K_j$ $(j \in J)$, *we have:*

$$\sum_{j=1}^{n} \left(\sum_{i=1}^{m} d_{ki} a_{ij} \right) x_j \leq b_k,\ k \in L.$$

We denote: $g_k(z) = \sum_{i=1}^{m} d_{ki} z_i$, $g_k(K_j) = \{g_k(z) : z \in K_j\}$, $k \in L$, $j \in J$,

$$g(K_j) = g_1(K_1) x \ldots x\, g_r(K_r),\quad b = (b_1, \ldots, b_r).$$

Then the problem I(D,b) is equivalent with the following linear programming problem with set-inclusive constraints:

$$\text{II}(b).\quad \sup\{cx : x_1 g(K_1) + \ldots + x_n g(K_n) \subseteq K(b),\ x \geq 0\}.$$

Let denote:

$$(2.6)\quad a'_{kj} = \sup\{g_k(z) : z \in K_j\},\ j \in J,\ k \in L.$$

By the equivalence between the problem I(D,b) and II(b) and from Theorem 2.1, it results the following theorem:

THEOREM 2.2 (/43/) If x' is an optimal solution for LP(A'), ***where the elements*** *a'_{kj} of the matrix A' are given by (2.6), then x' is an optimal solution for I(D,b).*

We mention that Soyster /33/ considered the same problem I(D,b) with the supplementary assumption that the matrix D is a nonnegative matrix.

Now, we present some nonlinear extensions of the problems (I), I(b) and II(b)given by Tigan /42/ (see, also /37/), for which it has been obtained deterministic equivalent problems via the support functionals of the activity sets.

Next we consider a nonlinear programming problem with monotone set-inclusive constraints, which includes as particular cases the mathematical programming with linear, polynomial and separable set-inclusive constraints (see, /42/ or /37/).

Let G_i ($i \in M = \{1, 2, \ldots, m\}$) be a real function defined on the subset $H_i \times T$ in $R^s \times R^n$:

$$G_i(A_i, x) = G_i(a_{i1}, \ldots, a_{is}, x_1, \ldots, x_n),$$

where $x_1, \ldots, x_n$ are the decision variables, while a_{ik} ($i \in M$, $k \in L' = \{1, 2, \ldots, s\}$) represent the constraint coefficients and A_i denotes the row vector $(a_{i1}, \ldots, a_{is})$. Let A be the matrix having the rows A_i ($i \in M$).

As in the case of the problem (I), we do not know exactly the values of the column coefficient vectors a_j ($j \in L'$) of the matrix A, but we know that they belong to some given domains of variation $K_j \subseteq R^m$ $(j \in L')$.

Let us denote, for every x in T :

$$G(K_1,\dots,K_s,x) = \{G(a_1,\dots,a_s,x) \in R^m : a_j \in K_j, \ (j \in L')\},$$

where $G(a_1,\dots,a_s,x)$ is a vector in R^m having the components $G_i(A_i,x)$, $i \in M$.

Next we suppose that:
h.1) For any point x in T, the function $G_i(.,x)$ $(i \in M)$ is a nondecreasing function;
h.2) For every $i \in M$, the row vector $A_i' = (a_{i1}',\dots,a_{is}')$ belong to H_i, where:
(2.7) $a_{ij}^{\bullet} = \sup\{a_{ij} : a_j \in K_j\}$, $(j \in L')$.

Let f be a real function defined on the set T and let K be a given subset in R^m.

Now we can state the nonlinear programming problem with monotone set-inclusive constraints:

$$\text{(MP)} \quad \sup\{ f(x) : G(K_1,\dots,K_s,x) \subseteq K, \ x \in T\}.$$

In the case when $K = K(b)$, we can associate to the problem (MP) with set-inclusive constraints the following deterministic problem:

$$\text{(MPA)} \quad \sup\{ f(x) : G_i(A_i',x) \leq b_i \ (i \in M), \ x \in T\}$$

<u>THEOREM 2.3</u> (/37/) *If the assumptions h.1) and h.2) hold and for every $x \in T$, $G_i(.,x)$ $(i \in M)$ are continuous functions, then the problems (MP) and (MPA) have the same feasible set and optimal solutions.*

Problem (I) has also been generalized by allowing uncertainty in the objective function. In Soyster /32/ the vector c of the objective coefficients is supposed to belong to a closed convex set C, for which if $c_1 \in C$ and $c_2 \geqslant c_1$, then $c_2 \in C$.

Soyster /32/ obtained some duality results for the following problem:

$$\text{II(C).} \quad \sup\left\{\inf_{c \in C} \ cx : x_1K_1 + \dots + x_nK_n \subseteq K(b), \ x \geqslant 0\right\}.$$

The results of Soyster /32/ are extended by Falk /12/, which has considered a stronger set of optimality conditions that leads **to a computationally implementable method of solution for II (C).**

By means of certain constraint qualifications, Pomerol /26/ completed the duality result of Soyster /32/. Other duality results for linear programming with set-inclusive constraints and generalized linear programming are obtained by Thuente /40/, Duca /10/, Rohn /27/ and Itbaev /14/.

In the case of the linear-fractional objective, duality theory has been developed by Singh /30/, Kaur /17/, Kaul, Kaur and Lyall /16/, Kaur and Bhatia /18/.

The mathematical programming with inexact objective function has been also studied by Bird and Chattergee /3/ (bounds of the optimal value for linear programming), Dumitru and Luban /11/ (deterministic equivalent problems for nonlinear programming with inexact objective function) and Stancu-Minasian and Tigan /38/ (generalized fractional max-min problems).

3. NOTIONS OF EFFICIENCY FOR MULTIOBJECTIVE PROGRAMMING WITH INEXACT DATA

In this section we shall present two types of efficiency for the multiobjective programming with inexact criteria.

As objective functions, we consider the real functions f_i defined on the subsets $M_i \times X$ in $R^s \times R^n$:

$$(3.1) \qquad f_i(C_i,x) = f_i(c_{i1},\dots,c_{is},x_1,\dots,x_n)$$

where $x = (x_1,\dots,x_n) \in X \subseteq R^n$ is the vector of decision variables, while c_{ik} $(i \in I = \{1,2,\dots,p\}$, $k \in J = \{1,2,\dots,s\})$ represent the objective coefficients and $C_i \in M_i \subseteq R^s$ denotes the row vector $(c_{i1},\dots,c_{is})$. Let C be the matrix having the rows C_i $(i \in I)$.

We suppose that the columns coefficient vectors c_j $(j \in J)$ belong to some given domains of variation $K_j \subseteq R^p$.

Next, we assume that:

H.1) K_j $(j \in J)$ are convex, compact and nonempty sets in R^p and the sets M_i $(i \in I)$ are given by:

$$(3.2) \quad M_i = \left\{ C_i = (c_{i1},\dots,c_{is}) \; : \; c_j \in K_j,\; j \in J \right\}.$$

H.2) The objectives f_i $(i \in I)$ are continuous functions on $M_i \times X$.

The assumption H.1) implies that M_i $(i \in I)$ are convex and compact sets. Then, by the continuity of the function f_i (see, H.2)), for any given point x' in X, the set :

$$(3.3) \quad f_i(M_i,x') = \left\{ f_i(C_i,x') \; : \; C_i \in M_i \right\}$$

is a bounded closed interval. Therefore, when the coefficients of the function f_i are perturbed in the set M_i, the values of the function f_i in a point $x' \in X$, lie in an interval.

Hence, it is natural that certain notions of efficient solution for the multiobjective programming with inexact objectives f_i $(i \in I)$, are related to a semiorder relation defined on the set

$$I(R) = \left\{ [a,b] \; : \; a,b \in R, \; a \leq b \right\},$$

of the bounded closed intervals in R.

To define a conservative notion of efficient solution, we consider the following semiorder relation on $I(R)$:

(3.4) $\quad [a,b] \leqslant_c [a',b'] \iff a \leq a'$,

and the strictly relation:

(3.5) $\quad [a,b] <_c [a',b'] \iff a < a'$.

Now, let $F : H \times F \longrightarrow R^p$ be the vectorial function having as components the functions f_i $(i \in I)$, where H denote the cartesian product $K_1 \times K_2 \times \ldots \times K_s$.

We consider a conservative notion of efficiency for the vectorial function F on the set X.

<u>DEFINITION 3.1</u> A point $x' \in X$ is said to be a c-efficient point for F on X if $f_i(M_i,x) >_c f_i(M_i,x')$, for some $x \in X$ and for some $i \in I$, implies that there exists at least one $j \in I - \{i\}$, such that $f_j(M_j,x) <_c f_j(M_j,x')$.

When the sets $K_j (j \in J)$ are singletons, then the notion of c-efficient point (or solution) becomes the usual notion of efficient point (see, Charnes and Cooper /7/, Benson and Morin /2/).

For any given $C \in H$, we denote by $E(F(C,.),X)$ the set of usual efficient points of the function $F(C,.)$ on X. For $x' \in E(F(C,.),X)$, $F(C,x')$ is said to be an efficient value.

Now we can state the conservative multiobjective problem with inexact data as:

(PMC). Find

(3.6) $\quad Vmax_c \left\{ F(H,x) \; : \; x \in X \right\}$,

which consists in the finding of the set of all c-efficient points of the function F on X, and where $F(H,x) = \{F(C,x) \; : \; C \in H\}$, for $x \in X$.

We denote this set by $c\text{-}E(F,X)$.

We associate to (PMC) the following mathematical programming problem with deterministic multiple objective functions and set-inclusive constraints:

(PMC1). Find

(3.7) $\quad Vmax \; y$

subject to

$$(3.8)\quad F(H,x) \subseteq K(y) = \left\{ z \in R^p \ :\ z \geqslant y \right\},$$

$$(3.9)\quad x \in X.$$

Concerning the multiobjective programming problems (PMC) and (PMC1) we can state the following property:

THEOREM 3.1 The point $x' \in X$ is a c-efficient solution for (PMC) if and only if there exists $y' \in R^p$ such that (x', y') is an efficient solution for the problem (PMC1).

PROOF: The problem (PMC) can be replaced by the following equivalent multiobjective problem:

(PMC2). Find $V\max \{ G(x) \ :\ x \in X \}$,

where $G : X \longrightarrow R^p$ is a vectorial function whose components $G_i : X \longrightarrow R$ $(i \in I)$, are defined by:

$$(3.10)\quad G_i(x) = \min \left\{ f_i(C_i, x) \ :\ C_i \in M_i \right\}.$$

By the assumptions H.1) and H.2), which imply the continuity of the functions $f_i(.,x)$ $(i \in I)$ on the compact set M_i, the function G_i is well defined.

It is easy to see that the problem (PMC2) has the same efficient values as the problem :

$$(PMC3).\quad V\max \left\{ y \,|\, y \in R^p,\ x \in X,\ F(C,x) \geqslant y \text{ for all } C \in H \right\}.$$

But the problem (PMC3) is equivalent to (PMC1). The theorem results by these equivalences.

We denote:

$$(3.11)\quad c'_{ij} = \inf \left\{ e_i c_j \ :\ c_j \in K_j \right\} \ (i \in I,\ j \in J)$$

where e_i is an unit row vector in R^p with 1 in the i-th component.

Next, we suppose that:

H.3) For any point $x \in X$, the function $f_i(.,x)$ $(i \in I)$ is a nondecreasing function.

H.4) The row vectors $C'_i = (c'_{i1}, \ldots, c'_{is})$ belong to $M_i (i \in I)$.

Now we associate to the problem (PMC1) (with set-inclusive constraints) the following deterministic problem:

(PMC4). Find $V\max \{ y \ :\ y \in R^p,\ F(C', x) \geqslant y,\ x \in X \}$,

where $C' = (c'_{ij})$ is a p x s matrix whose elements are obtained by (3.11).

From Theorem 2.3 we can easily get the following result.

THEOREM 3.2 If the assumptions H.1)-H.4) hold, then the problems (PMC1) and (PMC4) have the same feasible set and efficient solutions.

From Theorem 3.1 and 3.2, one obtains sufficient conditions for the multiobjective programming problem (PMC) with inexact data in the objectives can be solved via a deterministic multiobjective problem (PMC4). We mention that in the particular case when $f_i(.,x)$ and $f_i(C_i,.)$ $(i \in I)$ are linear functions, then (PMC4) is a linear multiobjective programming problem.

Now we consider a non-conservative notion of efficiency for the vectorial function F.

DEFINITION 3.2 A point $x' \in X$ is said to be a nc-efficient point for F on X if there exists $C' \in H$, such that the following conditions hold:

i) x' is an efficient point for $F(C',.)$ over X;

ii) if $f_i(C_i,x') > f(C_i',x')$ for some $C \in H$ and for some $i \in I$, then there exists at least one $j \in I$ such that $f_j(C_j,x') < f_j(C_j',x')$.

It can be easily shown that x' is a nc-efficient point for F on X, if and only if there exists $C' \in H$, such that the pair (C',x') is an efficient solution for the function F over the set $H \times X$.

Now we state the non-conservative multiobjective programming problem with inexact objectives as:

(PMNC).Find $V\max\{F(C,x) : C \in H, x \in X\}$,

which consists in the finding of the set of all nc-efficient points of the function F on X (or the set of all efficient points of F over $H \times X$). We denote this set by nc-$E(F,X)$.

When $f_i (i \in I)$ are linear (or linear-fractional) functions and X is a polyhedral convex set, it can be shown that problem (PMNC) can be reduced (as in the case of (PMC) problem) to a linear (or linear-fractional) multiobjective programming problem with generalized linear constraints.

For such problems, in the next section, we shall suggest an interactive approach to obtain nc-efficient solutions via goal programming.

Finally, we mention that another type of efficiency for linear multiple objective programming with interval coefficients was introduced by Bitran /4/.

A point $x' \in X$ is said to be efficient in the sense of Bitran if x' is an efficient point of the function $F(C,.)$ over X, for every $C \in H$.

This type of efficiency is very conservative and only under restrictive conditions on the set H, the multiple objective programming problem has efficient points.

4. GOAL PROGRAMMING WITH INEXACT DATA

In this section we present some extensions of the goal programming in order to include criterion functions with inexact data. These extensions establish a natural link between the goal programming, multiobjective programming, inexact programming with set-inclusive constraints and generalized linear and linear-fractional programming.

The classical goal programming problem (Charnes and Cooper /7/, Kornbluth and Steuer /19/) can be stated as follows:

(GP). Find the lexicographic minimum:

$$\text{Lex min}\left\{g^j(x,y,z) \ : \ j \in L\right\}$$

subject to:

(4.1) $f_i(C_i,x) + y_i - z_i = s_i$, for all $i \in I = \{1,2,\ldots,p\}$,

(4.2) $x \in X \subseteq R^n$, $y \geqslant 0$, $z \geqslant 0$,

where :

(i) $f_i (i \in I)$ are the criterion functions (see,(3.1)) and $C_i \in R^{s_i}$;

(ii) the numbers $s_i \in R$ $(i \in I)$ are the desired goal levels of achievement for the criterion functions f_i $(i \in I)$;

(iii) y_i and z_i are respectively underachievement and overachievement deviational variables;

(iv) $L = \{1,2,\ldots,q\}$ denotes the set of the priority levels;

(v) g^j $(j \in L)$ represents the deviational objective function of the j-th priority level, which can be taken, for instance, a weighted-sum of the deviational variables associated with the priority class j;

(vi) X is the feasible region of the problem.

We assume throughout that X is a polyhedral nonempty and bounded subset of R^n, defined by:

(4.3) $$X = \left\{x \in R^n \ : \ Ax = b, \ x \geqslant 0\right\},$$

where A is a given $m \times n$ real matrix and $b \in R^m$.

When the criterion functions $f_i(C_i,.)$ $(i \in I)$ are linear, we can solve the linear goal programming problem by solving a sequence of linear programming problems, one for each priority level.

Kornbluth and Steuer /19/ considered the case when the criterion functions f_i $(i \in I)$ are supposed linear-fractional and either each criteria is assigned to its own priority level or all the criteria are at the same level.

Stancu-Minasian /35/, /36/, /39/ considers also, that all the criteria are at the same priority level and a goal objective function of Chebyshev type (see, below (4.11)) must be minimized.

Charnes and Cooper /8/, for the linear case, and Peteanu and Tigan /25/, for the linear-fractional case, studied interval goal programming problems.

Now, we shall formulate an interval goal programming problem related to the conservative multiobjective programming problem (PMC).

We recall, that under the assumptions H.1) and H.2) (from§3), for any $x' \in X$, the set $f_i(M_i,x')$ (see, (3.3)) is a bounded closed interval. Then it is natural to associate to the criterion functions $f_i (i \in I)$ the interval goals $S_i = [s_i', s_i'']$, $(i \in I)$, where s_i', s_i'' are real numbers such that $s_i' \leqslant s_i''$.

We formulate the conservative interval goal programming problem, as follows:

(GPID). Find

$$(4.4)\quad \text{Lex min} \left\{ g^j(x,y,z) \ : \ j \in L \right\}$$

subject to:

$$(4.5)\quad f_i(M_i,x) \subseteq \left[s_i' - y_i, s_i'' + z_i \right], \quad i \in I,$$

$$(4.6)\quad x \in X, \ y \geqslant 0, \ z \geqslant 0.$$

In the constraints (4.5) the variables y_i and $z_i (i \in I)$ penalize the deviation of the interval $f_i(M_i,x)$ from the interval goal S_i, where M_i was defined by (3.2).

We remark that the interval goal programming problem (GPID) is a lexicographic optimization problem with set inclusive constraints.

Now, we suppose that the criterion functions $f_i (i \in I)$ in the problem (GPID) are linear-fractional functions, with uncertainty only in the numerator:

$$(4.7)\quad f_i(C,x) = F_i(C,x)/W_i(D,x), \quad i \in I$$

where

$$(4.8)\quad F_i(C,x) = \sum_{j=1}^{n} c_{ij}x_j + c_{io}$$

$$(4.9)\quad W_i(D,x) = \sum_{j=1}^{n} d_{ij}x_j + d_{io}.$$

In (4.8), $C = (c_{ij})$ is an arbitrary variable $p \times n$ matrix, whose columns c_j belong to the perturbations sets $K_j \subseteq R^p$ $(j=1,\dots,n))$ and $c_o = (c_{1o},\dots,c_{po})$ is a given real vector. We set $H=K_1 x \dots x K_n$. In (4.9) the $p \times n$ matrix $D = (d_{ij})$ and the vector $d_o = (d_{1o},\dots,d_{po})$ have fixed real elements.

Throughout this section we assume that:

(4.10) $W_i(D,x) > 0$, for all $x \in X$ and $i \in I$.

Also, we consider that all the criterion functions have the same priority level and the associated deviational objective function g has the form (see, /44/) :

(4.11) $$g(y,z) = \max_{i \in I} (\alpha_i y_i + \beta_i z_i)$$

where for each $i \in I$, the real numbers α_i and β_i are given intragoals weights.

Then, by (4.7), the constraints (4.5) of the problem (GPID) can be rewritten as:

(4.12) $$s_i' - y_i \leqslant F_i(C,x)/W_i(D,x) \leqslant s_i'' + z_i, \text{ for all } C \in H$$

and $i \in I$.

Using the assumption (4.10), we can transform (4.12) into the following constraints:

(4.13) $$(s_i' - y_i)W_i(D,x) \leqslant F_i(C,x) \leqslant (s_i'' + z_i)W_i(D,x) \text{ , for}$$

all $C \in H$ and $i \in I$.

Now the linear-fractional interval goal programming (GPID) can be restated as the following min-max problem with inexact nonlinear constraints:

(TP). Find $\min \max_{i \in I} (\alpha_i y_i + \beta_i z_i)$

subject to (4.6) and (4.13).

Using the Kornbluth-Steuer transformation /19/:

(4.14) $$u_i = y_i W_i(D,x), \quad v_i = z_i W_i(D,x), \quad i \in I ,$$

the problem (TP) becomes:

(FTP). Find $\min \max_{i \in I} (\alpha_i u_i + \beta_i v_i)/W_i(D,x)$

subject to:

(4.15) $$s_i' W_i(D,x) - u_i \leqslant F_i(C,x) \leqslant s_i'' W_i(D,x) + v_i, \text{ for all } C \in H$$

and $i \in I$,

(4.16) $$Ax = b, \ x \geqslant 0, \ u \geqslant 0, \ v \geqslant 0,$$

where $u = (u_1,\dots,u_p)$ and $v = (v_1,\dots,v_p)$.

Let d_j' and d_j'' ($j \in M \cup \{0\}$, $M = \{1,2,\dots,n\}$) be the vectors in R^p having the components $s_i'd_{ij}$, $s_i''d_{ij}$ $(i=1,2,\dots,p)$ respectively.

Then the constraints (4.15) can be transformed into the following constraints:

$$(4.17)\quad (c_1-d_1')x_1+\dots+(c_n-d_n')x_n+u \geqslant d_o'-c_o, \text{ for all } c_j\in K_j,\ (j\in M)$$

$$(4.18)\quad (c_1-d_1'')x_1+\dots+(c_n-d_n'')x_n-v \leqslant d_o''-c_o, \text{ for all } c_j\in K_j,\ (j\in M).$$

Therefore, the problem (FTP) can be reformulated, as follows:

(LFIP). Find $\min \max_{i\in I} (\alpha_i u_i+\beta_i v_i)/W_i(D,x)$

subject to (4.16), (4.17) and (4.18).

Let $G'=(g_{ij}')$ and $G''=(g_{ij}'')$ be p x n matrices, whose elements are defined by:

$$(4.19)\quad g_{ij}' = \min\left\{e_i(c_j-d_j') : c_j\in K_j\right\},\quad g_{ij}'' = \max\left\{e_i(c_j-d_j'') : c_j\in K_j\right\},$$

where $e_i \in R^p$ is the unit vector with 1 in i-th component and $K_j (j\in M)$ are supposed to be compact convex nonempty sets in R^p.

Let consider the following min-max linear-fractional programming problem with linear constraints:

(LFP). Find $\min \max_{i\in I} (\alpha_i u_i+\beta_i v_i)/W_i(D,x)$

subject to (4.16) and

$$G'x + u \geqslant d_o' - c_o,\quad G''x-v\leqslant d_o'' - c_o.$$

From Theorem 2.3 and by the equivalence of the problems (TP) and (LFIP) we can get the following result.

THEOREM 4.1 (/44/) If $f_i (i\in I)$ are linear-fractional functions (of the form (4.7)) which verify the condition (4.10), then (x',y',z') is an optimal solution for linear-fractional interval goal programming problem (TP) if and only if (x',u',v') is an optimal solution for the min-max linear-fractional problem (LFP), where:

$$u_i' = y_i' W_i(D, x'), \quad v_i' = z_i' W_i(D, x'), \quad i \in I.$$

By this theorem, we conclude that the inexact goal problem (TP) may be solved by means of a deterministic min-max problem with linear-fractional criteria and linear constraints. Such fractional min-max problems can be effectively solved by the numerical methods considered in /41/, /45/, /29/ or the procedures given in /8/, /20/.

Now we consider the case when both the numerator and the denominator of the fractional criteria (4.7) are with inexact data and we show that the corresponding interval goal programming problem can be solved by means of linear-fractional min-max programming problem.

Let suppose in addition that the matrix D (in (4.7) and (4.9)) is an arbitrary $p \times n$ matrix, whose columns d_j belong to some perturbation sets $H_j (j \in M)$ in R^p. Let $P = H_1 \times H_2 \times \ldots \times H_n$.

Now, we assume that:

(4.20) $W_i(D, x) > 0$, for all $D \in P$, $x \in X$ and $i \in I$,

(4.21) $F_i(C, x) \geqslant 0$, for all $C \in H$, $x \in X$ and $i \in I$,

(4.22) K_j and $H_j (j \in M)$ are convex nonempty and compact subsets in R^p.

In this case, we formulate the linear-fractional interval goal programming problem with inexact data, as the following min-max problem:

(TPO). Find $\min \max_{i \in I} g_i(y_i, z_i)$

subject to:

(4.23) $s_i' - y_i \leqslant \dfrac{F_i(C, x)}{W_i(D, x)}$, for all $C \in H$, $D \in P$ and $i \in I$,

(4.24) $s_i'' - z_i \geqslant \dfrac{F_i(C, x)}{W_i(D, x)}$, for all $C \in H$, $D \in P$ and $i \in I$,

(4.25) $Ax = b$, $x \geqslant 0$, $y \geqslant 0$, $z \geqslant 0$, *where* $g_i(y_i, z_i) = \max\{\alpha_i y_i, \beta_i z_i\}$, α_i *and* β_i $(i \in I)$ *being given intragoal real nonnegative weights.*

From the assumptions (4.20) and (4.21), it is reasonable to suppose that the goals $S_i (i \in I)$ *in problem (TPO) are nonnegative, that is:*

(4.26) $s''_i \geq s'_i \geq 0, \ i \in I.$

Since, by (4.20) and (4.21), if $y_i > s'_i$, *the constraints (4.23) are evidently verified, instead of the problem (TPO) we consider the following problem:*

(TP1). *Find* $\min \max_{i \in I} g_i(y_i, z_i)$

subject to (4.23)-(4.25) and:

(4.27) $y_i \leq s'_i, \ i \in I.$

We have the following property:

<u>THEOREM 4.2</u> *(/46/) If* (x', y', z') *is an optimal solution for (TP1) then it is an optimal solution for (TPO).*

Let associate to problem (TP1) the following min-max problem:

(TP2). *Find* $\min \max_{i \in I} g_i(y_i, z_i)$

subject to (4.25), (4.27) and

(4.28) $f_i(C', x) \geq (s'_i - y_i) W_i(D', x), \ i \in I,$

(4.29) $F_i(C'', x) \leq (s''_i - z_i) W_i(D'', x), \ i \in I,$

where $C' = (c'_{ij})$, $C'' = (c''_{ij})$, $D' = (d'_{ij})$ *and* $D'' = (d''_{ij})$ *are* $p \times n$ *matrices, whose elements are determined by:*

(4.30) $c'_{ij} = \min\{e_i c_j : c_j \in K_j\}, \quad c''_{ij} = \max\{e_i c_j : c_j \in K_j\}$
$i \in I, \ j \in J,$

(4.31) $d'_{ij} = \max\{e_i d_j : d_j \in H_j\}, \ d''_{ij} = \min\{e_i d_j : d_j \in H_j\},$
$i \in I, \ j \in J,$

where e_i *is a unit vector in* R^p.

From Theorem 2.3, we can easily get the following result:

THEOREM 4.3 (/46/) *If the assumptions (4.20)-(4.22) and (4.26) hold, then the problems (TP1) and (TP2) have the same feasible (optimal) solution sets.*

Now, using the change of variables:

$$(4.32) \qquad u_i = y_i\, W_i(D',x),\ v_i = z_i\, W_i(D'',x),\ i \in I,$$

the problem (TP2) becomes:

(LFP). Find

$$\min \max_{i \in I} \left(\max \left(\frac{\alpha_i u_i}{W_i(D',x)}, \frac{\beta_i v_i}{W_i(D'',x)} \right)\right)$$

subject to:

$$(4.33) \qquad F_i(C',x) \geqslant s_i' W_i(D',x) - u_i,\ i \in I,$$

$$(4.34) \qquad F_i(C'',x) \leqslant s_i'' W_i(D'',x) + v_i,\ i \in I,$$

$$(4.35) \qquad u_i \leqslant s_i'\ W_i(D',x),\ i \in I,$$

$$(4.36) \qquad Ax = b,\ x \geqslant 0,\ u \geqslant 0,\ v \geqslant 0.$$

From theorems 4.2 and 4.3, it results:

THEOREM 4.4. (/46/) *Let $f_i (i \in I)$ be linear-fractional functions of the form (4.7). Then, if the assumptions (4.20)-(4.22) and (4.26) hold and if (x',u',v') is an optimal solution for (LFP) problem then (x',y',z') is an optimal solution for (TPO) problem, where*

$$y_i' = u_i'/W_i(D',x'), \qquad z_i' = v_i'/W_i(D'',x'),\ i \in I.$$

Thus, by Theorem 4.4, it results that an optimal solution for the goal programming problem (TPO) with inexact criteria can be obtained by solving a deterministic linear-fractional min-max problem with linear constraints.

Also, when the criteria $f_i (i \in I)$ are linear functions it was shown (see, /44/) that the inexact linear goal problem is equivalent with a linear programming problem.

Now, in a similar manner to the conservative goal programming problem (GPID), we consider a goal programming problem associated to the non-conservative multiobjective programming problem (PMNC):

(GPNC). Find $\text{Lex min} \left\{ g^j(x,y,z) : j \in L \right\}$

subject to (4.6) and

$$(4.37) \qquad f_i(C_i,x) \in \left[s_i' - y_i,\ s_i'' + z_i \right],\ i \in I,$$

(4.38) $c_j \in K_j,\ j \in J = \{1, 2, \ldots, s\}$,

where c_j are the columns of the matrix C having the rows $C_i (i \in I)$.

In the constraints (4.37) the variables y_i and $z_i (i \in I)$ penalize the deviation of the value of f_i from the interval goal S_i.

When $f_i (i \in I)$ are linear-fractional functions of the form (4.7) with uncertainty only in the numerator, we formulate the non-conservative goal programming problem (GPNC), as follows:

$$\text{(TNC). Find } \min \max_{i \in I} (\alpha_i y_i + \beta_i z_i)$$

subject to (4.6) and

(4.39) $$s_i' - y_i \leq \frac{F_i(C, s)}{W_i(D, x)} \leq s_i'' + z_i, \quad i \in I,$$

(4.40) $c_j \in K_j,\ j \in M = \{1, 2, \ldots, n\}$,

By (4.10) and (4.14), from (TNC) one obtains the following linear-fractional min-max programming problem with generalized constraints:

$$\text{(FNC). Find } \min \max_{i \in I} (\alpha_i u_i + \beta_i V_i)/W_i(D, x)$$

subject to (4.16), (4.40) and

(4.41) $s_i' W_i(D, x) - u_i \leq F_i(C, x) \leq s_i'' W_i(D, x) + v_i, \quad i \in I$.

For (TNC) and (FNC) problems a similar result with Theorem 4.1, holds:

<u>THEOREM 4.5</u> If $f_i (i \in I)$ are linear-fractional functions of the form (4.7), under the assumptions (H.1) and (4.10), (x', y', z') is an optimal solution for the non-conservative linear-fractional interval goal programming problem (TNC) if and only if (x', u', v') is an optimal solution for the generalized linear-fractional problem (FNC), where $u_i' = y_i' W_i(D, x')$, $v_i' = z_i' W_i(D, x')$, $i \in I$.

This theorem provides an approach to solve the nonconservative interval goal programming problem by means of a fractional min-max problem with generalized linear constraints. Such generalized fractional optimization problems can be solved by the parametrical procedures considered in

/29/, /41/, /45/ and the generalized simplex algorithm /9/.

When the criteria $f_i (i \in I)$ are linear functions, it can be easily shown that the non-conservative linear interval goal programming problem is equivalent with a generalized linear programming problem.

5. FINAL REMARKS

Several applications of the mathematical programming with inexact constraints and of the interval goal programming to the multiobjective programming with inexact criteria were presented. It was considered two strategies.

In the case of the conservative strategy, it can be easily shown that by an appropiate selection of the interval goals and of the intragoals weights the optimal solution set of the interval goal problem (TC) contains at least a c-efficient point for the multiobjective programming problem (PMC). A similar assertion is also true for the non-conservative interval goal problem (TNC) and the multiobjective programming problem (PMNC).

This linear problem of multiobjective programming with inexact objectives (PMNC) can be also tackled by a strategy of maximin type (see, Soyster et al. /34/).

Using this approach, a solution for the multiple objective problem can be obtained as an optimal solution for a generalized bilinear max-min problem, which can be solved via generalized linear programming /9/.

We mention that Stancu-Minasian and Tigan /38/ have been presented a generalized linear programming approach for a related fractional bilinear max-min problem with inexact data, by employing Schaible's change of variable method /28/ for nonlinear fractional programming.

6. REFERENCES

1. *Bakan, G.M. and Ahmetzanova, Z.Zh. (1986) 'A problem of convex programming with multivalued parameters', (russian), Kibernet. i Vychisl.Tehn. 71, 104-108.*

2. *Benson, H. P. and Morin, T.L. (1977) 'The vector maximization problem: proper efficiency and stability', SIAM J. Appl. Math., 32, 1, 64-72.*

3. *Bird, J. F. and Chattergee, S. (1985) 'Objective function bounds for the inexact linear programming problem with generalized cost coefficients', Computing Oper. Res., 13 (5), 483-491.*

4. *Bitran, G.R. (1980) 'Linear MOP with interval coefficients', Manag. Sci., 26, (7), 694-706*

5. *Chadha, S. S. (1973) 'Duality theorems for a generalized linear and linear fractional program',*

Cahiers C.E.R.O., 15 (2), 167-173.

6. *Chadha, S. S. and Kaul, R. N. (1972) 'A linear fractional program with variable coefficients', J.Math.Sci., 7, 15-20.*

7. *Charnes, A. and Cooper W.W. (1961) Management Models and Industrial Applications of Linear Programming, John Wiley, New-York.*

8. *Charnes, A. and Cooper, W.W. (1977) 'Goal programming and multiple objective optimizations', European J. Oper. Res., 1, 39-45.*

9. *Dantzig, G.B. (1963) Linear Programming and Extensions, Princeton Univ. Press, Princeton, New-Jersey.*

10. *Duca, D. (1975) 'O teoremă de dualitate în programarea inexactă', Stud. Cerc. Mat. , 27 (4), 443-448.*

11. *Dumitru, V. and Luban, F. (1986) 'On some optimization problem under uncertainty', Fuzzy Sets and Systems, 18, 257-272.*

12. *Falk, E.J. (1976) 'Exact solutions of inexact linear programs', Oper.Res., 24 (4), 783-787.*

13. *Gould, F.J. (1972) 'Proximate linear programming : A variable extreme point method', Math. Programming, 3, 326-338.*

14. *Itbaev, M.N. (1985) 'On the duality problem of inexact and generalized linear programming', (russian), Izv. Akad. Nauk Kaz. SSR, Ser. Fiz. - Mat., 35-38.*

15. *Kaul, R.N. and Bhatia, D. (1974) 'Generalized linear fractional programming', Econom.-Mat.Obzor, 10 (3), 322-330.*

16. *Kaul, R.N., Kaur, S. and Lyall V. (1986) 'Duality in inexact fractional programming with set-inclusive constraints', J. Optim. Theor. Appl., 50 (2), 279-288.*

17. *Kaur, S. (1984) 'Inexact fractional programming with set-inclusive constraints,' Cahiers C.E.R.O., 26 (1-2), 33-41.*

18. *Kaur, S. and Bhatia, D. (1983) 'Duality theory for generalized fractional programming', Indian J. Pure Appl. Math., 14 (2), 257-264.*

19. *Kornbluth, J.S.H. and Steuer, R.E. (1981) 'Goal programming with linear-fractional criteria', European J. Oper.Res., 8 (1), 58-65.*

20. *Kornbluth, J.S.H. (1983) 'Max-min programming with linear fractional functions. Algorithms and examples', in*

P.Hansen (ed), Essays and Surveys on Multiple Criteria Decision Making, Springer, Berlin, pp.204-213.

21. Luhandjula, M.K. (1984) 'Fuzzy approaches for multiple objective linear fractional optimization', Fuzzy Sets and Systems, 13, 11-23.

22. Lyall, V. (1988) 'Duality theory in inexact multiobjective programming', J. Inf. Optim. Sci., (India), 9 (2), 199-206.

23. Negoiţă, C.V., Flondor, P. and Sularia, M. (1977) 'On fuzzy environment in optimization problems', Econom. Comput.Econom.Cybernet.Stud. Res., 1, 13-24.

24. Patkar, V.N., Saxena, P.C. and Parkash, O. (1979) 'Linear piecewise linear programs with variable coefficients', Pure Appl. Math. Sci. (India), 10 (1-2), 51-56.

25. Peteanu, V. and Tigan, St. (1985) 'The multiobjective linear-fractional programming and interval goal programming', Itinerant Seminar on Functional Equations, Approximation and Convexity, Iaşi (Romania), 40-45.

26. Pomerol, J. Ch. (1979) 'Constraint qualifications for inexact linear programs', Oper. Res., 27 (4), 843-847.

27. Rohn, J. (1980) 'Duality in interval linear programming', in Nickel K.L.E. (ed.), Interval Mathematics, Academic Press, pp.521-529.

28. Schaible, S. (1974) 'Nonlinear fractional programming,' Operations Res. Verfahren, 19, 103-115.

29. Schaible, S. (1985) 'Fractional programming with several ratios', Methods of Operations Research, 49, 77-83.

30. Singh, C. (1982) 'Convex programming with set-inclusive constraints and its applications to generalized linear and fractional programming', J. Optim. Theory Appl., 38 (1), 33-42.

31. Soyster, A.L. (1973) 'Convex programming with set-inclusive constraints and applications to inexact linear programming', Oper. Res., 21 (5), 1154-1157.

32. Soyster, A.L. (1974) 'A duality theory for convex programming with set inclusive constraints', Oper. Res. 22 (4), 892-898.

33. Soyster, A.L. (1979) 'Inexact linear programming with generalized resource sets', European J. Oper. Res., 3 (4), 316-321.

34. *Soyster, A.L., Lev, B. and Di Toof (1977) 'Conservative linear programming with mixed multiple objectives,' Omega, 5 (2), 193-205.*

35. *Stancu-Minasian, I.M. (1980) 'A survey of methods used for solving the linear fractional programming with several objective functions', Methods of Oper. Res. 40, 159-162.*

36. *Stancu-Minasian, I.M. (1984) Stochastic Programming with Multiple Objective Functions, Editura Academiei, Bucureşti and D. Reidel Publishing Company, Dordrecht, Boston, Lancester.*

37. *Stancu-Minasian, I.M. and Tigan, St. (1987) 'Inexact mathematical programming', Cluj-Napoca Univ., Seminar on Optimization Theory, Report no. 8, 99-116.*

38. *Stancu-Minasian, I.M. and Tigan, St. (1988) 'Generalized pseudofractional max-min problems', Itinerant seminar on functional equations approx., convexity, Cluj-Napoca Univ., 295-302.*

39. *Stancu-Minasian, I.M. and Tigan, St. (1988) 'A stochastic approach to some linear fractional goal programming problems', Kybernetika (Praha), 24 (2), 139-149.*

40. *Thuente, D. (1980) 'Duality theory for generalized programs with computational methods', Oper. Res., 28 (4), 1005-1011.*

41. *Tigan,St. (1975) 'Sur une méthode pour la résolution d'un probleme d'optimisation fractionnaire par segments', Analyse Numér. Théor. Approx., 4 (1), 87-97.*

42. *Tigan, St. (1976) 'Sur un problème de programmation mathematique nonlineaire avec constraintes de type inclusion', Analyse Numér. Théor. Approx., 5 (2), 219-224.*

43. *Tigan, St., (1976) 'Observaţii asupra programării matematice cu restricţii de tip incluziune cu aplicaţie la programarea matematică cu date inexacte', Al II-lea Simpozion de informatică şi Conducere, Cluj-Napoca, pp.182-186.*

44. *Tigan,St. (1985) 'Goal programming with inexact data', Cluj-Napoca University, Report no.5, 95-106.*

45. *Tigan,St. (1988) 'On some procedures for solving fractional max-min problems', Analyse Numér. Théor. Approx., 17 (1), 73-91.*

46. *Tigan, St. and Stancu-Minasian, I.M. (1989) 'Fractional goal programming with inexact data', Itinerant Seminar on Functional Equations, Approximation and Convexity, Cluj-Napoca Univ., pp.311-318.*

47. *Wadhwa, V. (1972) 'Linear fractional programs with variable coefficients', Cahiers C.E.R.O., 14 (4), 223-232.*

48. *Zimmermann, H.J. (1978) 'Fuzzy programming with several objective functions', Fuzzy Sets and Systems 1, 45-55.*

INDEX

A

α-cut, 242, 302, 311
α-level set, see also -cut, 194, 204, 214, 321, 326
aggregation, operator
    augmented minimax, 217
    consensory-like, 337
    general, 196, 206, 216, 332, 345
    lexicographic min, 406
    max, 321
    min, 278, 321
    minmax, 197, 207
    ordered weighted sum, 346
    partial, 333
alternative, 8
aspiration level, see also reference
    fuzzy, 250, 268, 280, 286, 373
associate deterministic problem, 23, 77, 105, 134, 139, 254, 273, 373, 380, 389

B

belief
    degree of, 335
    qualification, 337
bisection method, 198, 208
branch and bound, 110, 126

C

Chebyshev
    norm, 14, 110, 218, 258
    problem, 79
compromise
    programming, 230
    solution, 8, 91, 108, 112, 134, 194, 204, 215, 233, 243, 277, 287, 374, 382
consistency, 256, 331
constraint
    balance, 264
    fuzzy, 254, 275, 280, 301, 340
    set-inclusive, 396, 400, 403
    soft, 281, 285
credibility, level, 253
criterion space, 9

D

decision space, 9
decision variable, 8
    crisp, 264
    fuzzy, 241, 353
    random, 353
decomposition, 397
discretization method, 26
distribution problem (in stochastic programming), 77
domination
    α-possible, 303

E

efficiency, see solution
energy planning, 118
extension principle, 49, 51, 56, 62, 256, 282, 303

F

feasibility, 355
    α-possible, 303
    medium strong, 54
    medium weak, 54
    strong, 51
    very weak, 54
    weak, 51
feasible set, 258, 355, 371, 375, 378, 401, 412
fuzzy constraint, 45, 46, 250, 254, 275, 280, 301, 340
fuzzy decision, 47, 232, 340
fuzzy equal, 195, 205, 216, 253
fuzzy goal, 45, 46, 53, 54, 59, 194, 205, 215, 250, 261, 340
fuzzy interval order, 323
fuzzy logic, 331
fuzzy max, 56, 195, 205, 216
fuzzy min, 195, 205, 216
fuzzy number, 45, 192, 198, 214, 250, 279, 308, 367, 389
    comparison, 250, 269, 284, 321, 391
    cut of, see α-cut
    L-R type, 49, 52, 60, 251, 264, 281, 325, 373
        arithmetics, 255, 282
        flat, 49, 250, 254, 283, 325
        kernel of, 325
        middle (or mean) value of, 251, 269, 314
        reference function, 251, 269, 325, 373, 389
        spread of, 251, 269, 314, 389
        support of, 325
        trapezoidal, 326
        triangular, 48, 314, 357

fuzzy objective, 59, 250, 256, 260, 277, 285, 301, 340
fuzzy preference, 62, 196, 206
fuzzy relation
    binary, 323
        antisymmetric, 323
        Ferrers, 323
        irreflexive, 324
        transitive, 323
            max-min, 322
            negatively, 322
    inequality, 284, 324
fuzzy set, 302, 333, 362
    cartesian product, 363
    convex, 302, 308
    cut of, see α-cut
    intersection, 334, 363
    level set of, see α-cut
    normalized, 302
    support of, 302
    union, 334, 364
fuzzy solution, 241, 245

G

game
    against Nature, 354
goal, 354, 357
    fuzzy, 194, 250, 268, 277, 373, 376
    linguistic, 233, 244
    programming, 75, 136, 231, 236
        fuzzy, 230
        inexact, 405
        interval, 406, 414
        stochastic, 82
group decision-making, 85

I

importance degree, 337, 342
imprecision, 194, 267, 302, 319, 332, 395
interactive
    algorithm, 18, 92, 132, 146, 200, 211, 218, 258, 291, 365
    branch and bound, 110, 126
    computer program, 127, 221, 260, 267, 296
    phase, 107, 151, 377, 382, 391

L

Lagrangian function, 200, 211, 218
level, see also aspiration, 197
    plausibility, 366
linguistic quantifier, see quantifier
loss function, 163, 180
Lukasiewicz OR, 355

M

mathematical programming
    multiobjective, 8, 229, 236, 256, 308
        linear, 73, 193, 317, 380
            fractional, 202, 257, 273, 374
        fuzzy
            compromise, 230
            linear, 193, 249, 264, 279, 303, 353, 367
            fractional, 203
            goal, see goal programming
            nonlinear, 213
            parametric, 232, 241
        inexact, 395, 402
        integer, 110, 126
        nonlinear, 64, 104, 122
        possibilistic
            linear, 59, 63, 353, 357, 362
        stochastic
            linear, 24, 72, 79, 103, 117, 131, 353, 358, 362
    single objective
        flexible, 46
        fractional, 74, 95, 104, 113
        fuzzy, 45, 328
            with ambiguity, 48
            with vagueness, 46
            with vagueness and ambiguity, 60
        goal, see goal programming
        inexact, 396
            goal, 405
            linear proximate, 399
            nonlinear, 400
        linear, 199, 209, 289, 311, 328
            α-MOLP, 195
            fractional
                α-MOLFP, 204
                Gα-MOLFP, 206
            Gα-MOLP, 195, 199
            generalized, 396, 414

nonlinear
    α-MONLP, 216
    Gα-MONLP, 216
possibilistic, 45, 48, 51, 54, 59, 62, 63
        robust, 60
        semi-infinite, 305, 399
        stochastic, 76, 166
            with chance constrained approach, 25, 51, 56, 77, 94, 134, 143
            with recourse approach, 25, 77, 105, 126, 134, 141, 163
            fractional, 80
max-operator, 52, 321
mean dispersion approach, 139
mean value minimization, 24, 26
membership
    degree, 231, 333, 344
    function, 47, 49, 195, 205, 215, 250, 282, 321, 333, 358
minimax (or maximin) problem, 107, 197, 207, 288, 408
    augmented, 217
    bilinear, 414
minimum risk
    multiple objective problem, 89
    single objective problem, 78, 80
    vectorial solution, 91
min-operator, 47, 287, 321
multiobjective decision
    aid, 117, 127
    problem, 331, 348
        F-α-MODP, 196, 206
        under uncertainty, 103, 117, 196, 340
multiple fractile criterion model, 94

N

necessity, degree of, 51
non-domination
    α-possible, 304
    proper, 10
    weak, 10

O

operation, arithmetic
    extended, 255, 256, 282
optimization, see also mathematical programming
    farm structure, 264
    fuzzy, 287, 328
    soft, 333
    stochastic, 23, 163
ordered weighted average, 331, 345

P

parametric
    analysis, 108, 110, 120, 125, 146, 150, 156, 158, 383, 391
    programming, 18, 75, 232, 241
    representation, 181
Pareto-optimal, see also solution
    set, 317
pay-off, 354
    table, 12, 106, 137, 147, 153, 373, 381
penalty function approach, 24, 163
point
    efficient, see also solution, 259, 355, 392
        extreme, 313
    ideal, 12, 106, 138, 147, 258, 355, 375
    nadir, 12, 259, 375
    reference, 197
possibility, 354
    degree of, 51
    distribution, 354, 368
    measure, 51, 356
preference
    degree, 200
    information, 17
    relation
        fuzzy, 62, 322, 323
    parameters, 13
prescription, 7
programming, see mathematical programming

Q

quantifier, linguistic, 233, 331, 335, 340, 347
    absolute, 336
    fuzzy, 336, 338
    linear, 342
    monotonic, 338
    nondecreasing, 336
    proportional, 336

R

random coefficients, 23, 76, 103, 104, 133, 163
    with incomplete information, 131
reference
    function, see fuzzy number, L-R type
    level, 197, 207
    membership value, 197, 202, 203, 211, 212, 213, 221
    point, 13, 197

regression analysis, 168
representation theorem, 242
response surface methodology (RSM), 28
risk, see also safety, 105, 117, 254, 268, 389, 393

S

satisfaction
    degree of, 64, 136
    function, 234
safety, see also risk, 260, 390, 393
    optimistic, 251, 260, 373, 390
    parameters, 254, 260, 268, 373, 377
    perfect, 261
    pessimistic, 253, 260, 373, 390
scalarizing function, 13
scenario, 113, 121, 137, 163, 172, 366, 368, 388
sensitivity analysis, 199, 396
set coefficient, 395
set-valued function, 198, 208
simulation, 354
    Monte Carlo, 356
s-norm, 334
solution, compromise, 108, 112, 194, 204, 215, 233, 243, 287, 374, 382
    crisp, 243
    dominated, 240
    efficient, see also Pareto-optimal, undominated, 12, 74, 164, 166, 233, 258, 269, 308, 403
        β-possible, 59, 307, 309
        c-, 403
        C_m^J- , 176
        F-, 233
        M-, 233
        nc-, 405
        stochastic, 87, 88, 89, 94
    feasible, 355, 398
    fuzzy, 241, 245
    optimal, 199, 210, 229, 238, 305, 343, 398, 401
    Pareto-optimal, see also efficient, undominated, 193, 203, 213, 355
        α-, 194, 204, 214
        local, 215
        M-α-, 196, 199, 206
    satisfacing, or satisfactory, 194, 204, 215, 285
    satisfying
        $(\alpha,\beta)_1$-, 303, 305, 310, 314
        $(\alpha,\beta)_2$-, 307, 312, 315
    undominated (or nondominated), see also efficient and Pareto-optimal, 62, 354, 355, 357, 360

stochastic method (or approache)
    approximation, 27
    chance constrained, see mathematical programming
    gradient, 28
    with recourse, see mathematical programming
strategy, inexact programming
    conservative, 397, 414
    nonconservative, 398, 413
subjective probability, 103, 370, 389, 392
substituting problem, see associate deterministic problem

T

t-norm, 231, 237, 334, 354
    co-, 354
tradeoff information, 200, 211
transitivity, see fuzzy relation
triangular norm, see t-norm

U

uncertainty
    economic, 117, 124
    fuzzy, 250, 365, 396
    global, 369, 388
    modeling, 164, 264, 367, 371, 373, 388, 395
    possibilistic, 250, 264, 353
    probabilistic, 103, 117, 353, 365, 395
    separate, 369, 388
utility function, 17, 104

V

value function, 17

W

weighted sum, 14, 236

Z

zooming, 392

GPSR Compliance

The European Union's (EU) General Product Safety Regulation (GPSR) is a set of rules that requires consumer products to be safe and our obligations to ensure this.

If you have any concerns about our products, you can contact us on ProductSafety@springernature.com

In case Publisher is established outside the EU, the EU authorized representative is:

Springer Nature Customer Service Center GmbH
Europaplatz 3
69115 Heidelberg, Germany

Batch number: 10815974

Printed by Printforce, the Netherlands